ASPECTS OF HOMOGENEOUS CATALYSIS

Volume 1

ASPECTS OF HOMOGENEOUS CATALYSIS

A Series of Advances

EDITED BY

RENATO UGO

ISTITUTO DI CHIMICA GENERALE ED INORGANICA
MILAN UNIVERSITY

VOLUME 1

1970

CARLO MANFREDI, Editore - Milano

ISBN-13:978-94-010-3343-5   e-ISBN-13:978-94-010-3341-1
DOI: 10.1007/978-94-010-3341-1

CARLO MANFREDI EDITORE - MILANO

MONOTIPIA ROSSI-SANTI - MILANO 1970

Editorial Board

Contents of Volume 1

Dimerization and Co-dimerization of Olefinic Compounds by Co-ordination Catalysis
G. Lefebvre and Y. Chauvin

Selective Homogeneous Hydrogenation of Dienes and Polyenes to Monoenes
A. Andreetta, F. Conti and G. F. Ferrari

ASPECTS OF HOMOGENEOUS CATALYSIS

Volume I

Chapter 1

Recent Advances in Homogeneous Hydrogenation of Carbon-Carbon Multiple Bonds

R. S. COFFEY

I.C.I. Ltd., Heavy Organic Chemicals Division, Billingham, Teesside, U.K.

4

1. INTRODUCTION AND SCOPE

Research into homogeneous catalysis has been stimulated by the belief that homogeneous catalysts can be designed which are more efficient than their heterogeneous counterparts. It has proved notoriously difficult to reproduce active surfaces, and it is believed that greater reproducibility, better control of catalyst poisons, and more specific catalysts can be obtained in homogeneous systems by suitable control of the ligands surrounding the active centre. It should also be possible to design more active catalysts in homogeneous systems as each metal atom, or ion, is potentially a catalyst, whereas in heterogeneous systems only certain surface sites are catalytically active.

The use of homogeneous catalysts in the hydroformylation process, some dimerisations and polymerisations of olefins, and the oxidation of ethylene to acetaldehyde or vinylacetate confirm these beliefs. Another incentive to study homogeneous catalysts is that it is possible to obtain insight into the details of molecular processes and active intermediates, and this knowledge helps in our understanding of heterogeneous reactions. Although no large scale industrial homogeneous hydrogenation process has been developed, recent developments have produced catalysts which are highly specific and very active and are already finding uses in preparative chemistry. Consequently homogeneous hydrogenation is now established as a useful reaction in its own right, rather than only as an aid to understanding the heterogeneous counterpart.

Homogeneous hydrogenation dates from 1938 when Calvin noted that the hydrogen reduction of benzoquinone to quinol at 100 °C was catalysed by cupric ion [1]. Since then several general types of catalyst have been studied. There are reviews covering in detail hydrogenation catalysed by aqueous solutions of cobaltous cyanide [2, 3, 4,], aqueous and non-aqueous solutions of transition metal salts, e.g. halides [4] and carboxylates [5], and metal carbonyls, particularly of cobalt [6] and iron [7]. More general reviews include reference to organometallic compounds obtained by treating metal salts with aluminium alkyls or Grignard reagents [3, 8, 9,].

During the late fifties and early sixties studies in preparative and structural co-ordination chemistry established that transition metal hydrides, long thought to be key intermediates in hydrogenation and related catalytic reactions, are often stabilised by π bonding ligands, particularly tertiary phosphines [10]. It is in catalytic systems incorporating these ligands where the most significant advances in homogeneous hydrogenation have been made in the last few years.

The aim of this article is not to duplicate earlier reviews, but to supplement them with a review of the more recent literature not available to previous reviewers. The major advances have been made in hydrogenation of carbon-carbon multiple bonds and this will be dealt with most fully. As predicted the new homogeneous hydrogenation catalysts are highly specific. They usually hydrogenate olefins and acetylenes but they also show high specificity within this narrow range of substrates. However, specific hydrogenation of conjugated polyenes to monoenes will not be treated exhaustively as this is the subject of another article in the present series [11]. It is possible to divide a review into sections in various ways, the one selected here is to divide the catalysts into groups which emphasised the most important aspect of each catalyst system. As mentioned earlier homogeneous hydrogenation has now become a synthetic tool in organic chemistry, and so operating conditions are emphasised to help distinguish systems which have a practical use, from those which are of only academic interest.

2. BASIC REACTIONS IN HOMOGENEOUS HYDROGENATION

It is necessary to activate hydrogen (1), or olefin (or other substrate) (2), or both (3) before catalytic hydrogenation can occur. Transition metals form hydride complexes and olefin complexes, and generally operate by reaction (3). As will

$$L_xM \xrightarrow{H_2} L_xMH_y \xrightarrow{\text{olefin}} L_xM + \text{paraffin} \qquad (1)$$

$$L_xM \xrightarrow{\text{olefin}} L_xM(\text{olefin}) \xrightarrow{H_2} L_xM + \text{paratfin} \qquad (2)$$

$$L_xM \begin{array}{c} \xrightarrow{H_2} L_xMH_y \xrightarrow{\text{olefin}} \\ \\ \xrightarrow{\text{olefin}} L_xM(\text{olefin}) \xrightarrow{H_2} \end{array} L_xMH_y(\text{olefin}) \rightarrow L_xM + \text{paraffin.} \qquad \begin{array}{c} (3a) \\ \\ (3b) \end{array}$$

be seen later both routes (3a) and (3b) have been shown to occur. As the catalyst has to activate both hydrogen and olefin, and also favour the transfer of hydrogen onto the activated olefin, the intermediates in reactions (1) to (3) must have thermodynamic stability within quite narrow limits. If they are not sufficiently stable they will not form, whereas if an intermediate is too stable the full reaction sequence will not occur.

2.1. Coordinative unsaturation

An important feature of homogeneous catalysis in general is that active catalysts must have a free co-ordination site, so that a substrate, e.g. hydrogen or olefin, may become co-ordinated and activated [12, 13]. If a free site is not available it must be easily generated by dissociation of a ligand (4) or the complex must be "substitution labile" by a substrate molecule (5). In a catalytic cycle reaction (4) would precede reaction sequence (1) or (3a), whereas reaction (5) could replace the initial stage of reaction sequences (2) and (3b) or (3a).

$$L_{x+1}M \rightleftharpoons L_xM + L \tag{4}$$

$$L_{x+1}M + \text{olefin} \rightleftharpoons L_xM\,(\text{olefin}) + L \tag{5}$$

Thus 5 co-ordinate d^6 complexes or 4 co-ordinate d^8 complexe should be active catalysts, but 6 co-ordinate d^6 and 5 co-ordinate d^8 complexes will not be active unless they lose ligands. Long induction periods or the requirement of thermal stimulation are often indicative of ligand dissociation to give active catalysts. Catalyst poisons, e.g. phosphorus and sulphur compounds and carbon monoxide, act by occupying co-ordination sites and such sites can also be destroyed by bridging ligands. The free co-ordination site and the way it can be deactivated or poisoned is probably analogous to the poisoning of surface atoms of a heterogeneous catalyst.

2.2. Hydrogen activation

Three general ways of activating hydrogen have been recognised and they are exemplified by reactions (6)-(8) [4].

$$Ru^{III}Cl_6^{3-} + H_2 \rightleftharpoons Ru^{III}Cl_5H^{3-} + H^+ + Cl^- \tag{6}$$

$$2Co^{II}(CN)_5^{3-} + H_2 \rightleftharpoons 2Co^{III}(CN)_5H^{3-} \tag{7}$$

$$IrClCO(PPh_3)_2 + H_2 \rightleftharpoons IrH_2ClCO(PPh_3)_2 \tag{8}$$

Reaction (6) occurs most frequently and is a substitution reaction in which the metal is not oxidised. It depends on the ease of substitution of a ligand and the presence of suitable bases which may be the solvent or the substituted ligand itself, to prevent the reverse reaction.

Reaction (7) involves homolytic cleavage of the hydrogen molecule and the metal is oxidised. It has been suggested that this type of reaction may occur instead of reaction (6) in salts of transition metals, particularly carboxylates, where there are relatively short metal-metal distances [14].

Reaction (8) is an example of oxidative addition which has been extensively reviewed recently [12, 13, 15, 16].

The metal is formally oxidised from a d^8 to a d^6 configuration, for example. The energy change in reaction (8) approximates to (9), where E_{Ir-H} is the energy of the Ir–H bond, E_{H-H} is the energy of the hydrogen dissociation and P is the promotional energy in the ligand field associated with the electron transfer $5d_{z^2}$ to $6p_z$. As E_{H-H} is constant and E_{Ir-H}, or even

$$\Delta E = 2E_{Ir-H} - (E_{H-H} + P) \tag{9}$$

the general case E_{M-H}, varies within narrow limits, the value of ΔE depends mainly on P. Clearly electron withdrawing ligands will tend to increase P and electron donating ligands will have the reverse effect. This is demonstrated in Table 1 which contains the activation parameters for the forward step of reaction (8), and the free energy of activation can be regarded as a rough measure of the promotional process.

Table 1

ACTIVATION PARAMETERS FOR $IrXCO(PPh_3)_2 + H_2 \rightarrow IrH_2XCO(PPh_3)_2$

X	ΔH^*(kcal.mol.$^{-1}$)	ΔS^* (e.u.)	ΔF^*(kcal.mol.$^{-1}$)
Br[1]	12.0	— 14.0	16.2
Cl [1]	10.8	— 23.0	17.8
Cl [2]	11.5	— 20.0	17.6

[1] in benzene at 30° Ref. [17]
[2] in chlorobenzene at 30° Ref. [15]

Incidentally the entropy values indicate strict steric requirements for the reaction or a polar transition state. As the reaction goes faster in dimethylformamide than in benzene, the latter seems more likely [17]. The promotional energy within a given ligand field is also dependent on the metal, decreasing in going down a series e.g. Co > Rh > Ir. Thus the rhodium analogue $RhClCO(PPh_3)_2$ does not react with hydrogen under mild conditions.

The real significance of reaction (9) is that ΔE must be a small positive value for a complex to be a hydrogenation catalyst [18]. If it is negative, i.e. the promotional energy is high as in $RhClCO(PPh_3)_2$, the complex will not activate hydrogen. If it is positive but too large the hydride will be stable and hydrogen will not be transferred to the substrate as in $IrClCO(PPh_3)_2$, which is a very weak catalyst.

It has been suggested that coordination of hydrogen with a transition metal is brought about by donation of hydrogen bonding electrons to an empty metal orbital, or by donation of electrons from a metal orbital to the anti-bonding hydrogen orbital. In $d^8 - d^6$ oxidative additions it is more likely that the latter process occurs, as it is favoured by high electron density on the metal [18].

2.3. Olefin activaction and hydrogen transfer

Many olefin complexes of transition metals are known but the nature of the bonding in them is currently in dispute. Some equilibrium data and thermodynamic data is available on silver(I) [19] and copper(I) [19] complexes but very little is available on complexes of metals which are good homogeneous hydrogenation catalysts. Recently it has been shown that in the complexes Rhacac(olefin)$_2$ [20], IrXCO(PPh$_3$)$_2$(olefin) [15] and certain platinum(II) complexes [19] stability is decreased by alkyl substituents on the olefin bond but increased by electronegative substituents, e.g. halide, carboxy and cyanide. This emphasises the importance of back donation from the metal in stabilizing olefin complexes. In studying reaction (10) it was found that C$_2$F$_4$ reacted much more slowly than

$$\text{Rhacac}(C_2H_4)_2 + \text{olefin} \rightleftharpoons \text{Rhacac}(C_2H_4)\,(\text{olefin}) + C_2H_4 \qquad (10)$$

propylene, even though it formed a much more stable complex, and it was concluded that development of the σ bond and not of the π bond was crucial in forming the activated complex [20]. Clearly this could be of significance in all metal catalysed reactions of olefins. Recent ^{1}H n.m.r. studies on rhodium(I) [21] and platinum(II) [22, 23] complexes show that in solution the olefin is not static but rotates on the metal-olefin axis ($\Delta E = 10\text{-}15$ kcal.mol^{-1}) [21c, 23]).

The transfer of hydrogen to a co-ordinated olefin can be stepwise (11), or two hydrogen atoms can add simultaneously (12). (Reactions (11) and (12) are shown

$$\text{M (olefin)} \xrightarrow{\text{H}} \text{M alkyl} \xrightarrow{\text{H}} \text{M} + \text{paraffin} \qquad (11)$$

$$\text{M (olefin)} \underset{-\text{H}}{\overset{+\text{H}}{\rightleftharpoons}} \text{M alkyl} \underset{+\text{H}}{\overset{-\text{H}}{\rightleftharpoons}} \text{M (olefin}') \qquad (11a)$$

$$\text{M (olefin)} \xrightarrow{\text{H}_2} \text{M} + \text{paraffin} \qquad (12)$$

as an olefin complex reacting with hydrogen but the same result is achieved if a preformed hydride reacts with an olefin, and a hydrido-olefin complex may be an intermediate, see reactions (1)-(3)). Olefins can isomerise if the first step in (11) is reversible (11a). Many olefin isomerisations have been shown to go via (11a) which is known as "reverse alkyl formation". Olefin isomerisation has been reviewed recently [24] and work by Cramer and Lindsey is of particular significance [16, 25]. The occurrence of isomerisation during hydrogenation is regarded

as good evidence for stepwise addition of hydrogen and in catalytic deuteriogenation of olefins the formation of deutero olefins and paraffins with a variable
deuterium content (d_0 to $> d_2$) indicates reactions (11) and (11a) are taking place.

Recently, however, it has been shown that deuterium can be incorporated
into olefins via oxidative addition of hydrogen and vinyl groups (reaction 11b).
The hydrodimerisation of

$$Ru^{II+} + H_2C=CHCN \rightleftharpoons HRu^{IV}-CH-CHCN \tag{11b}$$

acrylonitrile in polar solvents in the presence of many ruthenium catalysts at 6-50
atmospheres of hydrogen and 100-150 °C gives a mixture of adiponitrile, dicyanobutene and acetonitrile [192-195]. In the presence of deuterium gas deuterium
is incorporated into the products and unchanged acrylonitrile and its distribution
is best accounted for by reaction (11b) rather than (11a) [195].

In deuterated hydroxylic solvents metal hydrides exchange with solvent deuterons via oxidative addition (13a) [16, 25]. This leads to exchange

$$M^{n+2}-H \underset{+H^+}{\overset{-H^+}{\rightleftharpoons}} M^n \underset{-D^+}{\overset{+D^+}{\rightleftharpoons}} M^{n+2}-D \tag{13a}$$

of hydrogen gas with solvent deuterons (13b) and also incorporation of

$$H_2 + D_2O \rightleftharpoons HD + HDO \tag{13b}$$

deuterium into olefins and paraffins via reactions (13a) and (11) and (11a). Reaction (13a) is general for transition metals, although it is not always certain which
valency states are involved [25b].

Hydrogenation of various substrates does not always require gaseous hydrogen. Hydrogen can be obtained from other organic molecules by intermolecular
hydrogen transfer (see section 10.2). In stepwise hydrogen transfer (11) the second
hydrogen atom can be derived from the solvent via a hydrolysis reaction (14).

$$M^n \text{ alkyl} + H^+ \rightarrow M^{n+1} + \text{paraffin} \tag{14}$$

Mechanistic schemes which have been proposed, and in some cases proved,
are made up of combinations of reactions (1) to (14).

3. TECHNIQUES FOR STUDYING HOMOGENEOUS HYDROGENATION

Kinetic studies, the study of reactive intermediates by nuclear magnetic
resonance, infrared, ultraviolet and visible spectra, electron spin resonance techniques and isotopic labelling studies have all been employed to good effect. One

of the problems in homogeneous catalysis is to ensure that reactions are truly homogeneous. In many systems, particularly where there are no ligands which stabilise reduced valency states or metal-hydrogen bonds, metal precipitation occurs and the onus is on the worker to prove he is dealing with a true homogeneous solution. It has been proved that the catalyst $(PtCl_2(C_2H_4)_2)$ [26] is truly homogeneous at low temperatures [27], but it has recently been shown that in hydrogenation of oleic acid at 200 °C and 250 atmospheres, catalysed by a mixture of copper and cadmium oleates, the actual catalyst is a copper metal colloid stabilised by a cadmium soap [28]. It had been assumed that this was a homogeneous catalyst system [3, 8].

Most transition metal complexes will catalyse hydrogenation of olefins under conditions vigorous enough to displace ligands. However, the severity of these conditions often causes metal formation and it is not clear that all high pressure systems are truly homogeneous. In the following account the catalysts have been shown to be truly homogeneous unless otherwise stated.

4. COBALTOUS CYANIDE AND RELATED SYSTEMS

Hydrogenations catalysed by the cobalt cyanide system have been reviewed [2, 3, 4]. It is generally used in aqueous solvent and at room temperature and pressure but it has been used at 40-90 °C and up to 100 atmospheres. It will not catalyse the hydrogenation of isolated olefins and acetylenes; conjugated dienes are reduced to monoolefins, and α-β olefins conjugated with aldehyde, carboxyl, keto and aromatic groups etc. may be hydrogenated, depending on the substituents of the double bond. The α-β unsaturated aldehydes give poor yields due to side reactions, e.g. aldolisation. Comprehensive lists of compounds which can be reduced have been drawn up [2a, 3, 8,] and only recent work will be discussed.

4.1. Cyano cobalt hydrides

It was thought that the same hydrido species $HCo(CN)_5^{3-}$ was obtained when aqueous solutions of cobaltous chloride and potassium cyanide (Co : CN=1 : 5) were treated with hydrogen (15), sodium borohydride, or allowed to stand under nitrogen (16) [2]. This has now been proved by ^{1}H.n.m.r.,

$$Co(CN)_5^{3-} + H_2 \rightleftharpoons 2HCo(CN)_5^{3-} \qquad (15)$$

$$2Co(CN)_5^{3-} + H_2O \rightarrow Co(CN)_5OH^{3-} + HCo(CN)_5^{3-} \qquad (16)$$

ultraviolet and visible spectroscopy, [29, 30] and a solid derivative $Cs_2Na[Co(CN)_5H]$ has at last been isolated and characterised [30]. Kinetic evidence had suggested that binuclear species, e.g. $Co_2(CN)_{10}^{6-}$ and $H_2Co_2(CN)_{10}^{6-}$ are intermediates in reaction (15) [2a, 31] but it now appears this is not true and only monomeric cobalt species are involved [29].

Aqueous solutions of $HCo(CN)_5^{3-}$ deteriorate on standing, particularly at high cobalt concentrations but 10% solutions of $Li_3[Co(CN)_5H]$ in ethanol appear to be stable indefinitely and this could have practical importance. Spectroscopic properties show that the hydride is the same as in aqueous solutions and a solvated salt has been isolated [32]. Hydrogenations in alcoholic solvents appear to be slower than the equivalent reaction in water [32, 33, 34].

Besides $HCo(CN)_5^{3-}$, low concentrations of other active species are present in solutions formed by reaction (15). It has been suggested they are $H_2Co(CN)_4^{3-}$ and $Co(CN)_6^{5-}$ formed by reaction (17), the former being favoured by CN : Co rations < 5 and the latter at CN : Co ratios > 5 [30]. On the basis of polarographic evidence it has been suggested that the cobalt(I) species

$$HCo(CN)_5^{3-} \rightleftharpoons H^+ + CN^- + Co(CN)_4^{3-}$$

$$Co(CN)_4^{3-} + H_2 \rightleftharpoons CoH_2(CN)_4^{3-} \qquad (17)$$

$$Co(CN)_4^{3-} + 2CN^- \rightleftharpoons Co(CN)_6^{5-}$$

is $Co(CN)_5^{4-}$ formed via reaction (18) particularly at pH > 10 [35]. Thus reduced cobaltous cyanide solutions contain at least three potential

$$HCo(CN)_5^{3-} + OH^- \rightleftharpoons H_2O + Co(CN)_5^{4-} \qquad (18)$$

reducing agents, $H_2Co(CN)_4^{3-}$, $HCo(CN)_5^{3-}$ and a cobalt(I) complex which may be $Co(CN)_5^{4-}$ or $Co(CN)_6^{5-}$, and the amount of each will vary with pH, solvent, other ions, etc.. Each active species will reduce substrates differently. The cobalt (I) complex is a redox catalyst and adds electrons followed by proton uptake and is probably responsible for redox reductions, e. g. reduction of benzoquinone [35b]. The monohydride transfers a hydride or hydrogen atom and appears to be responsible for the hydrogenations of activated olefins studied so far (see below), while the dihydride can be expected to transfer H_2 as other d^6 dihydrides (see section 6).

4.2 The mechanism of hydrogenation

Two distinct mechanisms for hydrogenation of activated olefins occur [2]. On the basis of preparative work and variation in products according to reaction conditions, Kwiatek had shown butadiene was converted to a mixture of butenes via the butenyl derivative (19),

$$HCo(CN)_5^{3-} + C_4H_6 \rightleftharpoons C_4H_7Co(CN)_5^{3-} \qquad (19)$$

$$C_4H_7Co(CN)_5^{3-} + HCo(CN)_5^{3-} \rightarrow C_4H_8 + 2Co(CN)_5^{3-} \qquad (20)$$

which reacted with more hydride (20). The cobalt hydride was regenerated by reaction (15). A detailed kinetic investigation of this reaction has proved this scheme and shown that two σ butenyl cobalt complexes and a π butenyl cobalt complex are in equilibrium [36] (See also reference [11]).

The kinetics of hydrogenation of cinnamic acid indicates a free radical mechamism via reactions (21) and (22), where S is $C_6H_5CH=CHCOO^-$, and the hydride is regenerated by (15) [2, 31c, 37]. The reversibility of reaction (21) is indicated by the deuterium content of products obtained in deuteriogenation [2]. Recently it has been reported

$$S + HCo(CN)_5^{3-} \rightleftharpoons \cdot SH + Co(CN)_5^{3-} \tag{21}$$

$$\cdot SH + HCo(CN)_5^{3-} \rightarrow SH_2 + Co(CN)_5^{3-} \tag{22}$$

that styrene is reduced by the same route [38]. The significance of this mechanism is that no intermediate organo-cobalt compounds are formed.

In fact when $HCo(CN)_5^{3-}$ reacts with activated olefins of general formulae $CH_2=CRX$ and $XCR=CRX$, where R is H, alkyl or substituted alkyl and X is aryl, CN, COR, COOR, and COO^- a variety of products are formed depending on R and X. The olefin may be hydrogenated (as above), or isomerised, or a stable organocobalt(III) complex may be formed as in reaction (23) [39]. The structure of the

$$CH_2=CRX + HCo(CN)_5^{3-} \rightarrow CH_3-CRX-Co(CN)_5^{3-} \tag{23}$$

cobalt alkyls suggest the cobalt hydrogen bond is polarised Co^+-H^- and it was thought they were formed via a polar mechanism [29]. However, two recent studies indicate that alkyl formation, hydrogenation, isomerisation and deuterium exchange during these reactions all go via a similar non-polar mechanism.

The rate of reaction (23) for a variety of olefins obeys the formula $r = = k\,[Co(CN)_5H^{3-}]$ [olefine]. The value of k for the following olefins decreases in the order butadiene, styrene, 2-vinylpyridine > acrylonitrile > acrylic acid and k increases on substituting alkyl for an α-hydrogen atom. This indicates an electrophilic rather than a nucleophilic attack on the olefin [33]. The reaction of $HCo(CN)_5^{3-}$ and $DCo(CN)_5^{3-}$ with a variety of unsaturated acids gives cobalt alkyls, hydrogenated products, isomerisation products, and deuterium substituted products. The relative rates of disappearance of the acids due to all these reactions indicates that $HCo(CN)_5^{3-}$ is an electrophilic reagent [39]. It was also shown that alkyl formation was a stereospecific *cis* addition. Thus $DCo(CN)_5^{3-}$ gives a threo derivative with fumarate and an erythro derivative with maleate. The deuterium content of the isomerised acids and saturated acid products indicated that hydrogenation and isomerisation were not stereospecific and that reaction (21) is reversible [39]. These reactions are now thought to go via hydride attack at the β carbon atom (24). If a stable radical can be formed, $Co(CN)_5^{3-}$ is eliminated (25). The radical can react with more hydride to give saturated product (22) or can

$$HCo(CN)_5^{3-} \; + \; \underset{}{>}C\overset{\beta}{=}\overset{\alpha}{C}< \;\rightarrow\; \left[(CN)_5Co...H...\underset{}{>}\overset{\beta}{C}\underset{\cdot}{---}\overset{\alpha}{C}<\right]^{3-} \qquad (24)$$

$$\left[(CN)_5Co...H...\underset{}{>}\overset{\beta}{C}\underset{\cdot}{---}\overset{\alpha}{C}<\right]^{3-} \;\rightleftharpoons\; Co(CN)_5^{3-} + H{-}\underset{}{>}\overset{\beta}{C}\underset{\cdot}{---}\overset{\alpha}{C}< \qquad (25)$$

loose hydrogen via the reverse of (21).

If the radical is not sufficiently stable to have a finite existence the intermediate formed in (24) rapidly rearranges to give a net *cis* addition of Co—H to the activated olefin [33, 39], (reaction 23).

So far no reductions via $H_2Co(CN)_4^{3-}$ have been described. However hydrogenation of cyclopentadiene in ethanol by solutions containing a CN : Co ratio of 4 and 5 appear to go via different mechanisms, and $H_2Co(CN)_4^{3-}$ may be involved at the low CN : Co ratio [32].

Hydrogenations catalysed by $HCo(CN)_5^{3-}$ have been likened to heterogeneous hydrogenations. The homolytic cleavage of hydrogen involving more than one metal atom is similar to hydrogen activation by a surface. The transfer of hydrogen atoms to the substrate is also reminiscent of heterogeneous hydrogenation. Although the formal oxidation state of cobalt is three in $HCo(CN)_5^{3-}$, polarographic measurements indicate that the system is really a hydrogen atom stabilised by cobalt (II) [2a, 31]. It has also been pointed out that deuteriogenation catalysed by cobalt cyanide or palladium give products with very similar deuterium distributions [2a].

The cobalt cyanide system is an example of catalytic hydrogenation in which the metal only activates hydrogen (reaction (1) section 2). The rates of reactions (21), (22) [39] and (23) [33,39] are independent of cyanide ion concentration indicating that no cobalt olefin complexes are formed. Other systems which react via reaction (1) are hydrogenations catalysed by non transition metals, e.g. the lithium aluminium hydride catalysed hydrogenation of acetylenes to *trans* olefins and of conjugated dienes to monoenes in tetrahydrofuran at 190 °C and 100 atmospheres [40]. These and related metal hydride catalysed reactions [41] go via metal alkyls which are cleaved directly by hydrogen (26).

$$MH_n \xrightarrow{\text{olefin}} M\,(alkyl)_n \xrightarrow{H_2} MH_n + \text{paraffin} \qquad (26)$$

4.3. Cyano-amine-cobaltate systems

Solutions of cobaltous cyanide in the presence of various chelating amines and at cyanide : cobalt ratios < 5 form hydrides which are more reactive than $HCo(CN)_5^{3-}$. Suitable amines are α, α'-dipyridyl (dipyr), ethylenediamine, propylenediamine, diethylenetetramine, triethylenetetramine, glyoxime etc. Replacement of cyanide ligands by amine ligands weakens the cobalt-hydrogen bond [42-46]. Piringer has described the reaction of such solutions with hydrogen

[42], the catalysed exchange of hydrogen with deuterium oxide [43] and the isolation of some amine complexes [44]. Wymore [45] has found that the systems are more active hydrogenation catalysts for activated olefins than is $HCo(CN)_5^{3-}$ and has exemplified the system cobalt $+ 2$ dipyridyl $+$ varying amounts of cyanide. Spectroscopic studies indicate that a mixture of products such as $Co(dipyr)_2(CN)_2$, $Co(dipyr) (CN)_3^-$ and $Co(CN)_6^{3-}$ exists with the latter predominating at CN : Co of > 6. The solutions absorb hydrogen, 10% at a CN : Co ratio of 2 and 90% of the theoretical amount at a CN : Co ratio of > 6. The most active hydrogenation catalyst appears to be $HCo(dipyr) (CN)_3^-$ and the kinetics of hydrogenation and exchange studies indicate the process goes via reactions (27) to (29) and involver organo cobalt intermediates. This resembles butadiene hydrogenation rather than styrene hydrogenation catalysed by $HCo(CN)_5^{3-}$.

$$2Co + H_2 \rightleftharpoons 2CoH \qquad (27)$$

$$CoH + CH_2{=}CH{-}C_6H_5 \rightleftharpoons CoCH_2CH_2C_6H_5 \qquad (28)$$

$$CoCH_2CH_2C_6H_5 + HCo \rightarrow 2Co + CH_3CH_2C_6H_5 \qquad (29)$$

4.4. Nickel cyanide systems

Solutions of $Ni_2(CN)_6^{4-}$ can be prepared by reducing $Ni(CN)_4^{2-}$ with sodium amalgam [46] or sodium borohydride [47]. These nickel(I) solutions appear to be more active than cobalt(II) cyanide solutions and will cause hydrogenation and isomerisation of isolated double bonds [47] as well as acetylenes [48], conjugated dienes [46], and olefins conjugated with other functional groups [49]. Experiments using deuterium oxide show that the hydrogen for hydrogenation is derived from water[49]. The nickel(I) is reoxidised to nickel(II) during the reaction and excess amalgam or borohydride is necessary to reduce it to the low valent state. As the reaction rate is increased by cyanide ion it is suggested that monomeric species, e.g. $Ni(CN)_4^{3-}$ are formed, which react with substrate to give $[Ni(CN)_3(substrate)]^{2-}$. It is not clear if nickel hydrides are involved [49].

5. BI-METALLIC SYSTEMS

The outstanding feature of this group is that the catalyst contains a metal-metal bond between a transition metal, usually platinum, and a main group element, usually tin. They should be distinguished from catalysts prepared from a transition metal complex and a main group hydride or alkyl, which generates a transition metal hydride (see section 8) and may contain bridging ligands between the main group and the transition element.

5.1. Platinum halide - tin halide systems

The first communication on this system described the hydrogenation of ethylene to ethane, and acetylene to a mixture of ethylene and ethane by a methanolic solution of $H_2PtCl_6(8 \cdot 10^{-3}M)$ and $SnCl_2 2H_2O(8 \cdot 10^{-2}M)$ at 25° and 1 atmosphere total pressure [50]. Since then several publications have appeared and it seems that in methanol there is an optimum ratio of Sn : Pt of about 10 : 1; terminal olefins are isomerised much rapidly than they are hydrogenated; internal olefins are hardly reduced at all; and on increasing the temperature platinum metal tends to precipitate out [25b, 51-53]. For example styrene is slowly hydrogenated, pent-1-ene is converted to pentane (4%) and pent-2-ene (96%, mainly the *trans* isomer), *trans* pent-2-ene is only isomerised to the *cis* isomer [53] and methyl linolenate (methyl octadeca-9, 12, 15,-trienoate) is slowly converted to a mixture of non-conjugated dienes, conjugated dienes, and trienes at 30-40° [54]. Under more vigorous conditions, (65° and 35 atmospheres), methyl linolenate gives monoene (32%) and conjugated diene (64%) [54]. At still higher temperatures and pressures (65-100 °C and 40 atmospheres) polyunsaturated esters are reduced to saturated esters but the results are variable, precipitates are sometimes formed, and it is not certain that true homogeneous hydrogenation is occurring [55, 56].

Mixtures of platinum and tin halides appear to be more effective catalysts in solvents other than methanol. A comprehensive study has shown that the rate of hydrogenation of cyclohexene in isopropanol depends upon platinum: tin ratio, concentration of water, and the rate is increased further by addition of hydrogen halides or lithium halides. Some selected results are given in Table 2. The opti-

Table 2

HYDROGENATION OF CYCLOHEXENE BY PLATINUM-TIN CATALYSTS IN ISOPROPANOL [1]

No.	Conc H_2O	Added halide (mole l.$^{-1}$)	Pseudo 1st order Rate constant (sec.$^{-1} \cdot 10^4$)
1	0.15 M	—	0.05
2	0.15 M	HCl (0.15 M)	3.3
3 [2]	0.15 M	HCl (0.15 M)	0.26
4	1.00 M	HCl (1.00 M)	0.42
5	0.15 M	HCl (0.05 M)	0.63
6	0.15 M	HCl (0.10 M)	2.3
7	0.15 M	HCl (0.23 M)	3.3
8	0.15 M	LiCl (0.15 M)	0.86
9	0.15 M	HBr (0.15 M)	2.3
10	0.50 M	HBr (0.15 M)	4.2
11	1.00 M	HBr (0.15 M)	10.0
12	5.00 M	HBr (0.15 M)	0.29
13	0.15 M	LiBr (0.15 M)	2.0

[1] Temp. 25°; H_2 1 atmos.; cyclohexene 0.17 M; $H_2PtCl_6 6H_2O$ 8 $\cdot$ 10^{-3} M; Pt : Sn = 1:6.
[2] Solvent methanol.

mum platinum: tin ratio was 1:6 and the benefit of added halide is clear. It seems bromide is more effective than chloride but the optimum platinum : halide ratio depends on the nature of the halide, the concentration of water, the acidity of the solution, and clearly the system is very complicated. Other alcohols were also better than methanol, but the conditions for them were not optimised [57].

Other substrates can be reduced in the isopropanol system and the order of activity of olefins was terminal > 1,2 disubstituted > trisubstituted. Some selectivity has been attained, e.g. acetylenes can be reduced to olefins, and 1-carboxy-2-alkylcyclohexa-2,5-diene can be reduced to 1-carboxy-2-alkylcyclohex-2-ene. The rate of isomerisation relative to hydrogenation is suppressed by hydrochloric acid [57].

Carboxylic acids (except formic acid), ketones, ethers and nitrobenzene are also suitable solvents for platinum halide-tin halide catalysts [58]. In acetic acid the rate of hydrogenation of hex-1-ene is increased by sodium chloride or bromide and the solutions can be used up to 80°, above which temperature metal precipitates out. The optium Pt : Sn ratio is 1 : 5, some olefin isomerisation occurs but it is much less than in methanol solutions.

5.2. Platinum-tin-phosphine and related systems

McQuillin et al. have reported that under ambient conditions solutions of the complexes $PtHCl(PPh_3)_2$ and $PtX_2(PPh_3)_2$ where X = Cl, Br or I ($5 \cdot 10^{-3}$ M) in the mixed solvent system benzene (3 parts) and methanol (2 parts) catalyse the hydrogenation of oct-1-ene ($4.5 \cdot 10^{-2}$ M) in the presence of $SnCl_2 2H_2O$ (10^{-2} M). The reaction is slow (< 20% hydrogenation in 100 minutes) and is accompanied by extensive isomerisation [59]. Only the hydride catalyses the hydrogenation of norbornadiene to norbornane and of 2,5-dimethylhex-3-yn-2,5-diol to 2,5-dimethylhexane-2,5-diol under similar conditions [60].

Bailar et al. have used similar catalysts to hydrogenate and isomerise olefins [61, 62] polyolefins [61-63], and unsaturated esters, e.g. soya bean oil methyl ester [55], methyl linoleate [56, 62], and its conjugated isomers [62], methyl oleate [56], and methyl linolenate [54]. Generally they have used much more vigorous conditions, the standard ones usually employed being 90-105 °C, 35-40 atmospheres of hydrogen, platinum complex conc. $5 \cdot 10^{-3}$ M — $1.5 \cdot 10^{-2}$ M, $SnCl_2 2H_2O$: platinum ratio 10:1, solvent benzene (3 parts) and methanol (2 parts), and substrate: platinum ratio 15:1 for the unsaturated oils and up to 150:1 for the olefins.

They have examined the effect of several variables on the rate of hydrogenation and isomerisation using methyl linoleate, methyl oleate [56], cycloocta-1,3-diene and cycloocta-1,5-diene [63] as substrates. Hydrogenation only goes at an appreciable rate around 90 °C and there is no advantage in operating at pressures higher than 40 atmospheres. Isomerisation occurs under milder conditions and the isomerisation of cycloocta-1,5-diene has been studied at 50 °C and 60 °C when no hydrogenation occurs [64]. Likewise isomerisation of hexa-1,5-diene has been studied at 40° and 1 atmosphere of hydrogen [62].

5.2.1. Effect of neutral ligand

The activity of platinum complexes of general formula $PtCl_2L_2$ depended greatly on the nature of L. Under the standard conditions (see above) it was found that the order of activity (based on conversion of methyl linoleate [56] or cyclooctadiene [63]) was $L = P(OPh)_3 > AsPh_3 > SePh_2 > SPh_2 > PPh_3 > > SbPh_3 > PBu_3$. After hydrogenation it was noted that precipitates were present in the systems $L = AsPh_3$ and $SbPh_3$ and although it was demonstrated the residues were inactive it does add a complicating factor in studying these homogeneous systems. The homogeneity of the PPh_3 system was confirmed by distilling off the hydrogenation product and recycling the soluble complexes which remained [56]. It has been possible to isolate various triphenylphosphine complexes from the reaction mixture but no definite phosphite, arsine, selenide, or sulphide complexes were isolated, (see later).

5.2.2. Effect of main group halide

The effect of metal salts other than $SnCl_2 2H_2O$ was examined under the standard conditions, which were the optimum conditions for $SnCl_2$, using $PtCl_2(PPh_3)_2$ as the source of platinum. The order of decreasing activity was $SnCl_2 2H_2O > SnCl_4 > GeCl_2 + HCl > GeCl_4 > PbCl_2 + HCl > SiCl_4 > SiHCl_3$ [56] while $AlCl_3$, $SbCl_3$ and $Hg(OAc)_2$ were completely ineffective [55]. The bromides $PtBr_2(PPh_3)_2 + SnBr_2$ gave a less active catalyst than the corresponding chlorides and solids were precipitated [56]. McQuillin et al. also found that the combination $PtBr_2(PPh_3)_2$ and $SnCl_2 2H_2O$ was a less active catalyst than $PtCl_2(PPh_3)_2$ and $SnCl_2 2H_2O$ for the hydrogenation and isomerisation of oct-1-ene under ambient conditions [59].

5.2.3. Effect of solvent

As in the $PtCl_2 + SnCl_2$ system the solvent has a large effect on the liganded platinum-tin system [63]. Hydrogenation of 3-ethylidene-cyclohexene under the standard conditions (see above), using $PtCl_2(PPh_3)_2$ as the source of platinum, gives a solvent activity of methylene dichloride, ethylene dichloride (100%) > > acetone (83%) > tetrahydrofuran = benzene (3 parts) + methanol (2 parts) (77%) > pyridine (0%), the figures in parenthesis being the amount of monoene formed after 5 hours reaction at 90 °C and 35 atmospheres hydrogen pressure. Similar results were obtained with 4-vinylcyclohexene and 1,4,9-decatriene. In acetone complications arise as mesityl oxide is formed via an aldol condensation and some of this is reduced to methyl isobutyl ketone. Polar substrates are reduced very efficiently in the absence of other solvents, e.g. mesityl oxide is reduced very efficiently when it is its own solvent. As with the $PtCl_2 + SnCl_2$ system changes in the solvent alter the relative rates of isomerisation and hydrogenation as well as altering the rate of both processes [63].

5.2.4. SPECIFICITY OF PLATINUM-TIN-PHOSPHINE SYSTEMS

During hydrogenation of non-conjugated polyenes [63] and esters of long chain unsaturated acids [55, 56] it was found that *cis-trans* isomerisation and double bond migration to give the thermodynamically more stable conjugated systems was much faster than hydrogenation. It was concluded that these systems were catalysts for the selective hydrogenation of polyenes to monoenes. However, hydrogenation of a wider range of substrates, see Table 3, and a more detailed

Table 3

HYDROGENATION OF OLEFINS (0.9 M) BY $PtCl_2(PPh_3)_2(5 \cdot 10^{-3}$ M), $SnCl_22H_2O$ $(5 \cdot 10^{-4}$ M), SOLVENT BENZENE (3)-METHANOL (2), $H_2 = 34$ atms., temp. 90°

Olefin	% isomerisation	% hydrogenation	
		to paraffin	to olefin
Ethylene	0	100	—
propylene	0	34.0	—
but-1-ene	46.7	11.3	—
cis but-2-ene [2]	27.3 [2]	1.6	—
trans but-2-ene [2]	9.8 [2]	0.6	—
hex-1-ene	74.5	12.0	—
hexa-1,5-diene [1]	77.7	2.5	19.8
hexa-2,4-diene [1]	3.1	0	0.9
methyl linoleate [1]	24.6	0	5.1
trans, *trans* methyl octadeca-9,11-dienoate	9.4 [2]	0	9.2

[1] conc = 0.2 M
[2] *cis-trans* isomerisation only

analysis of the products, Fig. 1 and 2, show that double bonds are hydrogenated in the order terminal olefins $\geqslant$ internal olefins > conjugated olefins and terminal olefins are isomerised faster than they are hydrogenated [62, 63]. It has also been shown that conjugated dienes, particularly butadiene slow down hydrogenation and isomerisation of other olefins [62]. Thus it would appear that during hydrogenation of long chain polyolefinic esters hydrogenation and isomerisation of isolated double bonds occurs until conjugated dienes are formed. These are preferentially complexed to the catalyst and are slowly reduced, and thus they slow down reduction of the remaining isolated double bonds.

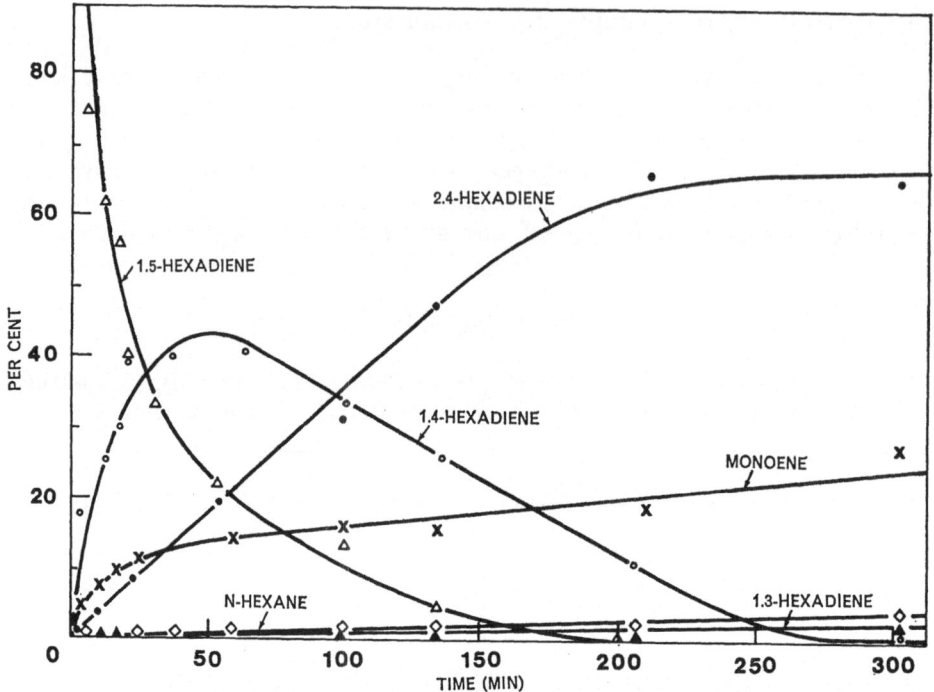

Figure 1. Catalytic hydrogenation of 1,5-hexadiene in benzene-methanol solvent under 34 atm of H_2 at 90°, using $PtCl_2(PPh_3)_2$-$SnCl_2$ catalyst.

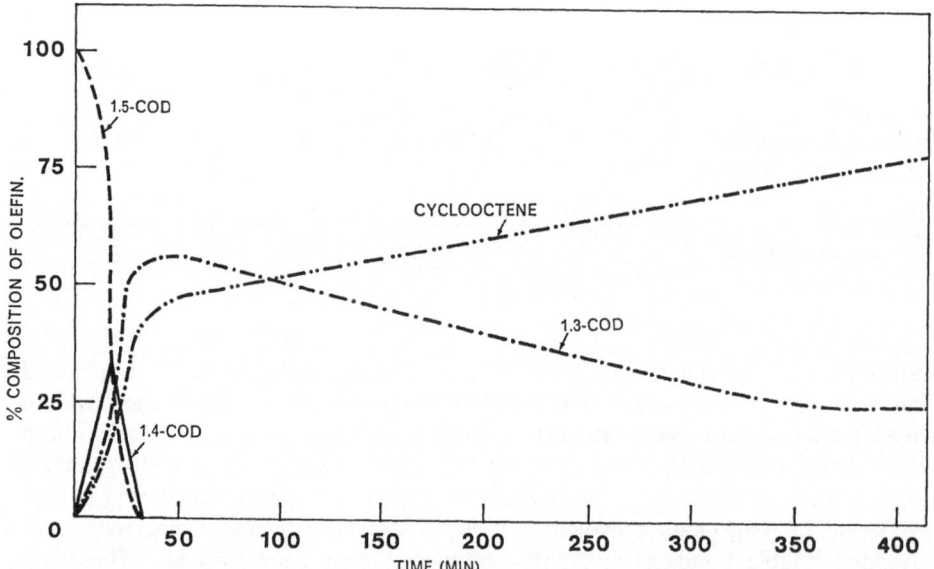

Figure 2. Catalytic hydrogenation of 1,5-cyclooctadiene in methylene chloride under 500 psi of hydrogen at 105°.

5.3. The mechanism of hydrogenation

It is not surprising that no detailed kinetic studies of such multi variable systems have yet appeared. However, a combination of the properties of platinum tin complexes, the comparative studies outlined above, studies on isomerisation of olefins and deuterium exchange reactions all point towards a general mechanism involving platinum hydrides and alkyls as intermediates.

5.3.1. PREPARATION OF PLATINUM TIN COMPLEXES

Mixtures of platinum halides and stannous chloride in vaious molar ratios in methanol give the anions $trans[PtCl_2(SnCl_3)_2]^{2-}$, $cis[PtCl_2(CnSl_3)_2]^{2-}$, [65] and $[Pt(SnCl_3)_5]^-$ [66] which can be isolated as salts using suitable cations. There is some doubt if $[Pt(SnCl_3)_5]^-$ exists to any great extent in solution [65] but an X-ray structure of one of its salts shows the platinum is surrounded by five tin atoms in a trigonal bipyramid arrangement, average platinum-tin bond length is 2.54Å [66].

In acetone $PtCl_2$ and $SnCl_2$ react to give the anion $[Pt_3Sn_8Cl_{20}]^{4-}$ thought to have structure I which reacts with cycloocta-1,5-diene to give a neutral complex $Pt_3Sn_2Cl_6(C_8H_{12})_3$ [67] now known to have structure II [68].

I   II

Triphenylphosphine complexes containing platinum-tin bonds can be prepared by adding PPh_3 to solutions of the ionic compounds described above [50, 65] or by adding $SnCl_2$ to platinum phosphine complexes $PtXCl(PP_3)_2$, where X = Cl [50, 65, 69], H [70] alkyl or aryl [70] and PR_3 is various phosphines (reaction (30))

$$PtClX(PR_3)_2 + SnCl_2 \rightarrow PtSnCl_3X(PR_3)_2 \qquad (30)$$

5.3.2. PROPERTIES OF PLATINUM TIN COMPLEXES

Infrared studies on $PtHSnCl_3(PR_3)_2$ and F^{19} n.m.r. studies on $PtRSnCl_3(PR_3)_2$ where R = meta or para C_6H_4F show that the $SnCl_3$ ligand has a strong $trans$ effect, is a weak σ donor and a strong π acceptor ligand [70]. The chemical properties of these complexes indicate that the $SnCl_3$ ligand assists hydrogenation in three ways:

(1) It increases the rate of coordinating olefins to platinum. This is demonstrated by the increased rate of formation of Zeise's salt (31) in the presence of 5 molar % of $SnCl_2$ [50, 71], and by the formation of the complexes

$$PtCl_4^{2-} + C_2H_4 \rightarrow PtCl_3C_2H_4^- + Cl^- \tag{31}$$

[(PtHSnCl$_3$(PPh$_3$)$_2$)$_2$ diene], where diene is cycloocta-1,5-diene or norbornadiene, and PtHSnCl$_3$(PPh$_3$)$_2$C$_8$H$_{12}$ and of some related complexes obtained by stirring mixtures of PtHCl(PPh$_3$)$_2$, $SnCl_2$ and the appropriate olefin [63].

(2) It stabilises the hydride ligand and increases the rate of hydride formation. This is shown by the ease of absorption of hydrogen by solutions of PtCl$_2$(PEt$_3$)$_2$ containing $SnCl_2$ compared to PtCl$_2$(PEt$_3$)$_2$ alone, and by reaction (32) which can be followed at 3 atmospheres and 30° [66].

$$[Pt(SnCl_3)_5]^{3-} + H_2 \rightleftharpoons [PtH(SnCl_3)_4]^{3-} + SnCl_2 + H^+ + Cl^- \tag{32}$$

(3) It increases the rate of formation of alkyl complexes. Thus the equilibrium (33) is set up at 25° and 1 atmosphere ethylene pressure in the presence of $SnCl_2$ [25b] but requires high ethylene pressures in the absence of $SnCl_2$ [72]. It is not clear if this is due to an increase in rate of

$$PtHCl(PEt_3)_2 + C_2H_4 \rightleftharpoons PtC_2H_5Cl(PEt_3)_2 \tag{33}$$

co-ordination of olefin or an increase in rate of hydrogen transfer from platinum to olefin.

5.3.3. POSSIBLE REACTIONS OCCURRING IN HYDROGENATION

There is abundant evidence from the reactions of platinum-tin complexes that activation of hydrogen occurs via reaction (34). (Compare reaction (6) section 2.2). In confirmation of this, the system

$$Pt^{2+} + H_2 \rightleftharpoons PtH^+ + H^+ \tag{34}$$

PtHCl(PPh$_3$)$_2$ + 2SnCl$_2$ is a better catalyst than PtCl$_2$(PPh$_3$)$_2$ + 2SnCl$_2$, but on adding a basic buffer to the dichloride system its activity becomes equal to that of the hydride, because the reverse of reaction (35) is suppressed [59].

$$PtClSnCl_3(PPh_3)_2 + H_2 \rightleftharpoons H^+ + Cl^- + PtHSnCl_3(PPh_3)_2 \tag{35}$$

Under much more vigorous conditions and in the absence of hydrogen gas hydrogen can be abstracted from some solvents (see section 10.2).

Cramer and Lindsey [25b] have proved conclusively that the system H$_2$PtCl$_6$–SnCl$_2$-H$_2$ in methanol at 25° catalyses the isomerisation and hydrogenation of butenes via the reverse alkyl mechanism or stepwise addition of 2 hydrogen atoms (36) and (37) (compare reactions (11) and (12) section 2.3). The extensive isomerisation occurring in all the platinum-tin catalysed hydrogenations suggests that stepwise addition of hydrogen always occurs in these systems. In deuterated methanol, CH_3OD, deuterium

$$PtH + (olefin) \rightleftharpoons PtH(olefin) \tag{36}$$

$$PtH(olefin) \rightleftharpoons Pt \text{ alkyl} \tag{37}$$

is incorporated into the residual olefin and hydrogenated product [25b, 73]. It has been suggested that this is because the second hydrogen is added via a hydrolysis step (38) (see reaction (14) section 2.3). Thus hydrogenation of

$$Pt^{2+} \text{ alkyl} + DCl \rightarrow Pt^{2+}Cl + d_1 \text{ paraffin} \tag{38}$$

ethylene in deuteromethanol or deuteriogenation of ethylene in methanol gives d_1 ethane as the major product. [73]. However, incorporation of deuterium in the hydrogen deuterated methanol experiment may be because a preformed hydride exchanges with deuterated solvent (39), a deuterated alkyl is formed via (36)

$$Pt^{2+}H + D^+ \rightleftharpoons Pt^{2+}D + H^+ \tag{39}$$

and (37) and then hydrogenolysis occurs (40). Reaction (40) could go via oxidative addition (8) or heterolysis (6). Reaction (39) has been well

$$Pt^{2+} d_1 \text{ alkyl} \xrightarrow{H_2} PtH + d_1 \text{ paraffin} \tag{40}$$

established by Cramer and Lindsey and along with reverse alkyl formation accounts for incorporation of deuterium into the olefin [25b]. In platinum systems reaction (39) can go via a platinum(IV) intermediate as proved by kinetic studies of reaction (41) [74], or via reduction to platinum(0) followed by

$$PtHCl(PEt_3)_2 + DCl \rightleftharpoons PtHDCl_2(PEt_3)_2 \rightleftharpoons PtDCl(PEt_3)_2 + HCl \tag{41}$$

reoxidation to platinum(II) [75]. It is not known how it goes in the platinum-tin systems. Although hydrolysis may occur in the second hydrogen addition in hydroxylated solvents (38), it is unlikely to occur in methylene dichloride when hydrogenolysis will be the major pathway (40).

In reactions (34) and (36)-(40) ligands other than the reacting ones have been omitted because it is not known what they are. They will be a mixture of solvent, halide, $SnCl_3$ and neutral ligands, olefin or added phosphine etc, depending on the composition and temperature of the catalyst solution. It is significant that complexes of general formula $[(PtHSnCl_3(PPh_3)_2)_2 \text{ diene}]$ and related complexes have been isolated from crude hydrogenation products [63].

5.4. Miscellaneous metal-tin-systems

Mixtures of Na_2IrCl_6 and $SnCl_2$ (optimum ratio Sn : Ir = 3.5) are catalysts for the slow hydrogenation of norbornadiene at room temperature and pressure. Under similar conditions mixtures of $PdI_2(PPh_3)_2(5 \cdot 10^{-3}$ M) and $SnCl_22H_2O(10^{-2}$ M) catalyse the slow hydrogenation of oct-1-ene in benzene methanol mixtures. The corresponding bromide and chloride are less active and cause isomerisation [51, 60]. Under more vigorous conditions 13-40 atmospheres of hydrogen and 60-90° mixtures of $PdCl_2(PPh_3)_2$ and $SnCl_22H_2O$(ratio 1 : 7) catalyse the hydrogenation of soya bean oil and other unsaturated esters. Hydrogenation appears to cease at the monoene stage and extensive isomerisation occurs. The corresponding arsine, stibine, etc. complexes are less good and the catalysts decompose. A feature of these reactions is that hydrogen for the hydrogenation can be derived from the solvent [154] (see section 10.2). No mechanistic studies have appeared on these systems.

6. CATALYSTS STABILISED BY TERTIARY PHOSPHINES

It is in this area where the most significant advances have been made in the last five years. Some of the most active homogeneous hydrogenation catalysts are to be found in this class of compounds and they are amenable to study by a variety of techniques. Studies by Osborn and Wilkinson and their co-workers on rhodium and ruthenium complexes are of particular importance. Although some minor aspects of their work have been shown to be incorrect and other aspects have been expanded, their work forms an excellent basis for further studies.

6.1. Chlorotris(triphenylphosphine)rhodium(I) and related systems

This system has proved exciting for the following reasons

(1) It is easy to use.

(2) When used correctly under mild conditions its activity on a metal weight basis is greater than many well known heterogeneous catalysts [18].

(3) It is highly specific for olefins, polyenes, conjugated polyenes (but not aromatics) and acetylenes, and shows some selectivity within this narrow range of substrates.

(4) It lends itself to detailed mechanistic studies. It has proved possible to isolate and study some of the intermediates which occur in the catalytic cycle and show how various inhibitors and poisons operate at a molecular level.

6.1.1. PREPARATION OF CATALYSTS

Several groups working independently appear to have synthesised RhCl(PPh₃)₃ during the period 1964-1965 [76-80], and some quickly noted it was an excellent hydrogenation catalyst at room temperature and pressure [76, 77, 81]. Previously RhCl(PHPh₂)₃ had been prepared from [RhCl(C₂H₄)₂]₂ via reaction (42) [82] and various phosphite analogues RhX(P(OR)₃)₃ had been known a long time [83].

Refluxing an ethanolic solution of RhCl₃3H₂O with six molar equivalents of triphenylphosphine gives RhCl(PPh₃)₃ in almost quantitative yield [18], the phosphine being the reducing agent [18, 79]. The corresponding bromide and iodide can be prepared similarly using the appropriate rhodium halide [77] or from RhCl₃3H₂O and an excess of lithium halide [18]. Only a few phosphine analogues can be prepared by this route, although some related complexes [RhCl(PR₃)₂]ₙ have been prepared. Reaction of RhCl₃3H₂O with other phosphines, arsines and stibines gives a variety of complexes of rhodium (III) [79, 84], rhodium (II), and rhodium (I) [85], depending on the ligand and the experimental conditions.

Table 4

PREPARATION OF RhXL₃, X = HALIDE, L = MONODENTATE PHOSPHINE, ARSINE OR STIBINE

Complex	Method of Synthesis	M.P. °C	Ref.
RhCl(P(C₆H₅)₃)₃	RhX₃ + excess L	157-158	[18]
RhBr(P(C₆H₅)₃)₃	RhX₃ + excess L	133-134	[18]
RhI(P(C₆H₅)₃)₃	RhX₃ + excess L	118-120	[18]
RhCl(P(C₆H₄-F-p)₃)₃	RhX₃ + excess L	——	[87]
Rh₂Cl₂(P(C₆F₅)₃)₄	RhX₃ + excess L	216-218	[106a]
Rh₂Cl₂(P(C₆H₅)(C₆F₅)₂)₄	RhX₃ + excess L	219-221	[106a]
Rh₂Cl₂(P(C₆H₅)₂(C₆F₅))₄	RhX₃ + excess L	175-185	[106a]
RhCl(As(C₆H₅)₃)₃	Rh₂Cl₂(C₂H₄)₄ + L	150-155	[93]
RhCl(Sb(C₆H₅)₃)₃	Rh₂Cl₂(C₂H₄)₄ + L	170 (d)	[93]
RhCl(PH(C₆H₅)₂)₃	Rh₂Cl₂(C₂H₄)₄ + L	173-175	[82]
RhCl(P(C₆H₅)(NC₅H₁₀)₂)₃	Rh₂Cl₂(C₂H₄)₄ + L	82	[89]
RhCl(P(C₆H₅)(NC₁₁H₈O)₂)₃C₆H₆	Rh₂Cl₂(C₂H₄)₄ + L	104-106	[89]
RhCl(P(C₆H₅)₂CH₃)₃	Rh₂Cl₂(C₂H₄)₄ + L	153	[92]
RhCl(PC₂H₅(C₆H₅)₂)₃	Rh₂Cl₂(C₈H₁₄)₂ + L	102-106	[87]
RhCl(P(C₆H₄-Cl-p)₃)₃	RhCl(C₄H₆)₂ + L	133-135	[77]
RhCl(P(C₆H₄-CH₃-p)₃)₃	RhCl(C₄H₆)₂ + L	———	[106b]

A better general procedure for preparing complexes RhXL₃ and RhXL₂, where X is halide and L is a tertiary phosphine, arsine or stibine, is to react a rhodium(I)

olefin complex with the appropriate amount of ligand, (42) and (43). This method is extremely important for synthesising catalysts in

$$(RhCl(olefin)_2)_2 + 6L \rightarrow 2RhClL_3 + 4olefin \qquad (42)$$

$$(RhCl(olefin)_2)_2 + 4L \rightarrow 2RhClL_2 + 4olefin \qquad (43)$$

situ without isolating the appropriate complex and has the advantage that the ratio of ligand to rhodium can be varied at will (see below). Suitable olefin complexes which have been used and some of the complexes $RhXL_3$ and $RhXL_2$ which have been characterised are summarised in Table 4.

6.1.2. Mechanism of hydrogenation

As a result of detailed studies Osborn, Wilkinson, et al [18, 86, 87] have proposed the mechanism for hydrogenation shown in Scheme 1. It is worthwhile

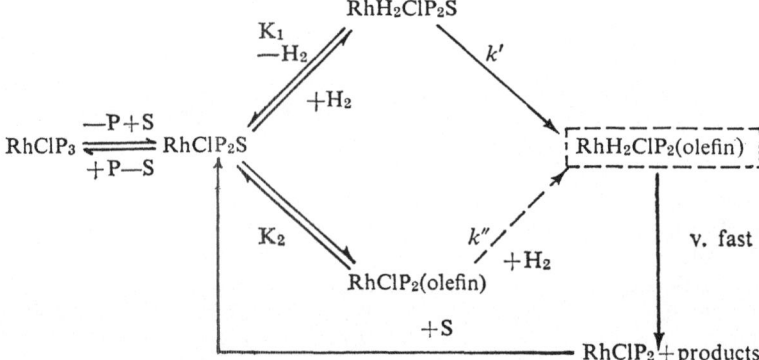

Scheme 1

assessing the data this is based upon and to discuss other relevant publications.

6.1.2.1. *Formation of a co-ordinatively unsaturated intermediate*

The evidence that monomeric $RhCl(PPh_3)_2$ and its analogues are important intermediates is overwhelming. It is often written as $RhCl(PPh_3)_2S$ where S is a solvent molecule easily replaced by phosphine or olefin. Studies on catalysts prepared in situ by treating rhodium(I) olefin complexes with varying amount of phosphine or arsine [87-90] show that the rate of hydrogenation depends on the ligand : rhodium ratio. The optimum ratio is slightly greater than 2 : 1 for trial-

kyl, triaryl or mixed alkylaryl phosphines and is quite sharpley defined-see Fig. 3
[87]. For some phenyl diamino phosphines and diphenyl amino phosphines the
optimum ratio of phosphine : rhodium is again between 2 : 1 and 3 : 1, but the
rate of hydrogenation is only marginally below the maximum at ratios as high as

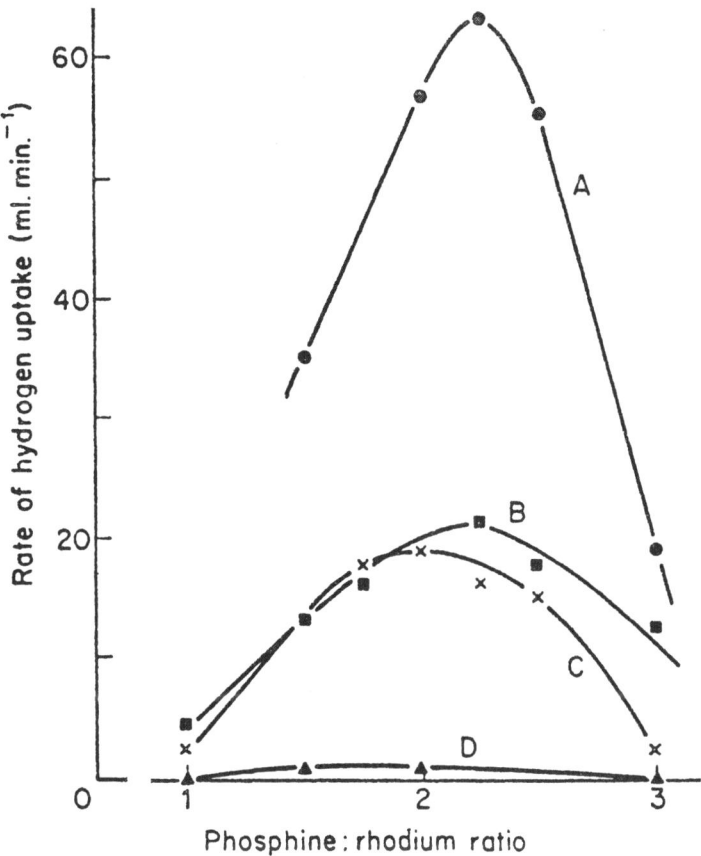

Figure 3. Hydrogenation of cyclohexene (0.6M in benzene) at 25° at catalyst concentration 1.25mM
with various phosphine: rhodium ratios; A, $(MeOC_6H_4)_3P$; B, Ph_3P; C, Ph_2EtP; D, PEt_3.

15 : 1 [89], see Table 5. This indicates that the position of the equilibrium (44)
concerning the coordinatively unsaturated intermediate depends greatly on the
ligand L.

$$RhClL_3 \rightleftharpoons RhClL_2 + L \qquad (44)$$

Solutions of $RhCl(PPh_3)_3$ and its analogues [87] are non-conducting, but
molecular weight measurements in benzene, chloroform and methylene dichloride
are approx. one half of theory in the concentration range $3 \cdot 10^{-3}$ M to $2 \cdot 10^{-2}$ M,
which is the range used for kinetic studies (see later). Furthermore attempted

Table 5

RATE (moles $\cdot$ l.$^{-1}$ $\cdot$ min.$^{-1}$) OF HYDROGENATION OF STYRENE BY THE SYSTEM $Rh_2Cl_2(C_2H_4)_4$
$+$ PHOSPHINE AT 40°, 1 ATM. OF H_2

Phosphine/rhodium ratio	2	3	4	5	15
Phosphine					
phenyl-dipiperidyl	1.12	1.11			1.10
phenyl-bisdimethylamino	0.658	0.605		0.584	
phenyl-bisdiethylamino	0.598	0.576		0.325	
triphenyl	0.437	0.392	0.051		

recrystallisation of $RhCl(PPh_3)_3$ leads to the dimeric compound $Rh_2Cl_2(PPh_3)_4$, which is converted back to $RhCl(PPh_3)_3$ with excess phosphine [18].

All the evidence listed above supports reaction (44) and indicates $RhCl(PPh_3)_3$ is completely dissociated in solution [18]. However, recent n.m.r. evidence does not support this. Phosphorus nuclear magnetic resonance studies on 10^{-3} M solutions of $RhCl(PPh_3)_3$ in benzene show that it is square planar, mutually *cis* and *trans* phosphine ligands can be distinguished, and there appears to be less than 5% dissociation to give free phosphine [91]. The ^{1}H n.m.r. spectrum of the methyl group of $RhCl(PMePh_2)_3$ indicates this complex is also square planar and relatively undissociated [92]. Further insight into these systems has been obtained by studying the ^{1}H n.m.r. spectrum of the methyl protons in mixtures of $RhCl(PPh_3)_3$ (0.1M) and tri-p-tolylphosphine (0.1M-1.0M) between —35° and 40 °C. There is a statistical distribution of liganded and free triphenylphosphine and tri-p-tolylphosphine within two minutes, so the tri-p-tolylphosphine can be regarded as labelled triphenylphosphine. It can be demonstrated that in the square planar complex, $RhClP_3$, the three phosphines exchange between *cis* and *trans* positions by an intramolecular process faster than the intermolecular exchange between liganded and non liganded phosphine. The authors suggest that in the complexes $RhClL_3$ one of the phosphines is loosely held in an outer co-ordination sphere, rather than being completely dissociated, but they are unable to explain the discrepancy between the n.m.r. studies and the molecular weight measurements which they have confirmed [91].

6.1.2.2. *The activation of hydrogen*

Solutions of $RhCl(PPh_3)_3$ are deep red. On bubbling hydrogen through they turn yellow-orange and the ^{1}H n.m.r. spectra show high field lines due to a rhodium hydride of structure (III) where S is solvent. A lot of the expected fine structure due to coupling between H-Rh, H-H and H-*cis* P is not seen due to line broadening caused by a trace of a paramagnetic impurity and possibly ligand exchange. The red colour is restored and the high field lines are lost on blowing nitrogen through the hydride solution and this reversible process can be repeated [18].

A variety of dihydrides can be isolated from solutions of $RhXL_3$ under hydrogen depending on the ligand, experimental conditions and the procedure used. Wilkinson et al obtained solids approximating to $RhH_2Cl(PPh_3)_2 S_x$ where S was a variety of solvents, e.g. $(RhH_2Cl(PPh_3)_2)_2CH_2Cl_2$ where methylene dichloride is a bridging ligand, and $RhH_2Cl(PPh_3)_2C_5H_5N$ [18]. The triphenylarsine analogue behaves similarly but no hydrides were isolated from $RhCl(SbPh_3)_3$ [93], while $RhCl(PEtPh_2)_3$ gave high yields of $RhH_2Cl(PEtPh_2)_3$ [87]. We obtained a solid analysing for $RhH_2Cl(PPh_3)_3$ rather than $RhH_2Cl(PPh_3)_2$ on treating $RhCl(PPh_3)_3$ in benzene with hydrogen but in acetonitrile $RhH_2Cl(PPh_3)_2(CH_3CN)_2$ was obtained [94]. In the presence of excess phosphine $RhH_2Cl(PPh_3)_3$ is obtained in high yield [79, 95] and the analogues $RhH_2X(PPh_3)_3$ where X is Br, I or $SnCl_3$ can be obtained similarly [94]. Hydrogen absorption (75-85% of theory) is complete within a few minutes when solutions of $RhCl(PR_3)_3$ are stirred under hydrogen (1 atm.) or solid is dropped into a stirred solvent [94]. The complex $RhH_2Cl(PPh_3)_3$ is isomorphous [94] with the iridium analogue known to have structure IV [96] and its [1]H n.m.r. spectrum would be similar to III. In view of

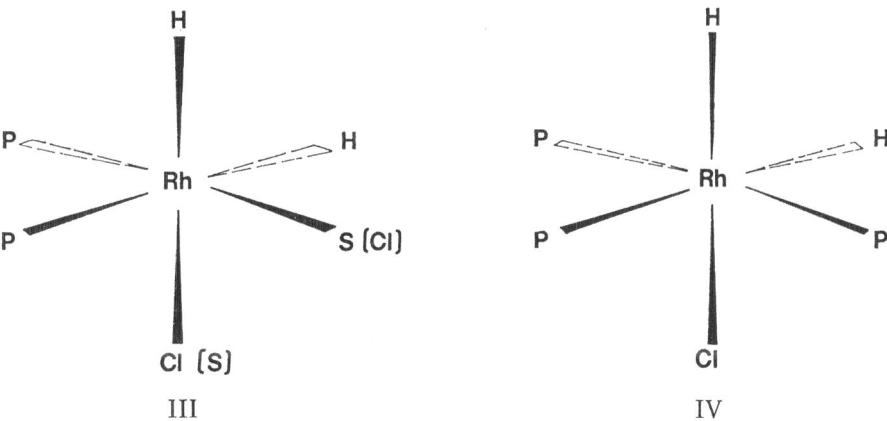

III   IV

the small degree of dissociation of $RhCl(PPh_3)_3$(see 6.1.2.1), solutions of the dihydride are best regarded as a mixture of species as represented by (45), although other species may be present.

$$RhCl(PPh_3)_3 \underset{-H_2}{\overset{+H_2}{\rightleftharpoons}} RhH_2Cl(PPh_3)_3$$
$$+ PPh_3 \updownarrow - PPh_3 \qquad\qquad + PPh_3 \updownarrow - PPh_3 \qquad (45)$$
$$RhCl(PPh_3)_2 \underset{-H_2}{\overset{+H_2}{\rightleftharpoons}} RhH_2Cl(PPh_3)_2 .$$

6.1.2.3 *The reaction of* $RhCl(PPh_3)_3$ *with olefins*

The complex reacts with ethylene to give $RhCl(PPh_3)_2C_2H_4$, a yellow compound characterised by an absorption band at 416 mμ. The equilibrium constant K_2, (scheme 1) of reaction (46) has been estimated at 100 mol^{-1}litre and

$$RhCl(PPh_3)_2S + C_2H_4 \rightleftharpoons RhCl(PPh_3)_2C_2H_4 + S \qquad (46)$$

20 mol^{-1}litre for the bromo analogue. Other simple olefins have a value of K_2 less than 0.05 mol^{-1} litre as no characteristic band in the 410 mμ region has been detected. The ^{1}H n.m.r. shows that there is rapid exchange between liganded and free ethylene at -50° and the average life time of ethylene on rhodium is less than 10^{-2} secs. [18].

6.1.2.4. *The overall kinetics of hydrogenation*

Osborn and Wilkinson and co-workers [18, 86, 87, 97] have conducted extensive kinetic studies in which they have studied the variables, temperature (15-30 ºC), hydrogen pressure (20-50 cm. of mercury), rhodium concentration (0.625 · 10^{-3} M to 2.5 · 10^{-3} M), olefine concentration (0.5 to 2.5M), olefine structure, variation of halide and phosphine ligand in the catalyst, the ratio of rhodium to phosphine, and solvent. The rate constants and thermodynamics data they derived are summarised in Table 6.

Table 6

RATE CONSTANTS AND THERMODYNAMIC DATA FOR THE HYDROGENATION OF OLEFINS.
CATALYST CONC. 1.25 · 10^{-3} M RhCl(PPh$_3$)$_3$
SOLVENT BENZENE UNLESS OTHERWISE STATED.

Olefin	k'_{298} ($\times 10^2$) [1] l.mol.$^{-1}$ sec.$^{-1}$	ΔH^* kcal. mol.$^{-1}$	ΔS^* e.u.	Ref.
Cyclopentene	34.3	16.7 ($\pm$ 0.4)	− 4.7	[86]
Cyclohexene	31.6	18.6 ($\pm$ 0.4)	1.3	[86]
	(25.0) [1]			[87]
	(28.0) [2]			[105]
Cyclohexene [2]	34.0	25.8	18.6	[87]
Cyclohexene [3]	29.6	16.6	− 5.3	[86]
Cyclohexene [4]	15.0	23.3	12.9	[18, 86]
Cyclohexene [4,5]	—	19.4	4.1	[86]
Cycloheptene	21.8	21.7 ($\pm$ 0.6)	11.0	[86]
Hex-1-ene	29.1	18.6 ($\pm$ 0.4)	1.1	[86]
Dodec-1-ene	34.3	16.9 ($\pm$ 0.4)	− 4.1	[86]
1-Methylcyclohexene	0.6	12.7 ($\pm$ 1.6)	−26.0	[86]
cis Pent-2-ene	23.2	12.6 ($\pm$ 0.7)	−20.2	[86]
2-Methylpent-1-ene	26.6	12.7 ($\pm$ 0.6)	−18.5	[86]
Styrene	93.0	11.1 ($\pm$ 0.5)	−21.5	[86]
cis 4-Methylpent-2-ene	9.9	16.1 ($\pm$ 0.8)	− 9.3	[86]
trans 4-Methylpent-2-ene	1.8	11.6 ($\pm$ 1.3)	−27.6	[86]

[1] Accuracy is $\pm$ 5.0% for values of $k' > 20 \cdot 10^{-2}$ l. mol.$^{-1}$ sec.$^{-1}$ but the errors are greater for the lower k' values.
[2] Catalyst is RhCl(P(C$_6$H$_4$–p–OCH$_3$)$_3$)$_3$ prep. in situ.
[3] Solvent is 1 : 1 hexane, methyl ethyl ketone
[4] Solvent is 1 : 1 hexane, benzene.
[5] Catalyst is RhBr(PPh$_3$)$_3$.

They have used a constant volume apparatus for these and related studies and calculated rates of hydrogenation from tangents to a plot of hydrogen partial pressure against time. Where possible the rate was measured as the pressure dropped below 50 cm. of mercury and only 1% of substrate had been reduced. In some experiments preformed catalyst $RhCl(PR_3)_3$ was used and in others the catalyst was prepared in situ from an olefin complex and phosphine, and it was shown these two methods gave identical results [87].

Using $RhCl(PPh_3)_3$ as a catalyst the full rate equation derived from scheme 1 is (47)

$$r = \frac{(K_1 k' + K_2 k'') [H_2] [\text{olefin}] [Rh]}{1 + K_1[H_2] + K_2[\text{olefin}]} \quad (47)$$

where a [ ] denotes a concentration, K_1 and K_2 are the equilibrium constants for reversible reactions and k' and k'' are the rate constants for the individual steps indicated in scheme 1 [18]. It is clear fron the preceding two sections that $K_1 > K_2$. Wilkinson et al. found that the only simple olefin complex which can be isolated, $RhCl(PPh_3)_2C_2H_4$, is not a hydrogenation catalyst (this is incorrect [95] see 6.1.2.7.) and concluded k'' is very small. Thus $K_2 k''$ is very small compared to $K_1 k'$ and can be ignored. The simplified rate expression can be transposed to (48) and (49).

$$\frac{1}{r} = \frac{C_1}{[\text{olefin}]} + C_2 \quad (48)$$

where $C_1 = \dfrac{1 + K_1[H_2]}{k' K_1 [H_2][Rh]}$ and $C_2 = \dfrac{K_2}{k' K_1 [H_2][Rh]}$

$$\frac{1}{r} = \frac{C_3}{[H_2]} + C_4 \quad (49)$$

where $C_3 = \dfrac{1 + K_2[\text{olefin}]}{k' K_1 [\text{olefin}][Rh]}$ and $C_4 = \dfrac{1}{k' [\text{olefin}][Rh]}$

The physical significance of the assumption $K_1 k' > K_2 k''$ is that hydrogenation proceeds exclusively via reaction of an olefin with a rhodium hydride and not via reaction of a rhodium olefin complex with hydrogen. This has been challenged [95] and will be discussed in section 6.1.2.7.

In favour of hydrogenation going via the rhodium hydride route of scheme 1, it was found that plots of $1/r$ against $1/[\text{olefin}]$ and $1/r$ against $1/[H_2]$ gave straight lines as required by (48) and (49) respectively. The rate constants for the slow rate determining step k', see Table 6, were calculated from C_1 on the assumption that $K_1[H_2]$ was much greater than one. The ratio C_1/C_2 gives K_2/K_1 as $1.6 \cdot 10^{-3}$ mole, which is probably lower than expected [18].

The main defficiency of the scheme is that a plot of catalyst concentration against rate is not a straight line passing through the origin as demanded by equation (47) [18]. Such plots are shown in Fig. 4 for two rhodium : phosphine ratios and although there is a linear dependence over a narrow concentration range it is clear the catalyst gets more active at lower concentrations [97]. Wilkinson has suggested that at low concentrations further dissociation occurs according to (50), and that RhCl(PPh3) is a more active

$$\text{RhCl(PPh}_3)_2 \; \rightleftharpoons \; \text{RhCl(PPh}_3) \; + \; \text{PPh}_3 \tag{50}$$

catalytic intermediate than RhCl(PPh3)2 [97]. In favour of further dissociation we have observed that solutions of RhCl(PPh3)3 (5 · 10⁻³ M) in tetrahydrofuran slowly deposit low yields of a solid analysing for [RhCl(PPh3)1.5C4H8O]$_n$ when kept under nitrogen [94].

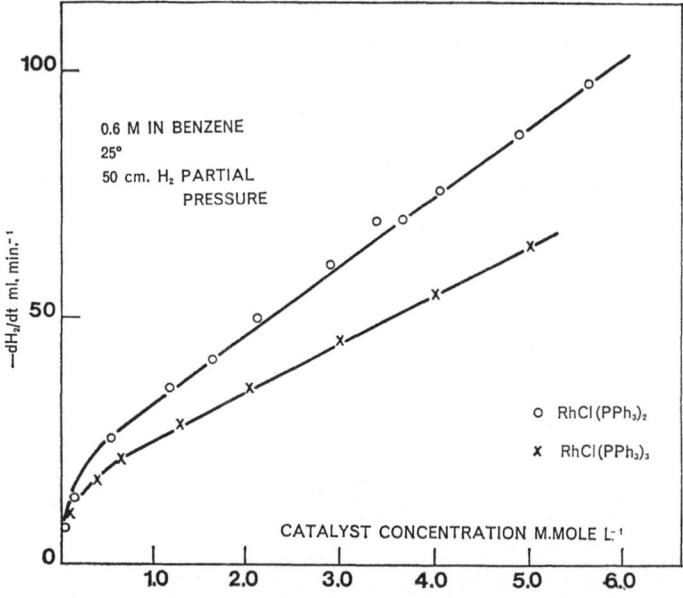

Figure 4. Hydrogenation of hex-1-ene.

Under rather different reaction conditions (amino phosphine ligands, styrene (4 parts), benzene (1 part) at 40 °C, and 1 atm. of hydrogen) Stern et al. have found the dependence of rate on catalyst concentration varies between zero and first order as the concentration decreases [89]. Thus it appears that a plot of rate against catalyst concentration is truly a curve and the order on rhodium decreases as the concentration gets higher due to incomplete dissociation of RhCl(PPh3)3 (see 6.1.2.1).

No other detailed kinetics have been reported but a value of k' for hydrogenation of cyclohexene in benzene agrees well with those quoted by Wilkinson, see Table 6.

In a study on hydrogenation of substituted cyclohexenes (0.7M) catalysed by RhCl(PPh₃)₃ (2.2 · 10⁻³ M) in benzene (3 parts) and ethanol (1 part) at 25° and 1 atm. of hydrogen the rate was zero order on olefin concentration [98] and a similar dependency was found under similar conditions in tetrahydrofuran [94] but the changes in experimental conditions would account for these changes.

A deuterium effect has been observed for reduction of cyclohexene and *cis* or *trans* 4-methylpent-2-ene where k'_H/k'_D falls between 1.0 and 0.8. This is now regarded as a secondary isotope effect and is not understood at the present time [87].

6.1.2.5. *The effect of phosphine on hydrogenation*

The effect of phosphine structure is very large, Fig. 3 shows that the order of activity is P(aryl)₃ > P(aryl)₂alkyl > P(alkyl)₃, and this has been confirmed by others [88-90]. In *para*-substituted aryl phosphines the order of activity is P(C₆H₄-*p*-OCH₃)₃ > P(C₆H₄-*p*-CH₃)₃ > P(C₆H₅)₃ > P(C₆H₄-*p*-F)₃ > P(C₆H₄-*p*--Cl)₃ [87, 90] and in dialkylamino phosphines the order appears to be P(C₆H₅)(NR₂)₂ ≈ P(C₆H₅)₂NR₂ > P(C₆H₅)₃ > P(NR₂)₃ [88, 89]. The ortho-substituted aryl phosphines are poor catalysts and although AsPh₃ is not active at 25° [97] it is active at 40° [88].

No doubt the phosphine ligand is important in determining the position of equilibrium (44) or more particularly the concentration of the active species RhH₂Cl(PR₃)₂ in the reaction represented by (45). All the good catalysts appear to contain an aryl group in the ligand and Wilkinson has suggested that these aryl groups act as electron sinks which can donate to, or withdraw electrons away from the metal at key stages in the hydrogenation process [87].

The order of activity within the *para*-substituted aryl phosphines parallels their order of basicity, but basicity cannot be the only important effect as replacing aryl groups by alkyl groups lowers the activity. Rough qualitative experiments show that RhH₂Cl(PEt Ph₂)₃ is more stable than its triphenylphosphine analogue and it may be that further alkyl groups stabilise the rhodium(III) state so much that hydrogen transfer within the complex RhH₂Cl(PR₃)₂(olefin) becomes rate determining [87].

To date the most active catalysts appear to be obtained from rhodium(I) olefin complexes plus 2 to 2.5 moles of the ligands P(C₆H₄-*p*-OCH₃)₃ [87, 90] or PC₆H₅(NC₅H₁₀)₂ [88, 89] per rhodium atom.

6.1.2.6. *Effect of substrate and solvent on hydrogenation*

The rate determining step was originally thought to be hydrogen transfer within the complex RhH₂Cl(PPh₃)₂(olefin) [18]. Later work showed that the nature of the substrate had a large effect on the activation parameters, see Table 6, and formation of RhH₂Cl(PPh₃)₂(olefin) is now considered rate determining (k' in scheme 1) [87]. The overall kinetics is the same for either rate determining step. As the olefin gets more substituted the rate decreases, the entropy of activation goes to more negative values, as expected on steric grounds, while the enthalpy of activation decreases. Styrene seems not typical of alpha olefine as the entropy of activation and the enthalpy of activation are very low. Substituted styrenes with either electron donating or electron withdrawing groups in the para position are hydrogenated faster than styrene itself. Clearly the aromatic ring introduces complicating factors [87]. The results in different solvents indicate that polar intermediates are favoured, but as the metal, hydride, and halide ligands can all be solvated this will be a very complicated effect.

The rate of hydrogenation of 1-methylcyclohexene in various solvents decreases in the order acetophenone > benzene-ethanol > nitrobenzene > cyclohexanone > ethylene dichloride > methylene dichloride > chlorobenzene while no reduction occurs in chloroform or benzonitrile [98].

Candlin and Oldham have provided further evidence that the nature of the substrate and the solvent is important [95]. They have quoted a list of relative rates of hydrogenation of various substrates based on an arbitrary rate of 1.0 for a terminal olefin. When a 1 : 1 mixture of a terminal olefine and another substrate is hydrogenated the relative rates of hydrogenation are not the same as those predicted from the relative rates of the hydrogenation of single substrates (see Table 7). Thus although alkynes and dienes are not reduced as fast as terminal olefins they are reduced faster than terminal olefins when mixed with them. The selectivity obtained however is not good in benzene as all the groups are

Table 7

RATE OF HYDROGENATION OF VARIOUS SUBSTRATES RELATIVE TO OCT-1-ENE WHEN REDUCED SINGLY OR IN COMPETITION WITH OCT-1-ENE.

SUBSTRATE (TOTAL) = 1.0 M, $RhCl(PPh_3)_3$ = 10^{-2} M, t = 22°, H_2 = 1 atm. Solvent = benzene.

Substrate	Rel. Rate of single substrates	Relative Rate in competition with oct-1-ene
Diethyl maleate	1.7	6.6
acrylonitrile	1.3	14.7
cyclopentene	1.0	1.2
C_6—C_{12} n-alk-1-enes	1.0	1.0
C_6—C_8 alk-1-ynes	0.85	1.7
2,4,4-trimethylpent-1-ene	0.70	0.41
cycloocta-1,3-diene	0.60	0.75
Cycloheptene	0.46	0.95
2-methylpent-1-ene	0.43	0.69
Cyclooctene	0.42	0.72
Methylmethacrylate	0.36	3.5
phenylacetylene	0.15	4.1
penta-1,3-diene	0.12	1.5

reduced together. Olefins containing a functional group, e.g. carboxylate, nitrile, ether, etc., are all reduced faster than terminal straight chain olefins. Terminal olefins from C_6 to C_{12} are hydrogenated at the same rate, and an increase in chain length imposes no steric constraints, which is in contrast to Wilkinsons results (see Table 6). This discrepancy may be due to trace impurities in individual olefins and errors in kinetic measurements ($\pm 5\%$ [87]) which are overcome in competition experiments. Candlin has also shown that the relative rates of hydrogenation of 1 : 1 mixtures of hex-1-yne and oct-1-ene depends greatly on the solvent (see Table 8). The greatly increased selectivity in changing from benzene to trifluoroethanol is due to an increase in the rate of hydrogenation of the alkyne and not to a reduction in the rate of hydrogenation of oct-1-ene, which is virtually the same in benzene or benzene trifluoroethanol mixtures [95].

Table 8

HYDROGENATION OF 1 : 1 MIXTURES OF HEX-1-YNE AND OCT-1-ENE IN VARIOUS SOLVENTS (50% BENZENE), CATALYST RhCl(PPh$_3$)$_3$

Solvent	Relative rates as single substrates		Relative rate $\dfrac{\text{hex-1-yne}}{\text{oct-1-ene}}$ in mixtures of substrates
	hex-1-yne	oct-1-ene	
benzene (only)	0.9	1.0	1.7
amyl acetate			1.7
tetrahydrofuran			1.8
dimethylformamide			1.8
methyl ethyl ketone			2.0
ethanol	0.9	1.4	2.6
ethyl acetate			3.1
chlorobenzene			> 5.0
phenol	2.4	1.0	Stepwise reaction
2,2,2-trifluoroethanol	12.0	0.9	Stepwise reaction

6.1.2.7. *The possibility of a rhodium(I) olefin complex as an intermediate in hydrogenation*

In contrast to Wilkinson and al., Candlin has found RhCl(PPh$_3$)$_2$C$_2$H$_4$ is a good hydrogenation catalyst for ethylene and other olefins [95]. The value given for the equilibrium constant of reaction (46) (see 6.1.2.3.) suggests that some free RhCl(PPh$_3$)$_2$ should be present in solution and it should at least catalyse hydrogenation of terminal olefins. One can only conclude that Wilkinson's results are incorrect on this point and it opens up the possibility that hydrogenation can actually progress via k'' in scheme 1. When the complex RhH$_2$Cl(PPh$_3$)$_3$ is added to an excess of 1 : 1 hex-1-yne and oct-1-ene under an atmosphere of nitrogen it stoichiometrically hydrogenates the substrates, but the relative amounts of each hydrogenated depends on the solvent system. In benzene the ratio of hexyne reduced : octene reduced is 1.9. This is very close to the competition figure obtained in catalytic hydrogenation in benzene (see Table 8). In benzene and ethanol mixtures the competition figure in the stoichiometric reaction is > 10 but only 2.6 in the catalytic reaction (see Table 8). Thus in benzene ethanol mixtures the preformed hydride is much more selective than the catalytic reaction. It follows they may go via different routes and in the catalytic hydrogenation an appreciable amount of product may be formed via the reaction of hydrogen with $RhCl(PPh_3)_2$(substrate), i.e. via k'' scheme 1 [95].

6.1.2.8. *The hydrogen transfer step*

Initially it was concluded that there was rapid simultaneous *cis* addition of two hydrogen atoms to the liganded olefin (51). It was thought the olefin rotated slightly on the rhodium olefin axis to assist this [18]. The alternative, stepwise addition of hydrogen atoms to give an intermediate

alkyl (see (11) section 2.3) was not favoured because (*a*) reaction of $RhH_2Cl(PPh_3)_2$ and ethylene gave ethane directly, no intermediate ethyl-rhodium was detected by H^1 n.m.r., whereas other rhodium hydrides have given ethyls on reaction with ethylene [16, 25, 99], (*b*) no isomerisation was observed in hydrogenation studies, (*c*) hydrogen exchange between deuterium and alcohols is slow suggesting the hydrogens react together or not at all [18]. Simultaneous *cis* addition was favoured because internal alkynes gave predominantly *cis* olefins and addition of deuterium to maleic acid and fumaric acid gave meso-1,2-dideutero-succinic acid and DL-sym-1,2-dideuterosuccinic acid respectively [18].

Since then a lot of evidence has accumulated which proves that the reverse alkyl mechanism actually does occur but of course it is impossible to prove that simultaneous addition does not take place to some extent. It may be that under certain conditions the second hydrogen adds so quickly that virtually simultaneous addition occurs. It is important to realise that this step in the reaction sequence does not affect the kinetic scheme.

The best evidence for reverse alkyl formation comes from reaction of olefins with deuterium and tritium. These reactions have been conducted under conditions very similar to those used in the kinetic studies (section 6.1.2.4.). Osborn and Wilkinson found that reduction of hex-1-ene with a 1 : 1 mixture of hydrogen and deuterium gave a mixture of products containing $C_6H_{13}D(27\%)$ [18], and partial reduction of cyclohexene with hydrogen containing tritium gives tritiated cyclohexene as well as cyclohexane [100]. A detailed study of the deuteriogenation of various substituted cyclohexene shows that d_0-d_4-cyclohexanes, HD, and some d_1-cyclohexenes are formed [98].

Occurrence of isomerisation during hydrogenation is also evidence for reverse alkyl formation. Bond has made a study of the reaction products in the $RhX(PPh_3)_3$, X=Cl, Br or I, catalysed hydrogenation of pent-1-ene, *cis* pent-2-ene and *trans* pent-2-ene at 25° in benzene at one atmosphere. It was found that no isomerisation of pent-1-ene was observed using the bromide or iodide as catalyst but traces of *trans* pent-2-ene were obtained with the chloride. All three catalysts cause isomerisation of the pent-2-enes, e.g. with $RhCl(PPh_3)_3$ the ratio of hydrogenation : isomerisation of *cis*-pent-2-ene is 1 : 0.6, the *trans* isomer being formed more readily than pent-1-ene. The actual product spectrum appears to depend on the halide, the catalyst concentration and is clearly very sensitive to reaction conditions [101]. Attempted hydrogenation of damsin gives isodamsin (52) and in the presence of

(52)

deuterium gas deuterated isodamsin is formed [102].

6.1.3. CATALYST POISONS

Many substances reduce the activity of these catalysts because they oxidise it to a form of rhodium(III) which cannot be reduced by hydrogen under the mild conditions employed. Thus carbontetrachloride and chloroform, give a series of rhodium trichlorides [94], allyl chloride gives $RhCl_2(C_3H_5)(PPh_3)_2$ [103] and acetyl chloride gives $RhCl_2COCH_3(PPh_3)_2$ [99].

Other substances act as poisons or inhibitors by simply competing with the substrate for a site on the metal and often complexes containing the inhibitor can be isolated. Pyridine and tetrafluoroethylene give complexes $RhCl(PPh_3)_2L$ [18], chelating olefins give $RhCl(PPh_3)(diene)$ where diene is cycloocta-1,5-diene or norbornadiene [78], and stable hydrides are formed in pyridine, methylene dichloride and acetonitrile (see 6.1.2.2.) None of these complexes are active catalysts under mild conditions, although they are catalysts under moderate hydrogen pressures or higher temperatures [18]. The low rate of hydrogenation of olefins in some solvents is a form of this site competition as is preferential reduction of one substrate before another. Clearly, the amount of inhibitor, or more particularly its ratio to catalyst is important in determining if a hydrogenation will proceed or not. Although dialkyl sulphides do not poison $RhCl(PPh_3)_3$ and allyl phenyl sulphide is readily reduced, traces of thiophenol (S : Rh = 2.5) reduce the catalysts activity and larger amounts (S : Rh = 42) render it virtually inactive [104]. Carbon monoxide, or organic compound from which it can be abstracted such as aldehydes, allylic alcohols (via isomerisation to the aldehydes), formic acid, etc., are strong poisons under mild conditions because the complex $RhClCO(PPh_3)_2$ is formed, and this is not a catalyst under ambient conditions [18].

Carboxylic acid groups do not favour hydrogenation, but it is not clear if this is a solvent effect or if acetates are formed. Oxygen, or a source of peroxides is a promoter at low levels, 0.2 to 1.0 moles per mole of catalyst [105], but above this level it is an inhibitor [18]. The products from the reaction of $RhCl(PPh_3)_3$ and oxygen are triphenylphosphine oxide, peroxy complexes such as $RhClO_2(PPh_3)_2$(solvent), and other rhodium(III) species [94]. Removal of phosphine as the oxide activates the catalyst system as the phosphine : rhodium ratio is reduced, but oxidation of the metal removes active catalyst. It would seem that at low oxygen or peroxide levels the net effect is slightly beneficial.

6.1.4. USEFUL APPLICATIONS OF $RhCl(PPh_3)_3$

So far this complex has been used for preparative purposes only. It is so much easier to use than its analogues that they will only be used in special cases. So far the compound has been used generally in unnecessarily large amounts and in benzene or benzene ethanol solvents. The complex only catalyses hydrogenation of olefinic and acetylenic groups and the order of activity is terminal olefin > disubstituted olefin > trisubstituted olefin $\gg$ tetrasubstituted olefin. Consequently some specific reductions have already been carried out, see Table 9. As more use is made of suitable solvent systems (see 6.1.2.6) this list will grow considerably. Several problems have arisen during attempted reductions. Although damsin is partly reduced to dihydrodamsin, much of it is isomerised to isodamsin (52) [102, 115] and coronopilin is converted to a mixture of dihydrocoronopilin and the isomer isocoronopilin [115]. Hydrogenation of allylic alcohols, e.g. geranial is accompanied by decarbonylation, presumably via rearrangement to an aldehyde [108]. Aldehydes always tend to interfere with hydrogenation because rapid decarbonylation occurs and the catalyst is converted to inactive $RhClCO(PPh_3)_2$ [18]. Wilkinson has shown that α-β-unsaturated aldehydes can be reduced to saturated aldehydes with little decarbonylation if the substrate is added to the reduced form of the catalyst, but conditions have to be carefully controlled [115]. Hydrogenation

Table 9

Substrate	Product	Ref.
1-methoxyocta-2,7-diene	1-methoxyoct-2-ene	[107]
ergosterol	5α,6-dihydroergosterol	[108]
pregna-5,16-dien-3β-ol-20-one acetate	pregnenolone acetate	[108]
(+) carvone	hydrog. of isopropylidene group	[108]
linalool	hydrog. of vinyl group	[108]
thebaine	8,14-dihydrothebaine	[108]
1-methoxycyclohexa-1,4-diene	cyclohexanone + 1-methoxycyclohexene	[108]
androsta-1,4-diene-3,17-dione ⎱ androsta-4,6-diene-3,17-dione ⎰	androsta-4-ene-3,17-dione	[109]
eremophilone	dihydroeremophilone	[110]
1,4-naphthoquinone	tetralin-1-4-dione	[111]
juglone	β-dihydrojuglone	[111]
2,3-dimethoxybenzoquinone	2,3-dimethoxycyclohexa-2-ene-1,4-dione (isomerises to 1,4-dihydroxy-2,3-dimethoxybenzene)	[111]
benzoquinone	quinol	[111]
1,4,5,8-tetrahydronaphthalene ⎱ 1,4-dihydrotetralin ⎰	9,10-octalin (80%) 1,9-octalin (20%)	[112] [112]
methyl-1,4,5,6-tetrahydro-1-naphthoate	methyl-9,10-octalin-1-carboxylate (96%) + methyltetralin-1-carboxylate. (4%)	[112]
γ-gurjunene	hydrogenation of isopropylidene group only	[113]

of allylic halides is also accompanied by hydrogenolysis of the halide [108] and the catalyst dies (see section 6.1.3), but allyl esters can be reduced without hydrogenolysis [94].

The catalyst is particularly good for hydrogenation of certain cyclic dienes, which isomerise or disproportionate with heterogeneous catalysts and give aromatic compounds. Thus dihydrotetralins and isotetralin can be reduced to octalins in very high yield [112] (see Table 9).

Because the catalyst can operate under conditions where little isomerisation occurs, and the addition of two hydrogen atoms occurs simultaneously or the second hydrogen adds extremely rapidly, RhCl(PPh3)3 is useful in preparing compounds labelled with deuterium at specific positions. With heterogeneous catalysts a wide spread of products is obtained, and often the deuterium is spread over more carbon atoms than the original double bond, due to isomerisation. Thus deuteriogenation of cyclohexene gives a virtually quantitative yield of 1,2 dideuterocyclohexane, methyl oleate gives methyl 9,10-dideuterostearate, and methyl linoleate gives methyl 9,10,12,13-tetradeuterostearate [108]. A variety of linear olefins have been deuterated and the mass spectral cracking patterns of the products confirm that deuterium has added specifically to the double bond

[116]. The deuteriogenation of various substituted cyclohexenes gives the expected product in 85-98% yield depending on the structure of the cyclohexene [98]. Olefins which on one double bond shift become tetrasubstituted olefins, e.g. damsin to isodamsin [102] (52), do not give high yields of dideuterated compounds and it appears that the specificity of dideuteriogenation decreases with increasing substitution of the double bond. Several specific deuteriogenations of unsaturated sterols have been recorded. With ergosterol 5α, 6-dideuteroergosterol is formed, the 7-ene position being uneffected [108, 117], androsta-1,4-diene-3,17-dione is converted to 1,2-dideuteroandrosta-4-ene-3,17-dione and androsta-4,6-diene-3,17-dione gives 6,7-dideuteroandrosta-4-ene-3,17-dione, the 4-ene position not being attacked in each case [109]. The various techniques used in preparing deuterated compounds has recently been reviewed and RhCl(PPh3)3 is regarded as important in this field [118].

A recent development is the use of $RhCl(PR^1R^2R^3)_{2 or 3}$ containing optically active phosphines to give catalytic asymmetric hydrogenations. Hydrogenation of α-ethylstyrene and α-methoxystyrene gives (S)-(+)-2-phenylbutane (7% optical yield) and (R)-(+)-1-methoxy-1-phenylethane (4% optical yeld) using a catalyst formed in situ from [RhCl(1, 5 hexadiene)]2 and (S)-(+) methylphenyl-n-propylphosphine (2.2 equivalents per rhodium) at room temperature and pressure [119]. Optically active hydrotropic acid (15% optical yield) from atropic acid and methyl succinic acid (3% optical yield) from itaconic acid have been prepared using RhCl3(PMePhn-Pr)3, and triethylamine as catalyst at 35-80° and 30 atmospheres of hydrogen [120].

6.1.5. SUMMARY

Osborn and Wilkinsons mechanistic scheme is basically correct but is too simplified. It appears that RhClP3 is dissociated to RhClP2 and also to RhClP. The n.m.r. data surprisingly suggests the degree of dissociation is quite small. Al the rhodium(I) species will give rhodium(III) hydrides and the catalytic activity of the solution will be the sum of the activity of all the species present. The relative amount of each species will depend on solvent, temperature, ligands, concentration of reactants, hydrogen pressure etc. There is evidence that hydrogenation can occur via attack of hydrogen on a rhodium(I) olefin complex and there is a great deal of evidence to suggest hydrogen addition is stepwise with di and tri substituted olefins but probably not with terminal olefins. The second addition is so fast that this is a minor point for most practical purposes as seen in section 6.1.4.

6.2. Hydridocarbonyltris(triphenylphosphine)rhodium(I)

This complex has been known since 1963 [121] and an X-ray study shows it to have structure (V) with Rh-H bond length of 1.7 Å [122]. It was known to catalyse hydrogenation of ethylene and hydrogen-deuterium exchange [137] but more recently Wilkinson has described more detailed studies on its preparation and properties [124] and its activity as a catalyst in hydrogenation [125], hydroformylation [126] and hydrogen-deuterium exchange.

$$
\begin{array}{c}
H \\
| \\
P \longrightarrow Rh \underset{\displaystyle |}{\overset{\displaystyle \diagup P}{\diagdown P}} \\
CO
\end{array}
$$

V

The complex is about as active as RhCl(PPh₃)₃ for hydrogenation of terminal olefins but it is much less active for internal olefins, conjugated dienes and acetylenes – see Table 10. It should thus have many useful applications in preparative chemistry.

Table 10

RATE OF HYDROGENATION OF 50 CC. OF BENZENE SOLUTIONS OF VARIOUS OLEFINS (0.7 M) CATALYSED BY RhHCO(PPh₃)₃ (1.25 · 10⁻³ M) AT 25° AND 50 CM. OF H₂ PRESSURE.

Olefin	Rate in cc. min⁻¹
hex-1-ene	16.7
hexa-1,5-diene	7.43
4-vinylcyclohexene	7.43
styrene	1.39
cis pent-2-ene, cyclohexene, penta-1,3-diene, 2-methylpent-1-ene, acrylic acid, 2-chloropropene, hex-1-yne	<0.1

6.2.1. THE MECHANISM OF HYDROGENATION AND HYDROGEN DEUTERIUM EXCHANGE

The properties of RhHCO(PPh₃)₃ and the kinetics of hydrogenation indicate that hydrogenations proceed via scheme 2 [125]. It is assumed k'' is almost zero

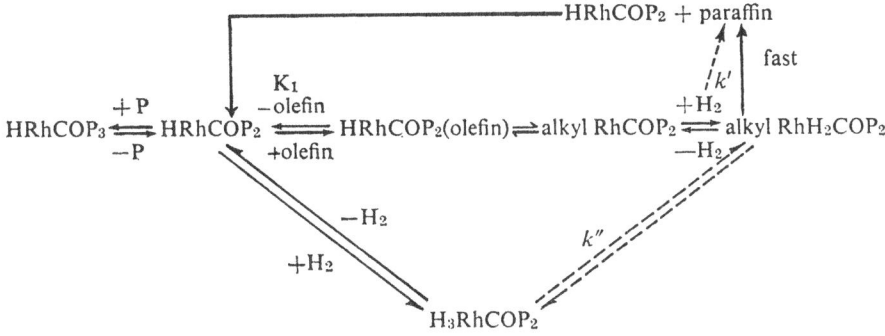

Scheme 2

under mild conditions as coordination of olefin would give a seven co-ordinate intermediate or direct formation of an alkyl. If this route is ignored the rate is given by (53) where k' is the rate determining constant,

$$r = \frac{k'K_1[H_2]\,[\text{olefin}]\,[\text{total Rh}]}{1 + K_1\,[\text{olefin}]} \tag{53}$$

K_1 is the equilibrium constant shown in scheme 2 and [ ] denotes a concentration. It is assumed all other reactions in scheme 2 are rapid. Equation (53) transposes to (54) and plots of $1/r$ against $1/[H_2]$ and $1/r$ against $1/[\text{olefin}]$ are straight, the former passing through the origin, as demanded by (54).

$$\frac{1}{r} = \frac{1}{[\text{olefin}]}\;\frac{1}{k'K_1[H_2]\,[\text{total Rh}]} + \frac{1}{k'[H_2]\,[\text{total Rh}]} \tag{54}$$

The rate constant k' is obtained from the intercept of the plot of $1/r$ against $1/[\text{olefin}]$. A deuterium effect was observed such that $k'_H/k'_D = 1.47$ which is of the order expected for oxidative addition of hydrogen as the rate determining step (k_H/k_D for addition of H_2 or D_2 to $IrClCO(PPh_3)_2$ is 1.22) [17].

6.2.2. SPECIFICITY OF THE CATALYST

Although no rhodium-alkyl derivatives have been prepared or detected by ^1H n.m.r. spectroscopy, the hydrogen deuterium exchange of $RhDCO(PPh_3)_3$ with olefins provides excellent evidence for the formation of an intermediate rhodium-alkyl [126]. The half life of the exchange reactions, as measured by following the formation of $RhHCO(PPh_3)_3$ by ^1H n.m.r., is 20 secs. for pent-1-ene and 1 hr. for pent-2-ene. The half life for isomerisation of pent-1-ene is greater than 1 hr. and all this shows that n-alkyl-rhodium groups are formed much more readily than isoalkyl groups. Wilkinson has proposed that the co-ordinately unsaturated intermediate $RhHCO(PPh_2)_2$ is square planar with *trans* phosphines and these occupy so much volume by rotations etc. that only a vinyl group can easily approach the rhodium atom, and in such a way that an n-alkyl rather than an iso-alkyl rhodium complex can subsequently form. This is demonstrated diagrammatically in structures VI and VII.

VI   VII

The strict steric requirements are reflected in the thermodynamic parameters particularly the entropy of activation obtained for hydrogenation of hex-1-ene, ΔH^* 10.6 kcals mol.$^{-1}$ and $\Delta S^* - 8.0$ e.u.. The corresponding values for the RhCl(PPh$_3$)$_3$ catalysed reaction are ΔH^* 18 kcal mol.$^{-1}$ and ΔS^* 1.3 e.u., reflecting less steric effects.

At concentrations below those used in the kinetic study the activity of the catalyst increases, see Fig. 5. This is explained by further dissociation of the

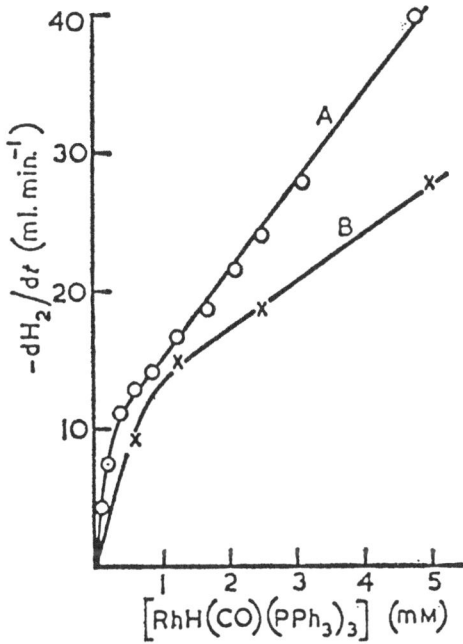

Figure 5. Rate of hydrogenation of alk-1-ene as a function of catalyst concentration in benzene at 25° with RhH (CO) (PPh$_3$)$_3$ at 50 cm. H$_2$ pressure: A, hex-1-ene; B. dec-1-ene.

catalyst according to (55) and the intermediate RhHCO(PPh$_3$) is more active

$$RhHCO(PPh_3)_3 \rightleftharpoons RhHCO(PPh_3)_2 \rightleftharpoons RhHCO(PPh_3) \qquad (55)$$

than RhH CO(PPh$_3$)$_2$ in hydrogenation. Evidence for this is that the specificity of hydrogenation drops at low catalyst concentrations, as would be expected, and molecular weight studies (cryoscopic and osmometric) in cyclohexane benzene and o-xylene between —29° and 38 °C support reaction (55). Dissociation is favoured by low concentrations, good coordinating solvents, and higher temperatures, whereas low temperatures, paraffin solvents, and high concentrations favour the left hand side of (55). The ^{1}H n.m.r. spectrum shows the complex is not dissociated and exchange of ligands is slow at -30° but at 25° extensive ligand exchange occurs [92, 124].

6.3. Miscellaneous phosphine complexes of rhodium

Several rhodium(III) phosphine complexes are active hydrogenation catalysts but they are best regarded as in situ preparations of $RhCl(PPh_3)_3$ ans its analogues.

The compound $RhCl_3(PPh_3)_3$, known to be a mixture containing some $RhCl(PPh_3)_3$, is a catalyst for hydrogenation of hex-1-ene [127] and a variety of acetylenes [128], $RhCl_3(PMePhn-Pr)_3$ has been discussed earlier (6.1.4.) and $RhHCl_2(PPh_3)_3$ catalyses isomerisation [129] and hydrogenation [95]. The isomerisation goes via formation of an alkyl intermediate [99, 129] and the hydrogenation most likely goes via $RhCl(PPh_3)_3$ as dissociation readily occurs (56) [99].

$$RhHCl_2(PPh_3)_3 \; \rightleftharpoons \; HCl \; + \; RhCl(PPh_3)_3 \qquad (56)$$

Mixtures of $RhCl_33H_2O$ and triphenylphosphine (2.25 to 6 moles) are catalysts for isomerisation and hydrogenation but these are the conditions in which $RhCl(PPh_3)_3$ and $RhHCl_2(PPh_3)_3$ are formed [88].

6.4. Phosphine complexes of iridium

Vaska's complex $IrClCO(PPh_3)_2$ and its analogues have provided valuable information on the basic concepts of oxidative addition and coordinative unsaturation [15] but it is only a very weak hydrogenation catalyst. A solution (10^{-3} to 10^{-2} M) in toluene at 40° and 1 atmosphere total pressure gives a 12% molar yield of ethane from ethylene in 24 hrs. A slower reaction occurs with acetylene, propylene [80] and but-1-ene which is converted to a mixture of isomers and butane at 60° [130]. Deuteriogenation of ethylene gives a mixture of deuterated ethane and ethylene which is consistent with reverse alkyl formation [130].

Hydrogenation in benzene is too slow for accurate kinetic work but the mechanism of hydrogenation of maleic and fumaric acid in dimethyl formamide has been studied at 80° [131]. The mechanism proposed is scheme 3 which has many similarities to the schemes proposed for $RhCl(PPh_3)_3$ and $RhHCO(PPh_3)_3$ (see Schemes 1, 2).

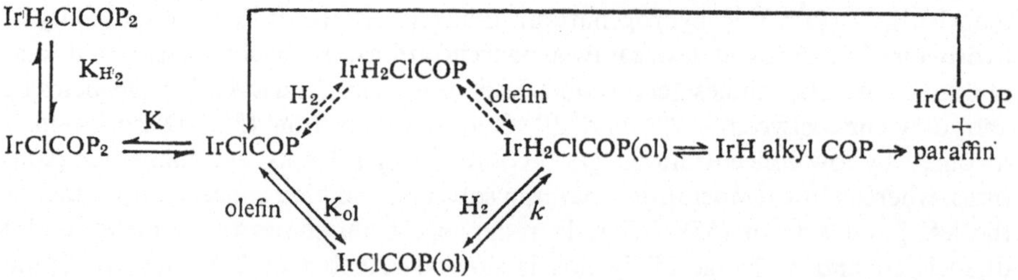

Scheme 3

Spectroscopic evidence shows that the equilibrium between $IrClCO(PPh_3)_2$ and $IrH_2ClCO(PPh_3)_2$ is set up quickly, and is followed by slow formation of a maleic anhydride complex. This stage and the hydrogenation reaction are slowed by the addition of excess phosphine. The hydrogenation goes up to a hundred times faster in the presence of traces of oxygen, which is most likely due to the removal of phosphine as phosphine oxide, so increasing the rate of formation of $IrClCOPPh_3$. The slow formation of $IrClCOPPh_3$ also accounts for the increase in rate of hydrogenation with time. The possible route via $IrH_2ClCOPPh_3$ is not favoured because no evidence was obtained for this intermediate, but evidence was obtained for the maleic acid complex.

Kinetic studies on this system have not been very rewarding [131]. Hydrogenation is first order on hydrogen, catalyst, and substrate at low concentrations tending to zero order as the concentration of each variable increases. The rate expression is

$$r = \frac{k\,K_{ol}[H_2]\,[\text{olefin}]\,[Ir]}{1 + K_{ol}[\text{olefin}] + [PPh_3]\,(1 + K_{H_2}[H_2])/K}$$

where [ ] denotes a concentration and the various equilibrium constants K and rate constants k are indicated in scheme 3. The expression is too complicated to check by a simple straight line plot.

The complex $IrHCO(PPh_3)_3(2.5 \cdot 10^{-3}\,M)$ slowly hydrogenates ethylene at 30° and one atmosphere of total pressure. The complex is recovered unchanged and solutions of it absorb hydrogen (70% of theory), and ethylene (28% theory) and it has been suggested the mechanism of hydrogenation goes via reactions (57) and (58) [132]. However, it is likely that dissocation of a phosphine is involved

$$IrHCOP_3 + C_2H_4 \rightleftharpoons IrHCOP_3C_2H_4 \rightleftharpoons IrC_2H_5COP_3 \tag{57}$$

$$IrC_2H_5COP_3 + H_2 \rightleftharpoons IrH_2C_2H_5COP_3 \rightleftharpoons C_2H_6 + IrHCOP_3 \tag{58}$$

as with $RhHCO(PPh_3)_3$ (Scheme 2).

The iridium(I) complexes containing chelating ligands $[Ir(Ph_2PC_2H_4PPh_2)_2]Cl$, $[IrCO(Ph_2PCH_2PPh_2)_2]Cl$ and $IrBr(Ph_2PC_2H_4SPh)_2$ are catalysts for hydrogenation of acetylenes and olefins at 150° and 100 atmospheres [133].

A comprehensive study on complexes of general formula $IrHX_2L_3$ has shown that the activity depends on X, L, temperature, pressure, solvent, substrate and added neutral ligands [134]. For hex-1-ene(1M) in toluene at 30 atmospheres of hydrogen the most active catalyst is $IrHCl_2(AsPh_3)_3$ which is active at about 35°. Other combinations of halide and neutral ligand require higher temperatures. The catalysts are more active after a pretreatment period and this suggests dissociation of ligands is an important step.

The complexes $IrH_3(PR_3)_2$ where PR_3 is PEt_2Ph and PPh_3 catalyse the hydrogenation of terminal olefins at room temperature and pressure in benzene [135] and dichloromethane [136], but the reaction is very slow. A monohydride $[IrH(PPh_3)_2(\text{solvent})]$ has been isolated from the products [136]. At 80° and 1 atmosphere pressure $IrH_3(PPh_3)_2(10^{-3}$ M in benzene) catalyses the hydrogenation of terminal and internal olefins, and extensive isomerisation occurs. The complex $IrH_3(PPh_3)_3$ is much more selective under these conditions and only causes hydrogenation of terminal olefins [135].

6.5. Phosphine complexes of ruthenium and osmium

Phosphine complexes of ruthenium chlorides are catalysts for the hydrogenation of olefins under ambient conditions, e.g. mixtures of $RuCl_3$ and triphenylphosphine (optimum ratio Rh : PPh_3 is 1 : 6), $RuCl_2(PPh_3)_4$ [137], $RuCl_2(PPh_3)_3$ [51, 137] and $[Ru_2(PPh_3)_6Cl_3]Cl$. Although $RuCl_2(PPh_3)_3$ is an active catalyst in benzene [51, 137] it is much more active in the presence of alcohols [137]. Wilkinson et al have shown that the reversible reaction (59) occurs and a base, e.g. alcohol, is required to drive the reaction to the right hand side. The true catalyst is $RuHCl(PPh_3)_3$ which can be prepared via reaction (59) [137-139].

$$RuCl_2(PPh_3)_3 + H_2 \rightleftharpoons RuHCl(PPh_3)_3 + HCl \qquad (59)$$

An X-ray structure shows it has structure VIII with benzene of crystallisation present in the lattice [140].

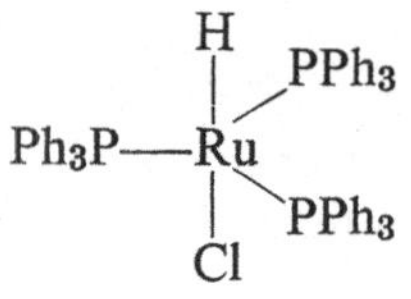

$$
\begin{array}{c}
\text{H} \\
| \quad \diagup \text{PPh}_3 \\
\text{Ph}_3\text{P}\!-\!\!\overset{|}{\underset{|}{\text{Ru}}} \\
\text{Cl} \quad \diagdown \text{PPh}_3
\end{array}
$$

VIII

In aromatic solvents $RuHCl(PPh_3)_3$ is the most active catalyst described so far for hydrogenating unsubstituted alk-1-enes. For example the rate of hydrogen uptake for 60 c.c. of a benzene solution of $RuHCl(PPh_3)_3$ $(8 \cdot 10^{-4}$ M) and hept-1-ene (1.2M) at 25° and 50 cm. hydrogen pressure is approx. 260 c.c. min^{-1}. The rate for $RhCl(PPh_3)_3$ under comparable conditions is 15 c.c. min^{-1} [139]. The rate is too fast to measure accurately and other factors prevent good kinetic data being obtained in this particular system. The complex has a low solubility, is slow to dissolve, and gives deeply coloured solutions making it difficult to prepare true homogeneous solutions. In addition the solution is very sensitive to minute traces of oxygen and peroxides and this increases the difficulty of obtaining reproducible results. Furthermore the catalyst dies off during hydrogenation. Extensive tests have shown that this is not due to solvent impurities such as water, thiophene etc., but to the slow formation of a complex with olefin, probably a π-allyl [139].

For practical purposes it is easier to use a ruthenium halide, e.g. $RuCl_2(PPh_3)_3$ and a suitable base, e.g. an alcohol benzene mixture, or benzene containing sodium phenoxide, sodium borohydride, caustic potash or triethylamine etc.,

Glacial acetic acid is not a strong enough base and pyridine is unsuitable as it forms inactive ruthenium complexes. The activity of solutions prepared this way depend on how quickly equilibrium (59) is set up and on the position of the equilibrium under the reacting conditions.

6.5.1. MECHANISM OF HYDROGENATION AND SPECIFICITY

Although good kinetic data have not been obtained in benzene solutions there is good evidence that the mechanism of hydrogenation goes according to scheme 4 [139]. This is very similar to scheme 2 for $RhHCO(PPh_3)_3$ which is isoelectronic with $RuHCl(PPh_3)_3$.

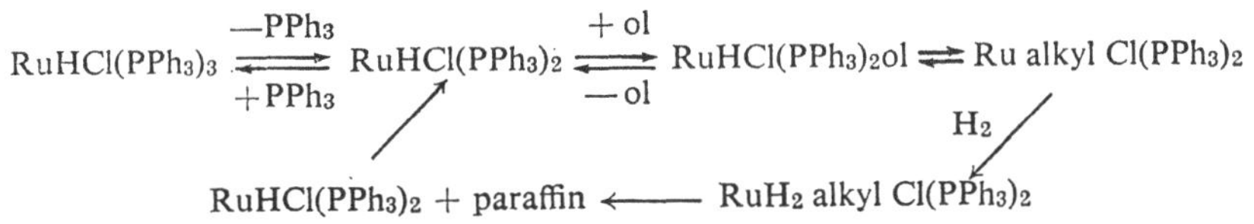

$$RuHCl(PPh_3)_3 \xrightleftharpoons[+PPh_3]{-PPh_3} RuHCl(PPh_3)_2 \xrightleftharpoons[-ol]{+ol} RuHCl(PPh_3)_2ol \rightleftharpoons Ru\ alkyl\ Cl(PPh_3)_2$$

$$RuHCl(PPh_3)_2 + paraffin \longleftarrow RuH_2\ alkyl\ Cl(PPh_3)_2 \quad H_2\nearrow$$

Scheme 4

The ruthenium hydride is not dissociated appreciably in solution but excess phosphine prevents hydrogenation suggesting a dissociative step is important. An alkyl derivative can be prepared by treating $RuHCl(PPh_3)_3$ with ethylene at 35 atmospheres pressure. Further evidence for reverse alkyl formation comes from exchange reactions between $RuDCl(PPh_3)_3$ and olefins. The half life for this reaction, as measured by ^1H n.m.r. spectra are ethylene, 15 seconds, alk-1-enes, 10 minutes, and alk-2-enes 30 minutes. Isomerisation of terminal olefins is slower than hydrogen deuterium exchange [129, 139] and all this suggests that n-alkyl-ruthenium complexes are formed much more readily than iso-alkyl-ruthenium derivatives. This results in very high selectivity for terminal olefins, see Table 11,

Table 11

COMPARATIVE RATES OF HYDROGENATION OF ALKENES RELATIVE TO HEPT-1-ENE CATALYSED BY $RuHCl(PPh_3)_3(8 \cdot 10^{-4}$ M) IN BENZENE AT 25° AND 50 CM. HYDROGEN

Alkene	Rate
hept-1-ene	1.0
hex-1-ene	1.6
cis + *trans* hex-2-ene	$<3 \cdot 10^{-4}$
hept-3-ene	$<3 \cdot 10^{-4}$
cyclohexene	$<1 \cdot 10^{-3}$
2-methylpent-1-ene	$<3 \cdot 10^{-4}$

and the reasons for this are the same as those advanced for the high selectivity of $RhHCO(PPh_3)_3$ (see 6.2.2.). The order of reactivity for cyclic olefins is cyclohexene > cyclopentene > cyclooctene [51], and acetylene and polyenes are slowly reduced by $RuCl_2(PPh_3)_3$ in benzene in the order diphenylacetylene > norbornadiene > stearolic acid [51]. In fact $RuHCl(PPh_3)_3$ and norbornadiene give a stable complex $RuHCl(PPh_3)_3 \cdot C_7H_8$ [139]; dienes and acetylenes are reduced first in mixtures with olefins.

There is no direct evidence that ruthenium(IV) hydrides are involved in the hydrogenolysis of the alkyl in scheme 4. Recently osmium(IV) hydrides OsH_4P_3 have been described [141] and by analogy ruthenium(IV) hydrides should also be formed, at least as transient intermediates. Ruthenium(IV) intermediates are likely to be involved in the exchange of Ru-H with deuterium (60) [139].

$$RuHCl(PPh_3)_2 \xrightleftharpoons[-D_2]{+D_2} RuD_2HCl(PPh_3)_2 \xrightleftharpoons[+HD]{-HD} RuDCl(PPh_3)_2 \qquad (60)$$

and in the deuteration of the ortho positions of the phenyl groups of the ligands, reaction (61), which can be brought about by deuterium derived from deuterium

$$RuDCl(PPh_3)_2 \rightleftharpoons \begin{array}{c} Cl \quad \overset{D}{|} \quad PPh_3 \\ \diagdown \; | \; \diagup \\ Ru \\ \diagup \; | \; \diagdown \\ H \quad | \quad PPh_2 \end{array} \rightleftharpoons RuHCl(PPh_3)(PPh_2C_6H_4 \; o \; D) \qquad (61)$$

gas [139] or deuteroolefins [129, 142].

The only report to date of a homogeneous osmium catalyst is that $OsHClCO(PPh_3)_3$ catalyses hydrogenation of acetylene to ethylene and ethane at $60°$ and 1 atmosphere of hydrogen [132].

6.6. Phosphine complexes of nickel, palladium and platinum

In benzene solution $NiX_2(PPh_3)_2 (5 \cdot 10^{-3} M)$ where X = Br or I slowly catalyses (10% conversion in 5 hours) the hydrogenation of oct-1-ene $(4.5 \cdot 10^{-2} M)$ at 20 °C and 1 atmosphere. No isomerisation was observed [59] but solutions of $NiBr_2(PEt_2Ph)_2 (0.15M)$ in benzene ethanol (1 : 1) at room temperature and 60 atmospheres of hydrogen catalyse the hydrogenation (15% yield) and isomerisation (85% yield) of hex-1-ene (3.5M) [143]. Under the more vigorous conditions of 40 atmospheres of hydrogen and 90 °C the complexes $NiX_2(PPh_3)_2$, X=halide $(1.2. \cdot 10^{-2} M)$ catalyse the hydrogenation of methyl linoleate $(6.8 \cdot 10^{-2} M)$ to a monoene and traces of the fully saturated methyl stearate. Isomerisation occurs to a lesser extent than hydrogenation and the yield of monoene is 84% for the iodide, 53% for the bromide and only 1% for the chloride. Benzene is a better solvent than tetrahydrofuran but there is evidence that hydrogen is transferred from the solvent [123] (see section 10.2). Similarly it is reported that $PdX_2(PPh_2)_2$, X=Cl or CN catalyses the partial hydrogenation of soya bean oil methyl ester [154] and cyclooctadienes [63] at 90° and 40 atmospheres of hydrogen. The analogous platinum complexes are less reactive [56]. The phosphine complexer $MCl_2(PBu_3)_2$, where M is Ni or Pd, are converted into catalysts by treatment with tri-isobutylaluminium. Thus, hex-1-ene is hydrogenated slowly over 19 hrs. at 25 °C and 3-4 atmospheres of hydrogen [166].

7. HYDROGENATION CATALYSED BY TRANSITION METAL SALTS IN POLAR MEDIA

Soon after Calvin's discovery [1], Japanese workers established that aqueous solutions of Pt(ethylenediamine)$_2$Cl$_2$ [144] and various amine derivatives of chodium(III) halides [145] catalysed the hydrogenation of quinones, fumaric acid and hydroxylamine. Since then the activation of hydrogen by transition metal salts have been extensively studied and reviewed [4, 5]. In aqueous solution, below 80 °C, the ions of manganese(VII), ruthenium(IV), ruthenium(III), osmium-(IV), rhodium(III), iridium(IV), palladium(II), copper(II), silver(I) and gold(III) catalyse the reduction by hydrogen of inorganic substrates such as iron(II), chronium(VI) and cerium(IV) [14].

Some of these systems also catalyse the hydrogenation of organic substrates, e.g. olefins and acetylenes. Two of these systems, the cobalt cyanide systems (section 4) and platinum-tin systems (section 5) have already been dealt with under more appropriate headings, and other classes are discussed below.

7.1. First row transition elements

The hydrogenation of cyclohexene and other simple olefins, catalysed by the carboxylates of the first row transition elements, scandium(III) to zinc(II), in ethanol at 20-60 °C and up to 100 atmospheres, has been extensively studied by Tupulov [5, 146]. The reactions are quite slow, depend only on hydrogen pressure, water is a poison and the order of activity is iron(III) > cobalt(II) > nickel-(II). The general mechanism proposed for all catalysts from d^0 to d^{10} configurations is shown in scheme 5 [146]. This differs quite radically from the mechanisms normally envisaged and composed from reactions (1) to (13) (see section 2), and is backed by extensive investigations of physical properties. In solution the carboxylates form dimers which have a "quasi aromatic" structure. This ring system reacts with olefins to give a compound in which the olefin is bound to two metal atoms. This is similar to one of the ways olefins and acetylenes are

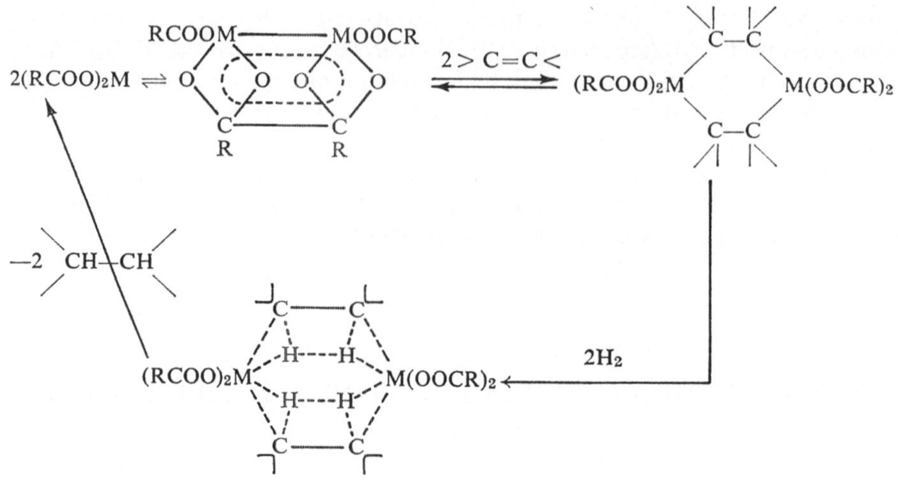

$$2(RCOO)_2M \rightleftharpoons O \underset{\underset{R}{\overset{|}{C}}---\underset{R}{\overset{|}{C}}}{\overset{RCOOM---MOOCR}{O\ O}} O \quad \underset{2> C=C <}{\rightleftharpoons} \quad (RCOO)_2M \overset{C-C}{\underset{C-C}{}} M(OOCR)_2$$

$$-2\ \diagdown CH\text{-}CH \diagup$$

$$(RCOO)_2M \overset{C---C}{\underset{C---C}{\overset{H---H}{\underset{H---H}{}}}} M(OOCR)_2 \leftarrow \quad 2H_2$$

Scheme 5

bound to heterogeneous surfaces. The olefin compound reacts with hydrogen, which is probably activated by simultaneous reaction with two metal atoms, i.e. a homolytic hydrogen cleavage as in reaction (7).

7.2. Noble metals in water, alcohols and amide solvents

Solutions of noble metal chlorides in aqueous hydrochloric acid react with hydrogen to give a hydride via reaction (6), which is an example of the general reaction (6a) [4]. In the absence of a suitable ligand to stabilise the hydride it decomposes and ultimately metal is precipitated. In catalytic hydrogenations the substrate reacts very rapidly with the hydride,

$$Ru^{III}Cl_6^{3-} + H_2 \rightleftharpoons HRu^{III}Cl_5^{3-} + H^+ + Cl^- \tag{6}$$

$$M^{x+} + H_2 \rightleftharpoons MH^{(x-1)+} + H^+ \tag{6a}$$

which is continuously regenerated until hydrogenation is complete, when metal deposition again occurs in the absence of stabilising ligands. Metal can be formed by a series of steps involving reactions (6a) and (62) or by

$$MH^{(x-1)+} \rightleftharpoons H^+ + M^{(x-2)+} \tag{62}$$

disproportionation. A similar sequence of reactions of metal halides with hydrogen occurs in amide solvents e.g. dimethylacetamide [147].

In the presence of acetate salts, aqueous solutions of palladium(II) catalyse the hydrogenation of activated olefins, e.g. α,β- unsaturated acids [148].

The hydrogenation activity of aqueous solutions of ruthenium(II) in hydrochloric acid has been thoroughly investigated by Halpern and co-workers [4, 149]. The mechanism for the hydrogenations is shown in scheme 6, where only the reacting ligands are shown, the others being a combination of water, halide etc.. The ruthenium(II) is generated by reaction of titanium(III) chloride

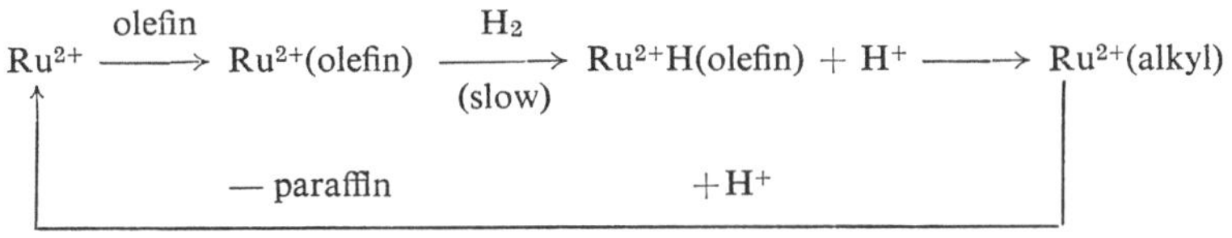

$$Ru^{2+} \xrightarrow{\text{olefin}} Ru^{2+}(\text{olefin}) \xrightarrow[\text{(slow)}]{H_2} Ru^{2+}H(\text{olefin}) + H^+ \longrightarrow Ru^{2+}(\text{alkyl})$$

$$- \text{paraffin} \qquad\qquad + H^+$$

Scheme 6

with ruthenium(III). Spectroscopic evidence shows that ethylene and α, β-unsaturated acids form complexes with ruthenium(II). However, only the activated olefins are catalytically hydrogenated, and the rate is given by $-d[H_2]/dt = k[H_2]$ · $[Ru^{II}(\text{olefin})]$ and is independent of the substrate, except at low concentrations. Experiments using deuterium, and or deuterium oxide show that hydrogen is added *cis* to maleic acid and the hydrogen is derived from the solvent. Thus the exchange reaction (63) is faster than alkyl formation.

$$Ru^{II}H + D^+ \rightleftharpoons Ru^{II}D + H^+ \qquad\qquad (63)$$

In dimethylacetamide ruthenium(III) is slowly reduced by hydrogen to ruthenium(II) at room temperature and pressure, but at 40 °C further reduction occurs to give ruthenium(I). The mechanism of this reduction is via a series of heterolytic splittings of hydrogen (6a). The ruthenium(I) solution catalyses hydrogenation of activated olefins at 80 °C and one atmosphere of hydrogen. Preliminary data indicate that ruthenium(I) is mainly dimeric in solution, but the hydrogenation of olefins is catalysed by a monomeric species [150].

Aqueous solutions of RhCl₃3H₂O do not catalyse hydrogenation of olefins [151] but ethanolic solutions are claimed to catalyse hydrogenation of hex-1-ene [152]. However, metal is precipitated during the process and it is not clear if it is a homogeneous hydrogenation.

However, solutions of rhodium(III) chloride in dimethylacetamide are catalysts for the hydrogenation of ethylene and some activated olefins, e.g. maleic acid, maleic anhydride, maleic esters etc.. [147] Typical conditions are 80 °C, 1 atmosphere of hydrogen, rhodium conc. approx. 10^{-3} M and substrate concentration 1 M. The mechanism is very similar to that proposed for hydrogenations catalysed by ruthenium(II) in water (see above). Initially rapid uptake of hydrogen occurs which corresponds to the formation of a rhodium(I) complex, reactions (64) and (65). In the absence of a suitable olefin substrate this rapidly gives rhodium metal. However in the presence of olefin a stable rhodium(I) complex is formed rapidly, (66) and this slowly reacts with hydrogen to give product and regenerate rhodium(I) (67). In the

$$Rh^{III} + H_2 \rightleftharpoons Rh^{III}H + H^+ \qquad (64)$$

$$Rh^{III}H \rightleftharpoons Rh^I + H^+ \qquad (65)$$

$$Rh^I + \text{olefin} \rightleftharpoons Rh^I(\text{olefin}) \qquad (66)$$

$$Rh^I(\text{olefin}) + H_2 \rightarrow Rh^I + \text{paraffin} \qquad (67)$$

presence of excess chloride ion the rate is given by $-d[\text{olefin}]/dt = k[\text{Rh}][\text{H}_2]$ which supports such a mechanism. In the absence of added chloride complicated kinetics are obtained due to bridged rhodium species [147]. The formula of the rhodium(I) olefin complex is not known and it is not known how hydrogen activation occurs in reaction (67) (see section 2.2.). The complex $RhCl_3(SEt_2)_3$ behaves kinetically like the system, $RhCl_3 3H_2O + Cl^-$, but is rather less active [153].

It appears that solutions of noble metal halides are much more active in dimethylformamide than in dimethylacetamide. Solutions of dicyclopentadiene (15%) in dimethylformamide are reduced at 25-80 °C and 1 atmosphere pressure by noble metal halides ($2 \cdot 10^{-2}$ M). The order of activity is $PdCl_2 > RhCl_3 > RuCl_3 > PtCl_2$ and the reactions appear to be homogeneous until most of the olefin has disappeared. The addition of thiophene, a heterogeneous catalyst poison, increases the activity of $PdCl_2$ proving the reaction is truly homogeneous [27].

A recent series of publications [155] has demonstrated that dimethylformamide solutions of rhodium(I) complexes with tyrosine, anthranilic acid or phenyl acetic acid are excellent catalysts for the hydrogenation of olefins. The structure of these complexes is not clear and they are best represented as H $[Rh_2(PhCH_2COO)_2Cl]$. Spectral evidence suggests the RCOO groups is a bridging ligand or a chelating ligand, rhodium interacting with the π electrons of the phenyl ring. Solutions of the complex absorb hydrogen to give rhodium hydrides which do not decompose to metal. The activity of the catalyst compares well with any other soluble rhodium catalyst. The nature of the intermediates during hydrogenation is not known but it has been shown by deuterium studies that reverse alkyl formation occurs. The catalysts also catalyse hydrogen deuterium exchange between deuterium and dimethylformamide (68).

$$RhD + OCHNMe_2 \rightleftharpoons Rh\text{---}O\text{---}CHDNMe_2 \rightleftharpoons RhH + OCDNMe_2 \qquad (68)$$

More recently a catalyst has been obtained by mixing $RhCl_3py_3$ with sodium borohydride in dimethylformamide. A solid analysing for $RhCl_2(BH_4)DMFpy_2$ has been isolated, and appears to be the actual catalyst [156]. The complex $RhCl_3py_3$ itself is a moderately active catalyst at room temperature and pressure in ethanol [152], but with borohydride in dimethylformamide has activity compared to the best rhodium catalysts. Presumably pyridine or dimethylformamide is easily displaced by olefin as hydrogenation is slowed down in the presence of an excess of pyridine.

The borohydride catalyst has been likened to heterogeneous catalysts because it resembles them in stereoselective hydrogenation of certain 3-oxo-Δ-4,5-steroids [157] and the hydrogenation of some cyclic olefins [158]. However, much more evidence is required to make these comparisons meaningful. The most important feature of this catalyst is that assymetric hydrogenations can be carried out in optically active amides. Thus, in (+) or (—) 1-phenylethyl-N-formamide methyl 3-phenylbut-2-enoate is hydrogenated to (+) or (—) methyl 3-phenylbutanoate in over 50% optical yield [159]. This is the best homogeneous asymmetric hydrogenation described to date and the catalyst should be of great value in this field.

Most of the catalysts described in this section contain so far the metal as an anion or a neutral complex. Recently cationic metal complexes which are efficient hydrogenation catalysts have been prepared. Salts of the cation $[RhH(NH_3)_5]^{2+}$ are easily prepared and are remarkably stable [160]. In aqueous solution an ammonia ligand is easily lost to give $[RhH(NH_3)_4H_2O]^{2+}$ and the hydride slowly forms an alkyl complex with olefins. At 60 °C and 1 atmosphere of hydrogen it appears to have some use as a catalyst for hydrogenation of water soluble olefins, e.g. acrylic acid [161].

More active cationic catalysts are complexes of general formula $[MH_2(PPh_3)_2S_2]^+$ where M is rhodium or iridium and S is a solvent molecule, acetone, ethanol, methoxy ethanol, dimethylacetamide etc.. These complexes are prepared via reactions (69) and (70), HX is a strong acid, e.g. perchloric.

$$[M(diene)Cl]_2 \xrightarrow[\text{ionic solvent}]{PPh_3, \; HX} [M(diene)(PPh_3)_2]^+ \qquad (69)$$

$$[M(diene)(PPh_3)_2]^+ \xrightarrow{H_2}{S} [MH_2(PPh_3)_2S_2]^+ \qquad (70)$$

The catalysts can be used in ionic solvents such as acetone and alcohols, and appear as active as $RhCl(PPh_3)_3$. They reduce acetylenes more readily than olefins, and the order of activity of olefins is terminal > *cis* > *trans* > trisubstituted as is always found [162].

Cationic catalysts can also be obtained by treating metal acetates containing metal-metal bonds, e.g. $Rh_2(CO_2CH_3)_4$, $Ru_2(CO_2CH_3)_4Cl$ and $Mo_2(CO_2CH_3)_4$ with a strong acid, e.g. $HBF_4(40\%)$ in methanol in the presence of two equivalents of triphenylphosphine per metal atom [163]. The nature of these catalysts has not been recorded but they may well be $[MH_2(PPh_3)_2S_2]^+$.

8. ORGANOMETALLIC CATALYSTS

The compounds described in this section are derived from Ziegler-Natta polymerisation catalysts. Natta proposed that addition of hydrogen controlled the chain length of the polymer by hydrogenolysis of the metal carbon bond [164]. These systems have been developed as hydrogenation catalysts by simply increasing the ratio of hydrogen to olefin. Some extremely active catalysts have been prepared and they show certain similar features. Generally the nature of the ligands round the metal are not known with certainty, all the systems are thought to operate by formation of a transition metal hydride, which reacts with olefin to give an alkyl (71), which

$$\text{MH} + \text{olefin} \rightleftharpoons \text{M-alkyl} \tag{71}$$

$$\text{M-alkyl} + \text{H}_2 \rightarrow \text{MH} + \text{paraffin} \tag{72}$$

in turn is hydrogenolysed (72). Several ways of producing the metal hydride have been used. Reaction of a transition metal salt with a metal alkyl, e.g. a Grignard reagent or a lithium alkyl, gives a transition metal alkyl, which cleaves with hydrogen, as in reaction (72), or eliminates olefin as in the reverse of (71), to give the intermediate hydride. The transition metal hydride can also be formed directly with various types of aluminium hydrides. Some of these catalysts have been used to reduce aromatic hydrocarbons as well as olefins.

8.1. Hydrogenation of olefins

Combinations of transition metal acetylacetonates (10^{-3} M) and aluminium alkyls, in paraffin solvents, are good catalysts for the hydrogenation of olefins (1 M) at 1-4 atmospheres and room temperature. Two independent groups of workers have come to the same general conclusions [165, 166]. The rate of hydrogenation is given by $-d[\text{H}_2]/dt = k[\text{cat}][\text{H}_2]$ and is generally independent from olefin concentration. There is an optimum ratio of aluminium to transition metal betwees 6 and 10 to 1 and there seems little difference which aluminium alkyl is used. The reason so much is required is that the aluminium alkyl is consumed by the acetylacetonate ligand, as well as by preparing a transition metal alkyl, and possibly reducing the transition metal to a suitable oxidation state. Generally the order of activity of various olefins is terminal > disubstituted > trisubstituted > tetrasubstituted, which are only reduced at 50 °C. Deuteriogenation gives mono, di and trideutero paraffins showing that alkyl formation is reversible [166]. Although acetylenes are hydrogenated they are also oligomerised to aromatic compounds. The activity of the metals is cobalt > nickel > iron > chromium > > titanium > manganese > vanadium [165, 166]. The metals are best introduced as oxygen containing salts e.g. carboxylates, alkoxides, phenoxides etc., halides are not suitable as the catalyst precipitates out [166].

These types of catalyst have been used under much more vigorous conditions, 150 °C and 150 atmospheres, to catalyse hydrogenation of soybean fatty esters, methyl linoleate and methyl linolenate to monoenes [167].

The reaction of Grignard reagents with cobalt stearate in ether has been studied in detail by Marko and co-workers. Treatment of the mixture with carbon monoxide and triphenylphosphine gives a variety of cobalt acyls [168] and cobalt hydrides [169], which proves that cobalt alkyl and cobalt hydrides are present in the mixture. Like the aluminium alkyl systems the nature of the Grignard is unimportant, but there is an optimum ratio of magnesium reagent: cobalt of about 3 to 8 : 1 for the hydrogenation catalysts [170]. It is thought that solutions of the cobalt alkyls and hydrides are stabilised by ligands which are generated in situ by the reaction of cobalt stearate with Grignard reagents. It was suggested that one such ligand is $C_{16}H_{33}$—CH=C(OMgCl)$_2$ which would revert to stearic acid on hydrolysis. Stearic acid is recovered quantitatively on hydrolysis of the mixture.

Lithium alkyls are also suitable for activating the transition metal [166]. In the hydrogenation of polybutadiene in cyclohexene at 40 °C and one atmosphere pressure by nickel salts, activated by lithium aluminium hydride or butyl lithium, the presence of an ether is essential and there is an optimum amount. The order of adding reagents is also important, the highest activity being obtained in the order, ether, butyl lithium, nickel salt [171]. This demonstrates the sensitivity of some of these organometallic catalysts.

A combination of titanium or zirconium cyclopentadienyl complexes, Cp_2MCl_2, and metal alkyls are also hydrogenation catalysts, but they are not as active as the Group VIII metal acetylacetonates [166]. The relationship between Cp_2TiCl_2 and the metal alkyl is quite complex. When the alkylating agent is an aluminium alkyl, the hydrogenation activity dies off rapidly, but does not die off in the presence of a small, critical amount of oxygen [172]. The optimum ratio of aluminium to titanium depends on the amount of oxygen used. These results indicate that titanium(IV) is the catalyst. In the presence of aluminium alkyls the titanium is rapidly degraded to titanium(III) and titanium(II) which have little activity. The oxygen is necessary to oxidise the titanium back to titanium(IV), but of course destroys aluminium alkyl in the process [172]. Indeed it has been reported that Grignard reagents and lithium alkyls are suitable for forming the catalysts, but aluminium alkyls are not [173]. The inactive titanium(II) and titanium(III) species are [$Cp_2TiCl \cdot Al(alkyl)_2Cl$] and [$Cp_2Ti \cdot Al \cdot (alkyl)_2Cl$] [173].

Under more vigorous conditions, viz $\cdot$ 40-50 °C and up to 60 atmospheres pressure, Cp_2TiCl_2 in combination with 4-7 equivalents of butyl lithium or a Grignard reagent catalyses hydrogenation of butadiene to butane but with Cp_2VCl_2 the hydrogenation stops at the butene stage. Monocyclopentadienyl complexes of nickel, cobalt and iron also reduce conjugated dienes to monoenes [174].

An important recent development in this field is the use of lithium tri-t butoxyaluminium hydride or di-t-butoxyaluminium hydride in place of aluminium alkyls etc.. The use of these reagents gives catalysts which are at least ten times more active than ones prepared from aluminium alkyls, the amount of aluminium alkoxy hydride added is not critical after a certain minimum (aluminium: transition metal ratio $\geqslant 4$), and the anion of the transition metal salt is not so critical. Cobalt salts are more active than nickel or iron salts [175], and Ti(OR)$_4$ has very little activity [176]. However, it appears that the titanium compound $Cp_2Ti(OPh)_2$ is the most active catalyst [176]. At 20 °C, and one atmosphere pressure and with titanium concentration of $2.5 \cdot 10^{-4}$ M the rate of hydrogenation of hex-1-ene(0.8M) is 0.572 mol.l^{-1}min.$^{-1}$. This highly active system $Cp_2TiX_2 + LiAlH_x(OR)_{4-x}$ is fairly sensitive to X, x and the ratio aluminium: titanium. Olefins are reduced in the order (relative rates in parenthesis) hex-1-ene(1.22), cyclopentene(0.75), cyclooctene(0.77), 2-methylpent-1-ene(0.61), hex-2-ene(0.31), cyclohexene(0.07), 2,3-dimethylbut-1-ene(0.025), 2,3-dimethylbut-2-ene(0). Little olefin isomerisation is observed in this rapid reaction [176].

8.2. Hydrogenation of aromatics

Under much more vigorous conditions, approx. 140-190 °C and approx. 60 atmospheres of hydrogen, mixtures of metal carboxylates and aluminium triethyl (ratio aluminium: metal 3 or 4 : 1) will reduce aromatic rings. The order of activity is nickel > cobalt > iron > chromium > copper. The catalyst solution is dark coloured but is truly homogeneous [177]. Similar results are obtained using metal acetylacetonates ($2 \cdot 10^{-4}$ M) and triethyl aluminium (ratio aluminium: metal 6 to 8 : 1) [178]. Other metal salts are not as effective, as in the hydrogenation of olefins. Substitution of alkyl groups in the aromatic ring lower the rate of hydrogenation and the activation energy for hydrogenation of benzene is 9.6 kcals[179]. As with olefins the rate of hydrogenation is given by $- d[H_2]/dt = = k[H_2][\text{catalyst}]$, the rate being independent of substrate. The active intermediate is presumably stabilised by arene ligands, which are reduced in a stepwise manner to cyclohexadiene (slow) to cyclohexene and cyclohexane (fast) [179].

Recently it has been shown that cobalt stearate and aluminium triethyl is active enough at 30 °C and 50 atmospheres hydrogen pressure to hydrogenate aromatics, provided the ratio of aluminium: cobalt is less than 2. The catalyst has no activity under these conditions at a ratio of 4. Xylenes are reduced to a mixture of *cis* and *trans* dimethylcyclohexanes, the *cis* isomer predominating In hydrogenations where there is a competition between benzene and another substrate for the catalyst it was found oct-1-ene was reduced whereas the benzene was not attacked, benzene was reduced twelve times faster than xylene and naphthalene was reduced about twenty times faster than benzene. The same catalyst also hydrogenates dienes (mainly to paraffins), furan, phenol, acetone and mesityl oxide, the latter two to alcohols [180].

9. METAL CARBONYLS

Hydrogenation of olefins to paraffins is a side reaction in hydroformylation and has recently been reviewed [6]. A minimum carbon monoxide pressure, depending on the temperature is required to keep the solution of catalyst stable, consequently hydroformylation always occurs. Metal carbonyls have recently been reported to catalyse hydrogenation of conjugated dienes to monoenes, sometimes with remarkable specificity. The use of iron pentacarbonyl to hydrogenate polyunsaturated fatty acid esters to monoene esters has been reviewed [7], and further advances in selective hydrogenation of conjugated dienes will be discussed elsewhere [11]. It has been reported that $Cp_2Ti(CO)_2$ ($2.5 \cdot 10^{-3}$ M) catalyses the hydrogenation of alk-1-ynes to alk-1-enes at 50 °C and 50 atmospheres pressure. The catalyst does not hydrogenate internal alkynes or conjugated dienes under similar conditions [181].

10. MISCELLANEOUS HYDROGENATIONS

More progress has been made in homogeneous hydrogenation of olefins than any other functional group. The hydrogenation of aromatics was discussed in section 8.2 and the reduction of acetylenes has been described throughout the preceding text. The cobalt cyanide system has been used to catalyse hydrogenation of allyl derivatives to propylene, epoxides to alcohols, benzil to benzoin, pyruvic acid and ammonia to alanine, oximes to amines, nitrobenzene to aniline, azoxybenzene, azobenzene, hydrazobenzene, and aliphatic nitro compounds to amines (full lists and details are given in Refs. 2a and 3).

Under an atmosphere of mixed gas ($H_2 : CO = 1 : 1$) cobalt hydrocarbonyl, which can be formed in situ from cobalt octacarbonyl or cobalt salts, catalyses hydrogenation of various substrates. Several aromatic compounds are partially reduced, e.g. anthracene to 9,10-dihydroanthracene, pyrene to 4,5-dihydropyrene, naphthalene to tetralin etc. Thiophene is reduced to thiolane, indoles to dihydroindoles and pyridine to N-methylpiperidine. Other functional groups can also be reduced, e.g. benzyl alcohol to toluene, ketones to secondary alcohols. aryl ketones to hydrocarbons and aldehydes to alcohols (see Ref. 3 for a full list).

10.1. Hydrogenation of aldehydes

The hydrogenation of aldehydes by cobalt hydrocarbonyl has been recently reviewed [6]. At 150 °C a minimum partial pressure of 32 atmospheres of carbon monoxide is required to prevent metal deposition. The true catalyst is thought to be the co-ordinatively unsaturated species $HCo(CO)_3$ and the mechanism proposed is via reactions (73) and (74), where M is cobalt. The rate of

$$R{-}CH{=}O + HM(CO)_3 \rightleftharpoons RCH{=}O \rightleftharpoons RCH_2COM(CO)_3 \qquad (73)$$
$$\downarrow$$
$$HM(CO)_3$$

$$RCH_2COM(CO)_3 + H_2 \rightarrow RCH_2OH + HM(CO)_3 \qquad (74)$$

reaction is given by $d[\text{ROH}]/dt = k[\text{RCHO}][\text{Co}][\text{H}_2][\text{CO}]^{-2}$. The inverse dependence on carbon monoxide is because its presence favours the formation of HCo(CO)_4 rather than HCo(CO)_3.

Rhodium carbonyls, prepared in situ from $\text{RhCl}_33\text{H}_2\text{O}$ and mixed gas, are also good catalysts for hydrogenation of aldehydes. At 150-200 °C and approx. 200 atmospheres of mixed gas the rate is given by $d[\text{ROH}]/dt = k[\text{RCHO}]$ [catalyst], where the catalyst concentration is given by the expression $[\text{Rh}]^{1/6} [\text{pH}_2]^{0.5} [\text{pCO}]^{-0.3}$, where p denotes a partial pressure. The power to one sixth on rhodium is because, under the conditions, the rhodium is thought to be mainly in the form of $\text{Rh}_6(\text{CO})_{16}$, which is inactive. The active species is thought to be HRh(CO)_3 and the reaction sequence is (73) and (74), where M is rhodium [182, 183].

Recently, phosphine stabilised cobalt carbonyls have been patented as hydrogenation catalysts [184]. Although several experimental procedures can be used it appears tha $\text{HCo(CO)}_3\text{PR}_3$ is formed in situ. This compound is much more stable than HCo(CO)_4 and it is useful in the area 100-200 °C and up to 25 atmospheres of hydrogen. Approximately 5 atmospheres partial pressure of carbon monoxide is required to prevent catalyst decomposition. A variety of substrates are reduced including olefins, aldehydes and ketones.

Phosphine complexes of other metals have been described as homogeneous catalysts for hydrogenation of aldehydes and ketones. For example in hydrocarbon or alcohol solvents mixtures of $\text{RhCl}_33\text{H}_2\text{O}$ triphenylphosphine are active at 110 °C and 60 atmospheres pressure [128] and $\text{IrH}_3(\text{PPh}_3)_3$ is active at 100 °C and 100 atmospheres of hydrogen [135]. The catalysts obtained by treating cobalt stearate with aluminium trimethyl (aluminium: cobalt ratio $\leqslant$ 2) also catalyses hydrogenation of ketones at 50-80 °C and 50-60 atmospheres of hydrogen [180].

All the systems which catalyse homogeneous hydrogenation of carbonyl groups mentioned above catalyse the hydrogenation of other functional groups. The order of activity for all these catalysts is terminal olefin > internal olefin > trisubstituted olefin > aldehydes > ketones, in other words the catalysts tend to be specific for olefins (or acetylenes) but not for carbonyl groups. However, in acetic acid at 50 °C and 1 atmosphere, $\text{IrH}_3(\text{PPh}_3)_3$ (10^{-3} to 10^{-2} M) catalyses the hydrogenation of aldehydes (0.5 to 1M) at a rate of approx. 0.5 mole l.^{-1} hr.^{-1}. Under the same conditions ketones, and oct-1-ene are not reduced, but some activated olefins, e.g. acrylic acid (but not acetonitrile) are slowly hydrogenated (0.1 mol. l.^{-1} hr.^{-1}) [94, 185]. This is the only example of specific hydrogenation of aldehydes under mild conditions. The reaction is not sensitive to excess phosphine (ratio excess PPh_3 : Ir = 4 : 1), and consequently it is unlikely

that a π bonded intermediate, such as $\left[\text{L}_x\text{Ir} \leftarrow \overset{\text{O}}{\underset{\text{CHR}}{||}} \right]$ is important. Intermediate

hydrido acetates of general formula $\text{IrH}_x(\text{OOCCH}_3)_y(\text{PPh}_3)_3$ have been isolated from the reaction mixture and appear to be the true catalyst. Their structure and exact composition is not known.

10.2. Some hydrogen transfer reactions

This section refers to some systems where a substrate is hydrogenated, but the source of hydrogen is other than hydrogen gas. This is not an extensive coverage, but lists some examples where a metal complex catalyses several hydrogen transfers per mole of catalyst.

The reaction of substituted cyclic ketones with aqueous isopropanol solutions of chloroiridic acid, containing various ligands, e.g. dimethylsulphoxide, trimethylphosphite or phosphorous acid gives very high yields of axial alcohol [186]. The reaction path does not appear to be simple as reductive etherification occurs to give isopropyl ethers, when the water content drops below 5%. From the sulphoxide systems a series of iridium complexes have been isolated, e.g. $IrCl_3(Me_2SO)_3$, $H[IrCl_4(Me_2SO)_2]_2(Me_2SO)$, and $IrHCl_2(Me_2SO)_3$ [186, 187]. The hydride, in refluxing aqueous (2%) isopropanol, containing a trace of hydrochloric acid, catalyses hydrogen transfer from the isopropanol to various substrate. For example 4-t-butylcyclohexanone gives 4-t-butylcyclohexanol (77% axial), various α-β ketones give saturated ketones [187] and diphenylacetylene gives cis-stilbene [188]. From benzylideneacetophenone and the hydride an intermediate was isolated [187], which was shown to have structure IX by X-ray analysis [189]. From the reduction of diphenylacetylene an intermediate X

was obtained which on hydrolysis gives cis-stilbene.

The phosphite systems are more complicated as the phosphite is oxidised to phosphate during the reaction and the hydrogen is derived from water. The formation of acetone is a minor side product and the reduction is stoichiometric on phosphite [190].

A true hydrogen transfer occurs between refluxing isopropanol and cyclohexanone in the presence of traces of $IrH_3(PPh_3)_3$, $IrH_3(PPh_3)_2$, $ReH_7(PPh)_3$ and $RuHBrCO(PEt_2Ph)_3$, when acetone can be steadily distilled off [94, 135]. With 4 substituted cyclohexanones equatorial alcohols are the major product [190]. This system cannot be used for reduction of aldehydes. Solutions of aldehydes (1M) in acetic acid are smoothly reduced to alcohols by formic acid at 50 °C in the presence of $IrH_3(PPh_3)_3$ (10^{-3}M), but saturated ketones are not reduced under these conditions [185].

It would appear that several reactions have to occur within a narrow range of conditions for these hydrogen transfers to occur. At some stage hydrogen has to be derived from the reductant, e.g. formic acid or isopropanol in the cases outlined above. This must be followed by addition of the metal hydride across the unsaturated system, e.g. formation of IX and X, and then the σ bond between the metal and the organic group must be cleaved by hydrolysis, which may or may not involve formation of another metal-hydrogen bond. Naturally the tendency for all these reactions to occur will depend on the metal, its ligands, the substrate, the hydrogen donor, and reaction conditions and this is why these hydrogen transfers tend to be very specific.

The high yield of axial alcohols obtained in reduction of cyclic ketones in sulphoxides and phosphites systems compared to the high yield of equatorial alcohols formed in phosphine systems is rather surprising, because both reactions appear to go via hydride intermediates and, to a large extent, the metal and its ligands must have similar steric sizes, unless strong solvation of the ligands plays an important part in the phospite and sulphoxide systems. An alternative explanation is that the metal hydrogen bond may be polarised differently in the two systems. In phoshine systems the metal hydrogen is polarised H^-—M^+

and reaction with a $\diagdown C = O$ group leads to an alkoxide intermediate, in which the bulky group projects away from the ring and hydrolysis of the O—M bond gives an equàtorial alcohol. With phosphite and sulphoxide ligands polarisation may be H^+—M^-, as indicated by the formation of intermediate IX, in which case a metal-carbon bond will be formed and this will project away from the ring, leaving a hydroxyl group in an axial position. Subsequent hydrolysis of the M—C bond will give an axial alcohol.

Hydrogen transfer to reduce olefins and dienes has also been recorded. Soya bean methyl ester, esters of linoleic and linolenic acid and other polyenes can be hydrogenated, mainly to monoenes, in the absence of hydrogen but in the presence of suitable solvents at about 100 °C. The compownds $MCl_2(PPh_3)_2$, where M is Pt [63] or Pd [154], combined with $SnCl_2$ are good catalysts in alcoholic media or tetrahydrofuran. The nickel complex $NiX_2(PPh_3)_2$, where X is halide brings about the hydrogenation in tetrahydrofuran or benzene [123]. The source of hydrogen in the latter case is not clear. It may come from the solvent, the ligand, via attack on the ortho position of the phenyl ring of the ligand, or from the substrate itself, i.e. the latter in effect disproportionates to monoene and a higher polyene. Evidence for this latter process is that homogeneous disproportionation of cyclohexa-1,4-diene to benzene and cyclohexene, catalysed by $IrClCO(PPh_3)_2$, has been described [191].

11. RECENT RESULTS

A review, which deals with homogeneous hydrogen-deuterium exchange, isomerisation and hydrogenation catalysed by bases, acids, non-transition metals, transition metals and enzymes, has recently appeared [197]. The complexes, $IrHCO(PPh_3)_2(ol)$, where ol = fumaronitrile (FUMN), cinnamonitrile, tetracyanoethylene and fumaric, maleic and cinnamic acids and their methyl esters, have been prepared by treating $IrHCO(PPh_3)_3$ with the appropriate olefin [198]. These are further examples of hydrido olefin complexes, thought to be intermediates in many hydrogenations and an X-ray structure of IrHCO-(PPh$_3$)$_2$(FUMN) shows that the hydrogen is *cis* to the olefinic ligand [199].

A recent review [234] shows some aspects of selective hydrogenation catalysis.

An investigation [235] on the kinetics of hydrogenation of ethylene on homogeneous platinum-tin chloride catalyst showed that the reaction rate, in different protic solvents, is $- d[C_2H_4]/dt = K_{eff}[Pt][C_2H_4](p_{H_2} > 200$ mmHg$)$.

This equation agrees with the already proposed mechanisms.

11.1 Catalysts stabilised by tertiary phosphines

An X-ray structure shows that RhCl(PPh$_3$)$_3$ is not square planar but distorted towards a tetrahedral structure [200]. Hydrogenation of vinylcyclopropane gives pent-2-ene which is reduced further to pentane. The stereochemical course of this reduction and related cyclopropylolefins indicates stepwise addition of hydrogen [201]. At 60 °C and 5 atmospheres the complex is useful for reducing the carbon-carbon double bond of α,β unsaturated ketones, acids, esters, nitro compounds and aldehydes. Decarbonylation of α,β unsaturated aldehydes can be diminished considerably if absolute ethanol is used as a solvent [202] and trichlorotris(4-biphenylyl-1-naphthylphenylphosphine)rhodium(III) is effective as a catalyst under these conditions [203].

Some very recent investigations [226] have shown that dissociation of RhCl(PPh$_3$)$_3$ occurs quite readily in benzene but it is completely inhibited by the presence of even small amounts of alcohol.

Moreover the same authors [227] have pointed out that the isomerisation of olefins over RhCl(PPh$_3$)$_3$ is dependent on the presence of oxygen in the system and on the solvent used. Also the products obtained on oxidation of RhCl(PPh$_3$)$_3$ are markedly dependent on the solvent in which the reaction is run [228]. These results show that oxygen and solvents play an important rôle in the catalytic activity and mechanism of action of RhCl(PPh$_3$)$_3$. They suggest that the rate enhancement of olefin hydrogenation over RhCl(PPh$_3$)$_3$, which is caused by oxygen, is not due to the oxidation of the dissociated phosphine, as previously suggested, but to the formation of a more active species.

In fact hydrogenation of the oxide species obtained from oxidation of RhCl(PPh$_3$)$_3$ in ethanol and benzene gives materials capable of olefin hydrogenation but isomerisation takes place during the reaction [229]. In the presence of PPh$_3$ these new catalytic species are more active and isomerisation is inhibited. These data serve to indicate why isomerization occurs readily during hydrogenation over RhCl(PPh$_3$)$_3$ in alcoholic solutions but not in benzene or benzene-hydrocarbon solvents even in the presence of oxygen.

The presence of alcohol prevent dissociation of RhCl(PPh$_3$)$_3$ to give PPh$_3$. Thus the oxidised species, which is formed in solution, is further hydrogenated in the absence of phosphine and isomerisation occurs. In benzene, however, phosphine is present and isomerisation is prevented.

Interestingly during the hydrogenation of the material obtained from the oxidation of $RhCl(PPh_3)_3$ in benzene, cyclohexane is formed by benzene hydrogenation, but only in small amounts [229].

Some of the rhodium carbonyls and hydrides, which are easily interconverted as shown in scheme 7, have been prepared, where L = PPh_3, PEt_2Ph, $PMePh_2$ and PEt_3. They are all good catalysts for hydrogenation of α-olegins under ambient conditions [204]. The nature of the inactive complex formed during

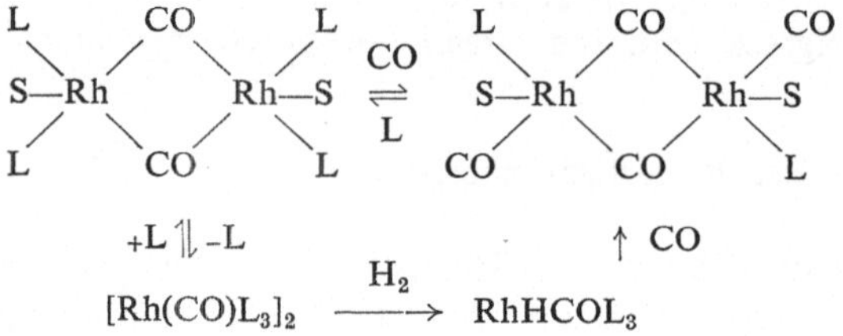

Scheme 7

hydrogenation catalysed by $HRhCO(PPh_3)_3$ has been studied. This complex is shown to be $[Rh(CO)(PPh_3)_2]_2$ [230]. The nitrosyl $RhNO(PPh_3)_3$ does not appear to give a hydride with hydrogen but it catalyses hydrogenation of simple olefins under ambient conditions and catalyses deuteriogenation of cyclohexene to $C_6H_{10}D_2$ (> 99 % isotopic purity) [205]. Studies of hydrogenation of hept-1-ene at 70° catalysed by $RhXCO(PR_3)_2$ show that isomerisation and hydrogenation occur at similar rates, but the relative rates of each depend on the nature of X and PR_3. The activity of the complexes decreases in the order X = Cl > Br > I and PR_3 = PPh_3 > $P(C_6H_{11})_3$ > $P(OPh)_3$, and the ratio rate of isomerisation rate of hydrogenation is greatest for the iodides [206].

The rate of the reaction of $IrXCO(PR_3)_2$ with hydrogen in toluene has been studied as a function of X and PR_3. The rate increases in the order X = I > > Br > Cl and PR_3 = $P(p\text{-tolyl})_3$ > PPh_3 > $P(OPh)_3$ > $P(alkyl)_3$. Aryl phosphine complexes are about 500-700 times more reactive than the alkyl phosphine complexes and iodides are about 500 times more active than chlorides. The activation energies lie between 10 and 13.6 Kcals, depending on the complex, and the values for $IrClCO(PPh_3)_2$(11.3 Kcals) and $IrBrCO(PPh_3)_2$ (11.9 Kcals) are in good agreement with those published earlier (see table 1) [207]. The hydrogenation of dimethylmaleate and maleic acid at 70 °C and 1 atmosphere catalysed by $IrXCO(PR_3)_2$ has been studied [208]. The sequence proposed is scheme 8 and the rate is given by (75).

$$IrH_2ClCOP_2 \underset{}{\overset{K_1}{\rightleftharpoons}} IrClCOP_2 \underset{}{\overset{K_2}{\rightleftharpoons}} IrClCOP_2ol \underset{}{\overset{K_3}{\rightleftharpoons}} IrClCOPol \xrightarrow[K]{H_2} IrClCOP + olH_2$$

Scheme 8

$$r = K\,[H_2][IrClCOPPh_3(\text{olefin})] \tag{75}$$

This differs in detail from scheme 3, which applies in dimethylacetamide.

A series of catalysts $RuH(OOCR)(PPh_3)_3$, where R is various alkyl aromatic and haloalkyl groups, have been prepared from $RuCl_2(PPh_3)_3$, the sodium salt of the appropriate acid and a reducing agent, e.g. hydrogen or sodium hypophosphite. They show similarities to $RuHCl(PPh_3)_3$, being specific for terminal and cyclic olefins, die off rapidly, sensitive to oxygen, inactive in pyridine, methylene dichloride and in the presence of excess phosphine, but they are easier to study than $RuHCl(PPh_3)_3$. Their activity is comparable to $RhCl(PPh_3)_3$ but the kinetics and mechanism is similar to $RhHCO(PPh_3)_3$ [209]. An X-ray structure of the acetate shows the ruthenium is at the centre of a distorted octahedron, the acetate is a chelating bidentate ligand [210]. Many phosphine complexes of ruthenium and osmium catalyse hydrogenation of a variety of functional groups, e.g. ketone and cyanide, at 40 atmospheres and t > 50 °C [211]. Also zerovalent platinum and palladium phosphine complexes catalyse hydrogenation of unsaturated organic substrates under pressure in homogeneous phase [233].

Recently it has been established there are two series of phosphine stabilised interconvertible cobalt hydride or cobalt nitrogen systems as shown in (76) and (77) [212]. The four complexes catalyse hydrogenation of ethylene under ambient

$$
CoHN_2(PPh_3)_3 \underset{+N_2}{\overset{+H_2}{\rightleftharpoons}} CoH_3(PPh_3)_3 \tag{76}
$$

$$
CoN_2(PPh_3)_3 \underset{+N_2}{\overset{+H_2}{\rightleftharpoons}} CoH_2(PPh_3)_3 \tag{77}
$$

conditions, presumably via the di and trihydrides [213]. A variety of phosphine stabilised cobalt carbonyls and hydrido carbonyls are hydrogenation catalysts at moderate temperatures and pressures. Thus $CoH(CO)_2(PBu_3)_2$ and $CoH(CO)(PBu_3)_3$ [214] catalyse hydrogenation of aldehydes, under hydroformylation conditions and acetylenes and olefins at 60 °C and 30 atmospheres. At 30 °C $CoH(CO)(PBu_3)_3$ selectively hydrogenates linear olefins [215]. The paramagnetic complex $(Co(CO)_2PBu_3)_3$ and $Co(\pi\text{-}C_4H_7)(CO)_2PBu_3$ catalyse hydrogenation and and isomerisation of olefins at 60 °C and 15 atmospheres of hydrogen. As isomerisation is faster than hydrogenation these complexes do not show a great deal of selectivity between different olefins [216]. The phosphite complexes $CoX(P(OEt)_3)_4$ where X = halide are catalysts for the hydrogenation of acetylenes and activated olefins under mild conditions. When C_2H_5OD is used as solvent deuterium is not incorporated into the substrate [217].

It was reported that certain soluble rhodium complexes of formula $[RhH_2L_2S_2]^+$ (L = PPh_3, PPh_2Me, $PPhMe_2$, PMe_3) can function also as homogeneous catalysts for the hydrogenation of ketones, when promoted by small quantities of water [231]. Aldehydes are also initially reduced but a rapid decrease in catalytic activity is soon observed. These catalysts reduce olefins, diolefins and acetylenes but the reduction of olefins is inhibited by the addition of water.

64

11.2 Organometallic Catalysts

Catalysts formed from nickel and cobalt carboxylates and $AlEt_3$ (> 1 mole) or lithium alkyls (8 moles) are useful for hydrogenation of polythene and considerable isomerisation occurs [218]. Both the black solid and the black solution formed from metal halides and $AlEt_3$ are active catalysts [219]. An extensive study [220] on the catalyst formed from $Co(acac)_2$, $Ni(acac)_2$ and $Fe(acac)_3$ and $Al(i.C_4H_9)_3$ show that cobalt is the most active catalyst and the ratio of aluminium to transition metal is very critical for optimum activity (5 for Ni and Co and 8-10 for Fe). The deleterious effect of excess aluminium can be minimised by adding $Al(iC_4H_9)_3$ as a complex with a Lewis base, e.g. dioxane, or by adding tertiary butanol, which gives a slightly less active catalyst that is insensitive to tertiary butanol. Isomerisation occurs and the catalyst shows great selectivity; hex-1-ene is reduced before cyclohexene in 50 : 50 mixtures and phenylacetylene (7 parts) is reduced before hex-1-ene (93 parts). Zirconium tetrachloride reacts with $PhCH_2MgCl$ to give $ZrCl_x(PhCH_2)_{4-x}$ where $x = 1$ to 3 and these compounds are good catalysts for hydrogenation of olefins at 0 °C and aromatic rings at 50 °C [221].

11.3 Complexes Active in Polar Media

Dimethylformamide solutions of complexes of rhodium with aromatic acids, particularly amino acids, are active for the hydrogenation of aromatic rings (anthacene > benzene > naphthalene). The L-tyrosine derivatives of rhodium catalyse an asymmetric hydrogenation of acetoacetic ester to ethyl β-hydroxybutyrrate and it has been suggested these catalysts resemble the enzyme hydrogenase [222]. Aromatic ligands, e.g. duroquinone, mesitylene etc. also stabilise reduced solutions of rhodium, ruthenium, iridium and molybdenum in dimethyl formamide and they are good olefin hydrogenation catalysts [222].

Hydrochloric acid aqueous solutions of the anion $[RuCl_4(bipyr)]^{2-}$ catalyse homogeneously the hydrogenation of maleic acid to succinic acid at 80 °C and 1 atmosphere of hydrogen pressure [232].

The mechanism of hydrogenation by this complex does not involve first a π complex, in contrast to the $[RuCl_4]^{2-}$ system [149].

The catalyst system obtained by $NaBH_4$ reduction of $RhCl_3py_3$ in dimethylformamide hydrogenates diphenylacetylene to *trans* stilbene. Deuteriogenation studies show that *cis* addition occurs with most substrates but during deute-

XI

riogenation of diphenylacetylene the ortho position of the benzene ring is deuterated. It is suggested 1 : 4 addition occurs to give XI followed by a 1 : 3 hydrogen (or deuterium) shift [223].

11.4 Hydrogen Transfer between Organic Molecules

The disproportionation of cyclohexadienes to cyclohexene and benzene is catalysed by $[Rh(\pi\text{-}C_5Me_5)(C_6H_8)]$ [224]. Unlike the iridium catalysts [191] this reaction goes via the cyclohexa-1,3-diene which disproportionates. Some experimental details of the Henbest reduction $(H_2IrCl_6/P(OMe)_3/isopropanol)$ of 3-oxo steroids have been reported. The oxo groups at 6, 7, 11, 12, 17, 17a (D-homo) and 20 are not noticeably reduced [225]. The hydrides, $RhHCO(PR_3)_3$ catalyse the hydrogen transfer (78) at 130 °C, the reaction being favoured by the more basic phosphines [204].

$$C_6H_5CH_2OK + C_6H_{13}CH=CH_2 \longrightarrow C_6H_5COOK + C_8H_{18} \qquad (78)$$

12. SUMMARY

Some of the expectations of homogeneous hydrogenation have certainly been met. For example several systems have appeared which are of comparable activity to a heterogeneous catalyst containing an equivalent weight of metal. The word comparable is used advisedly because most comparisons are made under conditions which are the optimum operating conditions for the homogeneous systems. Homogeneous systems are not rugged, in that they operate best within quite narrow experimental conditions. Furthermore, special care in preparation will give heterogeneous catalysts of a very wide range of activities and the most active forms are not always employed when comparing them with homogeneous systems.

The most active homogeneous catalysts described so far appear to be those formed by interaction of transition metal salts and organoaluminium compounds, particularly lithium tri-t-butoxyaluminium hydride (See section 8). Other highly active catalysts under the appropriate conditions are $RuHCl(PPh_3)_3$ (see Section 6.4), dimethylformamide solutions of $H[Rh_2Cl(RCOO)_2]$, dimethylformamide solutions of $RhCl_3pyr_3/NaBH_4$, (see Section 7.2) and phosphine complexes of rhodium of general formula $RhXL_x$, where X is halide, L is certain tertiary phosphines and x is usually 2 to 3 (see Section 6.1).

High degrees of specificity have also been demonstrated. In hydrocarbon/alcohol solvents all homogeneous catalysts reduce olefins in the order terminal olefin > disubstituted olefin > trisubstituted olefin but some of the selectivity within this range is quite striking. The most selective is $RuHCl(PPh_3)_3$, which only reduces terminal olefins at an appreciable rate.

The system which has been used most is $RhCl(PPh_3)_3$. Besides hydrogenating the less hindered double bonds preferentially, this complex is useful for controlled deuterium labelling and is useful in hydrogenation of double bonds where side reactions occur with heterogeneous catalysts, e.g. hydrogenolysis and aromatisation (see Section 6.1.4.). The specificity of some phosphine stabilised catalysts can be altered quite drastically by altering solvents. Thus, in trifluoethanol or phenol, $RhCl(PPh_3)_3$ becomes completely specific for hydrogenation of acetylenes in the presence of olefins (see section 6.1.2.6), and in acetic acid $IrH_3(PPh_3)_3$ is specific for hydrogenation of aldehydes (see Section 10.2). Recently the stereospecific hydrogenation of olefins to optically active products has been accomplished using $RhClL_3$ containing optically active phosphines, and even more promising results have been obtained using $RhCl_3pyr_3/NaBH_4$ in optically active amides (see Section 7.2).

Future developments in selectivity can be expected in homogeneous hydrogenation. For laboratory purposes, there seems no virtue in developing more active catalysts than those already described, unless the activity goes hand in hand with increased selectivity. The problem of separating product from catalyst and recovering the latter for further use will always be a detraction from using homogeneous catalysts, particularly those containing rare metals, on a large scale.

There are numerous instances of similarities between homogeneous and heterogeneous catalysis and hence an understanding of the former will automatically increase our understanding of the latter. Both types of systems are poisoned by neutral ligands such as carbon monoxide, excess phosphine or pyridine, thiols etc.. Hydrogenation of olefins is slowed down in both types of system by the presence of conjugated dienes or acetylenes (see Sections 5.2.4, 6.1.2. and 6.4.1) because they are more strongly co-ordinated than olefins. It is clear that on a molecular level the nature of the bonding between the metal atom and various ligands, substrates, poisons etc. is similar in both homogeneous and heterogeneous systems. The similarity between bonding in metal complexes and in molecules absorbed on the surface of metals has been pointed out [196] and reviewed [17]. The metal atoms in a heterogeneous catalyst are regarded as surrounded by ligands (which happen to be other metal atoms), and so the electrons fall into two discrete bands, the t_{2g} and e_g bands, due to the crystal field, as in complexes of transition metals.

The greatest advantage obtained in studying homogeneous systems is that it is much easier to gain insight into the nature of the reactive intermediates. Thus in phosphine complexes the formation of metal hydrides, metal olefin complexes, and metal alkyls can be demonstrated by spectroscopic techniques, and the nature of the other ligands on the metal atom is known with more certainty than in other systems. Preliminary observations suggest the $RhCl_3pyr_3/NaBH_4$ system in amide solvents will also be amenable to similar studies.

There are still some aspects of the best studied system, $RhCl(PPh_3)_3$, which are not fully understood, e.g. the true degree of dissociation of the catalyst, and the relative importance of reaction of a rhodium hydride with an olefin and a rhodium olefin complex with hydrogen (see Section 6.1.2.7). It has been suggested that higher specificity is shown by RhCl(PPh₃)₃ if the hydride route is taken. Heterogeneous hydrogenation of acetylenes by palladium gives olefins when the catalyst is pretreated with hydrogen but paraffinic products are obtained in the early stages, if the catalyst is pretreated with the acetylene [60]. In certain homogeneous systems the rate of hydrogenation is a maximum at substrate: catalyst ratio of 1 : 1, falls to a minimum at a ratio of 2, and then rises again as the ratio increases. The minimum rate at a ratio of substrate: catalyst of 2 probably corresponds to complexing of all the free sites by olefin and hence hydrogen cannot be activated and this has parallels in heterogeneous catalysis [60].

The cobalt cyanide systems appear to function by the metal activating hydrogen only, whereas other transition metal systems appear to activate both the substrate and hydrogen. It would appear that stepwise addition of hydrogen to an olefin is universal but in some cases, notably RhCl(PPh₃)₃ and related system the second addition is so fast it and is virtually simultaneous addition of two hydrogen atoms. The occurrence of deuterated olefins during hydrogenation is regarded as evidence for reverse alkyl formation but recently it has been shown that deuterated olefin can be formed via oxidative addition of an olefin to give a hydridovinyl complex (see section 2).

13. REFERENCES

[1]   M. CALVIN, "Trans. Far. Soc.", *34*, 1181, (1938).

[2]   (a) J. KWIATEK, "Catalysis Rev.", *1*, 37, (1967).
      (b) J. KWIATEK, I. L. MADOR, and J. K. SEYLER, "Advan. Chem. Ser.", *37*, 201, (1963).
      (c) J. KWIATEK, and J. K. SEYLER, "Advan. Chem. Ser.", *70*, 207, (1968).

[3]   C. W. BIRD, *Transition Metal Intermediates in Organic Synthesis*, p. 248, (1967), Logos Press, (London).

[4]   (a) J. HALPERN, "Ann. Rev. of Phys. Chem.", *16*, 103, (1965).
      (b) J. HALPERN, "Chem. Eng. News", *44*, Oct, p. 68, (1966).
      (c) J. HALPERN, "Advan. Chem. Ser.", *70*, 1, (1968).

[5]   V. A. TUPULOV, "Russ. J. Phys. Chem." (English Translation), *39*, 1251, (1965).

[6]   A. J. CHALK, and J. F. HARROD, *Advan. Organometallic Chem.*, vol. 6, p. 119, (1968) Ed. F.G.A. Stone and R. West Academic Press, (New York).

[7]   E. N. FRANKEL, T. L. MOUNTS, R. O. BUTTERFIELD, and H. J. DUTTON, "Advan. Chem. Ser.", *70*, 177, (1968).

[8]   M. M. JONES, *Ligand Reactivity and Catalysis* p. 192. (1968), Academic Press, (New York).

[9]   J. C. LAUER, "Annalen", *10*, 301, (1965).

68

[10] (a) J. CHATT, "Proc. Chem. Soc.", 318, (1962).
     (b) A. P. GINSBERG, *Transition Metal Chemistry*, Vol. 1, p. 111. (1965) Ed. R. L. Carlin. Dekker, (New York).
     (c) M. L. H. GREEN, and D. J. JONES, *Advan. Inorg. and Radiochem.*, Vol. 7, p. 114, (1965) Ed. H. J. Emeleus and A. G. Sharpe, Academic Press (New York).

[11] A. ANDREETTA, F. CONTI, and G. F. FERRARI, *Aspects Homog. Cat.*, Vol. 1, p. 193, (1970), Ed. R. Ugo, Manfredi, (Milan).

[12] S. CARRÀ, and R. UGO, "Inorg. Chim. Acta. Rev.", *1*, 49, (1967).

[13] J. P. COLLMAN, "Acc. Chem. Res.", *1*, 136, (1968).

[14] A. B. FASMAN, and Zh. A. IKHSANOV, "Kin. and Cat.", *8*, 53, (1967).

[15] L. VASKA, "Acc. Chem. Res.", *1*, 335, (1968).

[16] R. CRAMER, "Acc. Chem. Res.", *1*, 186, (1968).

[17] P. B. CHOCK, and J. HALPERN, "J. Amer. Chem. Soc.", *88*, 3511, (1966).

[18] J. A. OSBORN, F. H. JARDINE, J. F. YOUNG, and G. WILKINSON, "J. Chem. Soc. A", 1711, (1966).

[19] *See reference 20 for appropriate references.*

[20] R. CRAMER, "J. Amer. Chem. Soc.", *89*, 4621, (1967).

[21] (a) R. CRAMER, "J. Amer. Chem. Soc.", *86*, 217, (1964).
     (b) R. CRAMER, "J. Amer. Chem. Soc.", *89*, 5377, (1967).
     (c) R. CRAMER, J. B. KLINE, and J. D. ROBERTS, "J. Amer. Chem. Soc.", *91*, 2519, (1969).

[22] A.R. BRAUSE, F. KAPLAN, and M. ORCHIN, "J.Amer.Chem. Soc.", *89*, 2661, (1967).

[23] C. E. HOLLOWAY, G. HULLEY, B. F. G. JOHNSON, and J. LEWIS, "J. Chem. Soc. A", 53, (1969).

[24] M. ORCHIN, *Advan. in Catalysis*, Vol. 16, p. 1, (1966).

[25] (a) R. CRAMER, "J. Amer. Chem. Soc.", *88*, 2272, (1966).
     (b) R. CRAMER, and R. V. LINDSEY, "J. Amer. Chem. Soc.", *88*, 3534, (1966).

[26] J. H. FLYNN, and H. M. HULBURT, "J. Amer. Chem. Soc.", *76*, 3393, (1954).

[27] P. N. RYLANDER, N. HIMELSTEIN, D. R. STEELE, and J. KREIDL, "Engelhardt Industries Inc., Tech. Bull.", *3*, 61, (1962).

[28] J. D. RICHTER, and P. J. VAN DEN BERG, "J.Amer. Oil Chem. Soc.", *46*, 155, (1969).

[29] M. G. BURNETT, P. J. CONNOLLY, and C. KEMBALL, "J.Chem. Soc. A", 800, (1967).

[30] (a) R. G. BANKS, and J. M. PRATT, "J. Chem. Soc. A", 854, (1968).
     (b) R. G. BANKS, P. K. DAS, H. A. O. HILL, J. M. PRATT, and R. J. P. WILLIAMS, "Disc. Far. Soc.", *46*, 80, (1968).

[31] (a) B. DE VRIES, "J. Catalysis", *1*, 489, (1962).
     (b) O. PIRINGER, and A. FARCAS, "Z. Phys. Chem." (Frankfurt), *46*, 190, (1965).
     (c) L. SIMANDI, and F. NAGY, *Proc. Symposium on Co-ordination Chemistry, Tihany, Hungary, 1964*, p. 83, (1965), Akad. Kiado, (Budapest).

[32] G. PREGAGLIA, D. MORELLI (the late), F. CONTI, G. GREGORIO, and R. UGO, "Disc. Far. Soc.", *46*, 110, (1968).

[33] J. HALPERN, and LAI-Y. WONG, "J. Amer. Chem. Soc.", *90*, 6665, (1968).

[34] K. TARAMA, and T. FUNABIKI, "Bull. Chem. Soc. Japan", *41*, 1744, (1968).

[35] (*a*) J. HANZLIK, and A. A. VLCEK, "J. Chem. Soc. D", 47, (1969).
     (*b*) J. HANZLIK, and A. A. VLCEK, "Inorg. Chem.", *8*, 669, (1969).

[36] M. G. BURNETT, P. J. CONNOLLY, and C. KEMBALL, "J. Chem. Soc. A", 991, (1968).

[37] L. SIMANDI, and F. NAGY, "Acta Chim. Acad. Sci. Hung.", *46*, 137, (1965).

[38] W. STROHMEIER, and N. IGLAUER, "Z. Phys. Chem" (Frankfurt), *51*, 50, (1966).

[39] L. M. JACKMAN, J. A. HAMILTON, and J. M. LAWLOR, "J. Amer. Chem. Soc.",
     90, 1914, (1968).

[40] L. H. SLAUGH, "Tetrahedron", *22*, 1741, (1966).

[41] L. H. SLAUGH, "J. Org. Chem.", *32*, 108, (1967).

[42] O. PIRINGER, and A. FARCAS, "Z. Phys. Chem." (Frankfurt), *46*, 190, (1965).

[43] (*a*) O. PIRINGER, and A. FARCAS, "Nature", *206*, 1040, (1965).
     (*b*) O. PIRINGER, A. FARCAS, and U. LUCA, *Proc. 9th I. C.C.C. St, Moritz-Bad*,
     p. 202, (1966), Ed. W. Schneider.
     (*c*) A. FARCAS, U. LUCA, N. MORAR, and. O. PIRINGER, "Z. Phys. Chem." (Frank-
     furt), *58*, 87, (1968).

[44] R. RIPON, A. FARCAS, and O. PIRINGER, "Zeit. fur Anorg. und allg. Chemie",
     346, 211, (1966).

[45] C. E. WYMORE, "Chem. Eng. News", *46*, p. 52, April, (1968).

[46] M. G. BURNETT, "Chem. Comm.", 507, (1965).

[47] T. MIZUTA, H. SAMEJIMA, and T. KWAN, "Bull. Chem. Soc. Japan", *41*, 727,
     (1968).

[48] M. S. SPENCER, and D. A. DOWDEN, U. S. Pat. (to Imperial Chemical Industries
     Ltd.) 2,966,534, (1960).

[49] W. H. DENNIS Jr., D. H. ROSENBLATT, R. R. RICHMOND, G. A. FINSETH, and
     G. T. DAVIS, "Tetrahedron Lett.", 1821, (1968).

[50] R. D. CRAMER, E. L. JENNER, R. V. LINDSEY Jr., and U. G. STOLBERG, "J. Amer.
     Chem. Soc.", *85*, 1691, (1963).

[52] G. C. BOND, and M. HELLIER, "Chem. and Ind." (London), 35, (1965).

[51] I. JARDINE, and F. J. McQUILLIN, "Tetrahedron Lett.", 4871, (1966).

[53] G. C. BOND, and M. HELLIER, "J. Catalysis", *7*, 217, (1967).

[54] E. N. FRANKEL, E. A. EMKEN, H. ITATANI, and J. C. BAILAR Jr., "J. Org. Chem.",
     32, 1447, (1967).

[55] J. C. BAILAR Jr., and H. ITATANI, "J. Amer. Oil Chem. Soc.", *43*, 337, (1966).

[56] J. C. BAILAR Jr., and H. ITATANI, "J. Amer. Chem. Soc.", *89*, 1592, (1967).

[57] H. VAN BEKKUM, J. VAN GOGH, and G. VAN MINNEN-PATHRIS, "J. Catalysis",
     7, 292, (1967).

[58] L. P. VAN'T HOF, and B. G. LINSEN, "J. Catalysis", *7*, 297, (1967).

[59] P. ABLEY, and F. J. McQUILLIN, "Disc. Far. Soc.", *46*, 31, (1968).

[60] I. JARDINE, R. W. HOWSAM, and F. J. McQUILLIN, "J. Chem. Soc. C", 260, (1969).

[61] R. W. ADAMS, G. E. BATLEY, and J. C. BAILAR Jr., "Inorg. Nucl. Chem. Lett.", *4*,
     455, (1968).

[62] R. W. ADAMS, G. E. BATLEY, and J. C. BAILAR Jr., "J. Amer. Chem. Soc.",
     90, 6051, (1968).

70

[63] H. A. TAYIM, and J. C. BAILAR Jr., "J. Amer. Chem. Soc.", *89*, 4330, (1967).

[64] H. A. TAYIM, and J. C. BAILAR Jr., "J. Amer. Chem. Soc.", *89*, 3420, (1967).

[65] J. F. YOUNG, R. D. GILLARD, and G. WILKINSON, "J. Chem. Soc.", 5176, (1964).

[66] R. D. CRAMER, R. V. LINDSEY Jr., C. T. PREWITT, and U. G. STOLBERG, "J. Amer. Chem. Soc.", *87*, 658, (1965).

[67] R. V. LINDSEY Jr., G. W. PARSHALL, and U. G. STOLBERG, "Inorg. Chem.", *5*, 109, (1966).

[68] L. J. GUGGENBE, "Chem. Comm.", 512, (1968).

[69] M. C. BAIRD, "J. Inorg. Nucl. Chem.", *29*, 367, (1967).

[70] R. V. LINDSEY Jr., G. W. PARSHALL, and U. G. STOLBERG, "J. Amer. Chem. Soc.", *87*, 658, (1965).

[71] R. CRAMER, "Inorg. Chem.", *4*, 445, (1965).

[72] J. CHATT, R. S. COFFEY, A. GOUGH, and D. T. THOMPSON, "J. Chem. Soc. A", 190, (1968).

[73] A. P. KRUSHEK, A. A. TOKINA, and A. E. SHILOV, "Kin. and Cat.", *7*, 793, (1966).

[74] C. D. FALK, and J. HALPERN, "J. Amer. Chem. Soc.", *87*, 3523, (1965).

[75] F. CARIATI, R. UGO, and F. BONATI, "Inorg. Chem.", *5*, 1128, (1966).

[76] J. F. YOUNG, J. A. OSBORN, F. H. JARDINE, and G. WILKINSON, "Chem. Comm.", 131, (1965).

[77] (a) R. S. COFFEY, and J. B. SMITH, British Pat. (to Imperial Chemical Industries Ltd.) 1,121,642.
      (b) K. C. DEWHIRST, U.S. Pat. application, (to Shell Oil Co.). Ser. No. 417,482.

[78] M. A. BENNETT, and P. A. LONGSTAFF, "Chem. and Ind." (London), 846, (1965).

[79] A. SACCO, R. UGO, and A. MOLES, "J. Chem. Soc. A", 1670, (1966).

[80] L. VASKA, and R. E. RHODES, "J. Amer. Chem. Soc.", *87*, 4970, (1965). (see footnotes).

[81] F. H. JARDINE, J. A. OSBORN, G. WILKINSON, and J. F. YOUNG, "Chem. and Ind." (London), 560, (1965).

[82] R. G. HAYTER, "Inorg. Chem.", *3*, 301, (1964).

[83] L. VALLARINO, "J. Chem. Soc.", 2473, (1957).

[84] J. CHATT, and B. L. SHAW, "J. Chem. Soc.", 2508, (1964).

[85] M. A. BENNETT, R. BRAMLEY, and P. A. LONGSTAFF, "Chem. Comm.", 806, (1966).

[86] F. H. JARDINE, J. A. OSBORN, and G. WILKINSON, "J. Chem. Soc. A", 1574, (1967).

[87] S. MONTELATICI, A. VAN DER ENT, J. A. OSBORN, and G. WILKINSON, "J. Chem. Soc. A", 1054, (1968).

[88] R. STERN, Y. CHEVALLIER, and L. SAJUS, "Compt. Rend.", *264*, Series C, 1740, (1967).

[89] Y. CHEVALLIER, R. STERN, and L. SAJUS, "Tetrahedron Lett.", 1197, (1969).

[90] L. HORNER, H. BÜTHE, and H. SIEGEL, "Tetrahedron Lett.", 4023, (1968).

[91] D. R. EATON, and S. R. STUART, "J. Amer. Chem. Soc.", *90*, 4170, (1968).

[92] K. C. DEWHIRST, W. KEIM, and C. A. REILLY, "Inorg. Chem.", *7*, 546, (1968).

[93] J. T. MAGUE, and G. WILKINSON, "J. Chem. Soc. A", 1736, (1966).

[94] R. S. COFFEY, unpublished result.

[95] J. P. CANDLIN, and A. R. OLDHAM, "Disc. Far. Soc.", *46*, 60, (1968).

[96] R. C. TAYLOR, J. F. YOUNG, and G. WILKINSON, "Inorg. Chem.", *5*, 20, (1966).

[97] C. O'CONNOR, and G. WILKINSON, "Tetrahedron Lett.", 1375, (1969).

[98] A. S. HUSSEY, and Y. TAKEUCHI, "J. Amer. Chem. Soc.", *91*, 672, (1969).

[99] M. C. BAIRD, J. T. MAGUE, J. A. OSBORN, and G. WILKINSON, "J. Chem. Soc. A", 1347, (1967).

[100] A. L. ODELL, J. B. RICHARDS, and W. R. ROPER, "J. Catalysis", *8*, 393, (1967).

[101] G. C. BOND, and R. A. HILLYARD, "Disc. Far. Soc.", *46*, 20, (1968).

[102] J. F. BIELLMANN, and M. J. JUNG, "J. Amer. Chem. Soc.", *90*, 1673, (1968).

[103] D. N. LAWSON, J. A. OSBORN, and G. WILKINSON, "J. Chem. Soc. A", 1733, (1966).

[104] A. J. BIRCH, and K. A. M. WALKER, "Tetrahedron Lett.", 1935, (1967).

[105] H. VAN BEKKUM, F. VAN RANTWIJK, and T. VAN DE PATTE, "Tetrahedron Lett.", 1, (1969).

[106] (*a*) R. D. W. KEMMITT, D. I. NICHOLS, and R. D. PEACOCK, "J. Chem. Soc. A", 1898, (1968).
(*b*) M. A. BENNETT, and P. A. LONGSTAFF, *Proc. 9th I.C.C.C., St. Moritz-Bad*, p. 349, (1966), Ed. W. Schneider.

[107] S. TAKAHASKI, H. YAMAZAKI, and N. HAGIHARA, "Bull. Chem. Soc. Japan", *41*, 254, (1968).

[108] A. J. BIRCH, and K. A. M. WALKER, "J. Chem. Soc. C", 1894, (1966).

[109] C. DJERASSI, and J. GUTZWEILLER, "J. Amer. Chem. Soc.", *88*, 4537, (1966).

[110] M. BROWN, and L. PISZKIEWICZ, "J. Org. Chem.", *32*, 2013, (1967).

[111] A. J. BIRCH, and K. A. M. WALKER, "Tetrahedron Lett.", 3457, (1967).

[112] J. J. SIMS, V. K. HONWAD, and L. H. SELMAN, "Tetrahedron Lett.", 87, (1969).

[113] J. F. BIELLMANN, and H. LIESENFELT, "Bull. Soc. Chim. France", 4029, (1966).

[114] H. RÜERSH, and T. J. MABRY, "Tetrahedron", *25*, 806, (1969).

[115] F. H. JARDINE, and G. WILKINSON, "J. Chem. Soc. C", 270, (1967).

[116] J. R. MORANDI, and H. B. JENSEN, "J. Org. Chem.", *34*, 1889, (1969).

[117] A. J. BIRCH, and K. A. M. WALKER, "Tetrahedron Lett.", 4939, (1966).

[118] M. FETIZON, "Bull Soc. Chim. France", 651, (1969).

[119] L. HORNER, H. SIEGEL, and H. BÜTHE, "Angew. Chem. Inter. Ed.", *7*, 942, (1968).

[120] W. S. KNOWLES, and M. J. SABACKY, "Chem. Comm.", 1445, (1968).

[121] S. S. BATH, and L. VASKA, "J. Amer. Chem. Soc.", *85*, 3500, (1963).

[122] S. J. LA PLACA, and J. IBERS, "Acta Cryst.", *18*, 511, (1965).

[123] J. C. BAILAR, Jr., and H. ITATANI, "J. Amer. Chem. Soc.", *89*, 1600, (1967).

[124] D. EVANS, G. YAGUPSKY, and G. WILKINSON, "J. Chem. Soc. A", 2660, (1968).

[125] C. O'CONNOR, and G. WILKINSON, "J. Chem. Soc. A", 2665, (1968).

[126] D. EVANS, J. A. OSBORN, and G. WILKINSON, "J. Chem. Soc. A", 3133, (1968).

[127] J. A. OSBORN, G. WILKINSON, and J. F. YOUNG, "Chem. Comm.", 17, (1965).

72

[128] G. Wilkinson, British Pat., 1,138,601.

[129] B. Hudson, P. C. Taylor, D. E. Webster, and P. B. Wells, "Disc. Far. Soc.", 46, 37, (1968).

[130] G. G. Eberhardt, and L. Vaska, "J. Catalysis", 8, 183, (1967).

[131] B. R. James, and N. A. Memon, "Canad. J. Chem.", 46, 217, (1968).

[132] L. Vaska, "Inorg. Nucl. Chem. Lett.", 1, 89, (1965).

[133] K. A. Taylor, "Advan. Chem. Ser.", 70, 195, (1968).

[134] M. Yamaguchi, "Kogyo Kagaku", 70, 675, (1967).

[135] R. S. Coffey, British Pat. (to Imperial Chemicals Industries Ltd.), 1,135,979.

[136] M. Giustiniani, G. Dolcetti, M. Niccolini, and U. Belluco, "J. Chem. Soc. A", 1961, (1969).

[137] D. Evans, J. A. Osborn, F. H. Jardine, and G. Wilkinson, "Nature", 208, 1203, (1965).

[138] P. S. Hallman, D. Evans, J. A. Osborn, and G. Wilkinson, "Chem. Comm.", 305, (1967).

[139] P. S. Hallman, B. R. McGarvey, and G. Wilkinson, "J. Chem. Soc. A", 3143, (1968).

[140] A. C. Skapski, and P. R. Troughton, "Chem. Comm.", 1230, (1968).

[141] P. G. Douglas, and B. L. Shaw, "J. Chem. Soc. D", 624, (1969).

[142] D. E. Webster, and P. B. Wells, "Platinum Metals Review", 13, 104, (1969).

[143] G. Wilkinson, British Pat. 1,130,749.

[144] Y. Shibata, and E. Matsumoto, "Nippon Kagaku Zasshi", 60, 1173, (1939), Chem. Abst., 34, 1582.

[145] M. Iguchj, "Nippon Kagaku Zasshi", 60, 1787, (1939).

[146] V. A. Tulupov, "Russ. J. Phys. Chem.", (English Translation) 41, 456, (1967).

[147] B. R. James, and G. L. Rempel, "Disc. Far. Soc.", 46, 48, (1968).

[148] E. B. Maxted, and S. M. Ismail, "J. Chem. Soc.", 1750, (1964).

[149] J. Halpern, J. F. Harrod, and B. R. James, "J. Amer. Chem. Soc.", 88, 5150, (1966).

[150] B. Hui, and B. R. James, "J. Chem. Soc. D", 198, (1969).

[151] B. R. James, and G. L. Rempel, "Canad. J. Chem.", 44, 233, (1966).

[152] R. D. Gillard, J. A. Osborn, P. B. Stockwell, and G. Wilkinson, "Proc. Chem. Soc.", 284, (1964).

[153] B. R. James, F.T.T.Ng, and G. L. Rempel, "Inorg. Nucl. Chem. Lett.", 4, 197, (1968).

[154] H. Itatani, and J. C. Bailar Jr., "J. Amer. Oil Chem. Soc.", 44, 147, (1967).

[155] N. V. Borunova, L. Kh. Friedlin, M. L. Khidekel, S. S. Danielova, V. A. Avilov, and P. S. Chekrii, "Zvest. Akad. Nauk. S.S.S.R. Ser Khim", 434, (1968). V. A. Avilov, Yu. G. Brod'ko, V. B. Panov, M. L. Khidekel, and P. S. Chekrii, "Kin. and Cat.", 9, 582, (1968). O. N. Efimov, M. L. Khidekel, V. A. Avilov, P. S. Chekrii, O. N. Eremenko, and A. G. Ovcharenko, "J. Gen. Chem. U.S.S.R.", 38, 2581, (1968).

[156] I. JARDINE, and F. J. McQUILLIN, "J. Chem. Soc. D", 477, (1969).

[157] I. JARDINE, and F. J. McQUILLIN, "J. Chem. Soc. D", 503, (1969).

[158] I. JARDINE, and F. J. McQUILLIN, "J. Chem. Soc. D", 502, (1969).

[159] P. ABLEY, and F. J. McQUILLIN, "J. Chem. Soc. D", 477, (1969).

[160] K. THOMAS, J. A. OSBORN, A. R. POWELL, and G. WILKINSON, "J. Chem. Soc. A", 1801, (1968).

[161] A. R. POWELL, "Platinum Metal Review", 11, 58, (1967).

[162] J. R. SHAPLEY, R. R. SCHROCK, and J. A. OSBORN, "J. Amer. Chem. Soc.", 91, 2816, (1969).

[163] P. LEGZKINS, G. L. REMPEL, and G. WILKINSON, "J. Chem. Soc. D", 825, (1969).

[164] G. NATTA, G. MAZZANTI, P. LONGI, and F. BERNARDINI, "Chimica e Industria" (Milan), 41, 519, (1959).

[165] V. KALECHITS, and F. K. SHMIDT, Kin. a nd Cat.", 7, 541, (1966).

[166] M. F. SLOAN, A. S. MATLACK, and D. S. BRESLOW, "J. Amer. Chem. Soc.", 85, 4014, (1963).

[167] Y. TAJIMA, and E. KUNIOKA, "J. Amer. Oil Chem. Soc.", 45, 478, (1968).

[168] P. SZABO, and L. MARKO, "J. Organometal. Chem.", 3, 364, (1965).

[169] J. PALAGYI, G. PALYI, and L. MARKO, "J. Organometal. Chem.", 14, 238, (1968).

[170] F. UNGVARY, B. BABOS, and L. MARKO, "J. Organometal. Chem.", 8, 329, (1967), and B. BABOS, F. UNGVARY, L. PUPP, and L. MARKO, "Magyar Kem. Foly.", 75, 127, (1969).

[171] E. W. DUCK, J. M. LOCKE, and Ch. J. MALLINSON, "Annalen", 719, 69, (1969).

[172] I. V. KALECHITS, V. G. LIPOVICH, and F. K. SHMIDT, "Kin. and Cat.", 9, 16, (1968).

[173] K. SHIKATA, K. NISHINO, K. AZUMA, and Y. TAKEGAMI, "Kogyo Kagaki Zasshi", 68, 358, (1965).

[174] Y. TAGIMA, and E. KUNIOKA, "J. Org. Chem.", 33, 1689, (1968).

[175] R. STERN, and L. SAJUS, "Tetrahedron Lett.", 6313, (1968).

[176] R. STERN, G. HILLION, and L. SAJUS, "Tetrahedron Lett.", 1561, (1969).

[177] S. J. LAPPORTE, and W. R. SCHUETT, "J. Org. Chem.", 28, 1947, (1963).

[178] V. G. LIPOVICH, F. K. SHMIDT, and I. V. KALECHITS, "Kin. and Cat.", 8, 812, (1967).

[179] V. G. LIPOVICH, F. K. SHMIDT, and I. V. KALECHITS, "Kin. and Cat.", 8, 1099, (1967).

[180] C. BRESSAN, and R. BROGGI, "Chimica e Industria" (Milan), 50, 1194, (1968).

[181] K. SONOGASHIRA and N. HAGIHARA, "Bull. Chem. Soc. Japan", 39, 1178, (1966).

[182] B. HEIL, and L. MARKO, "Chem. Ber.", 99, 1086, (1966).

[183] B. HEIL, and L. MARKO, "Acta Chim. Acad. Sci. Hung.", 55, 107, (1968).

[184] (a) French Pat. (to Montecatini Edison), 1,509,863.
(b) British Pat. (to Shell Chemicals), 942, 435.

[185] R. S. COFFEY, "Chem. Comm.", 923, (1967).

[186] Y. M. Y. HADDAD, H. B. HENBEST, J. HUSBANDS, and T. R. B. MITCHELL, "Proc. Chem. Soc.", 361, (1964).

[187] J. TROCHA-GRIMSHAW, and H. B. HENBEST, "Chem. Comm.", 544, (1967).

74

[188] J. TROCHA-GRIMSHAW, and H. B. HENBEST, "Chem. Comm.", 757, (1968).

[189] M. McPARTLIN, and R. MASON, "Chem. Comm.", 545, (1967).

[190] H. B. HENBEST, and T. R. B. MITCHELL, *Ann. Meeting of the Chem. Soc. Nottingham*, paper 4.16, (1969).

[191] J. E. LYONS, "J. Chem. Soc. D", 564, (1969).

[192] S. KUNICHIKA, Y. SAKAKIBATA, and T. NAKAMMA, "Bull. Chem. Soc. Japan", *41*, 390, (1968).

[193] A. MISONO, M. HIDAI, I. INOMATA, and Y. UCHIDA, "Chem. Comm.", 704, (1968).

[194] J. D. McCLURE, R. OWYANG, and L. H. SLAUGH, "J. Organometal. Chem.", *12*, p 8-p 12, (1968).

[195] E. BILLIG, C. B. STROW, and R. L. PRUETT, "Chem. Comm.", 1307, (1968).

[196] C. E. BOND, "Disc. Far. Soc.", *41*, 200, (1966).

[197] M. E. VOL'PIN, and I. S. KOLOMNIKOV, "Russian Chem. Rev.", *38*, 273, (1969).

[198] W. H. BADDLEY, and M. S. FRASER, "J. Amer. Chem. Soc.", *91*, 3661, (1969).

[199] K. W. MUIR, and J. IBERS, "J. Organometal. Chem.", *18*, 175, (1969).

[200] P. B. HITCHCOCK, M. McPARTLIN, and R. MASON, "J. Chem. Soc. D", 1367, (1969).

[201] C. H. HEATHCOCK, and S. R. POULTER, "Tetrahedron Lett.", 2755, (1969).

[202] R. E. HARMON, J. L. PARSONS, D. W. COOKE, S. K. GUPTA, and J. SCHOOLENBERG, "J. Org. Chem.", *34*, 3684, (1969).

[203] R. E. HARMON, J. L. PARSONS, and S. K. GUPTA, "J. Chem. Soc. D", 1365, (1969).

[204] G. GREGORIO, G. PREGAGLIA, and R. UGO, "Inorg. Chim. Acta", *3*, 89, (1969).

[205] J. P. COLLMAN, V. W. HOFFMAN, and D. E. MORRIS, "J. Amer. Chem. Soc.", *91*, 5659, (1969).

[206] W. STROHMEIER, and W. REHDER-STIRNWEISS, "J. Organometal. Chem.", *18*, P28; *19*, 417, (1969).

[207] W. STROHMEIER, and T. ONODA, "Zeit. Naturforsch.", 24B, 515, (1969).

[208] (a) W. STROHMEIER, and T. ONODA, "Zeit. Naturforsch.", 24B, 461, (1969).
(b) W. STROHMEIER, and T. ONODA, *Colloques Internationaux, du Centre National de La Recherche Scientifique*, No. 191, Paris, Oct. (1969).

[209] D. ROSE, J. D. GILBERT, R. P. RICHARDSON, and G. WILKINSON, "J. Chem. Soc. A", 2610, (1969).

[210] A. C. SKAPSKI, and F. A. STEPHENS, "J. Chem. Soc. D", 1008, (1969).

[211] K. C. DEWHIRST, U.S. Pat. (to Shell Oil Co.), 3,454,644.

[212] G. SPEIER, and L. MARKO, "Inorg. Chim. Acta", *3*, 126, (1969).

[213] A. MISONO, Y. UCHIDA, T. SAITO, and K. M. SONG, "Chem. Comm.", 419, (1967); A. MISONO, Y. UCHIDA, M. HIDAI, and T. KUSE, "Chem. Comm.", 981, (1968).

[214] G. PREGAGLIA, A. ANDREETTA, and R. UGO, "Chimica e Industria" (Milan), *50*, 1332, (1968).

[215] G. PREGAGLIA, A. ANDREETTA, and G. FERRARI, *IV Int. Conf. on Organometallic Chem., Bristol*, July (1969), paper L3.

[216] G. Pregaglia, A. Andreetta, G. Ferrari and R. Ugo, "J. Chem. Soc. D", 590, (1969).

[217] See Ref. [197], p. 279.

[218] D. R. Witt, and J. P. Hogan, "*Amer. Chem. Soc. Div. of Polymer Chem.*", Polymer Preprints, Vol. 10, No. 1, p. 255, (1969).

[219] Y. Takegami, T. Ueno, and T. Fujii, "Bull. Chem. Soc. Japan", *42*, 1663, (1969).

[220] W. R. Kroll, "J. Catalysis", *15*, 281, (1965).

[221] U. Zucchini, U. Giannini, E. Albizzata, and R. D'Angelo, "J. Chem. Soc. D", 1174, (1969).

[222] See Ref. [197], p. 280.

[223] P. Abley, and F. J. McQuillin, "J. Chem. Soc. D", 1503, (1969).

[224] K. Moseley, and P. M. Maitlis, "J. Chem. Soc. D", 1156, (1969).

[225] P. A. Browne, and D. N. Kirk, "J. Chem. Soc C", 1653, (1969).

[226] R. L. Augustine, and J. F. van Peppen, "J. Chem. Soc. D", 497, (1970).

[227] R. L. Augustine, and J. F. van Peppen, "J. Chem. Soc. D", 495, (1970).

[228] R. L. Augustine, and J. F. van Peppen, "J. Chem. Soc. D", 497, (1970).

[229] R. L. Augustine, and J. F. van Peppen, "J. Chem. Soc. D", 571, (1970).

[230] M. Yagupsky, C. K. Brown, G. Yagupsky, and G. Wilkinson, "J. Chem. Soc. A", 937, (1970).

[231] R. R. Schrock, and J. A. Osborn, "J. Chem. Soc. D", 567, (1970).

[232] B. C. Hui, and B. R. James, "J. Inorg. Nucl. Chem. Lett.", *6*, 367, (1970).

[233] British Pat. (to Imperial Chemical Industries Ltd.), 1,154,937.

[234] J. E. Lyons, L. E. Rennick and J. L. Burmeister, "Ind. Eng. Chem. Prod. Res. Dev." *9*, 2, (1970).

[235] A. P. Krushek, N. F. Shvetsova and A. E. Shilov, "Kinet. Kataliz.", *10*, 1226, (1970).

Chapter 2

Stereoselectivity in Organic Syntheses Involving Nickel-co-ordinated Intermediates

G. P. CHIUSOLI

Centro Ricerche di Chimica Organica, Montecatini-Edison, Novara, Italy

1. INTRODUCTION

The aim of this review is to present an outlook of the stereoselective reactions* of unsaturated species which occur on nickel complexes, giving rise to *cis* or *trans* isomers. The synthetic potentialities of these reactions will also be put forward.

(*) According to Eliel [1] a reaction is considered stereoselective if one product is formed predominantly from either of two stereoisomeric reactants, whereas it is considered stereospecific if each stereosomeric reactant gives rise to two sterically different products.

In the interest of clarity the subject is divided into three classes:

a) reactions in which a triple bond is converted into a *cis* or *trans* double bond;

b) reactions in which a vinylic double bond is converted into a *trans* double bond;

c) reactions in which an allylic double bond reacts, giving rise to a *cis* or *trans* double bond.

Reactions of classes *a*) and *b*) may occur together with reactions of class *c*), an allyl group almost always being present in the reagents. In order to distinguish classes *a*) and *b*) from class *c*), however, the examination of the behaviour of allyl groups containing substituents that can give rise to stereoisomers will be omitted in treating the first two classes.

2. REACTIONS OF THE TRIPLE BOND

2.1. Conversion into a cis double bond

The first reaction on nickel which was shown to be stereoselective was the synthesis of acrylic esters from acetylene, tetracarbonylnickel and alcohols [2]. Working with higher acetylenes [3] it could be established that this reaction gives *cis* products (1):

$$C_6H_5-C\equiv CH + CO + ROH \xrightarrow{\ Ni\ } \begin{array}{c} H \\ \diagdown \\ C_6H_5 \diagup \end{array} C=C \begin{array}{c} H \\ \diagup \\ \diagdown COOR \end{array} \qquad (1)$$

Later, another synthesis on nickel was found, in which an allyl group and carbon monoxide were added to the two ends of the acetylene molecule giving rise to a *cis* double bond [4]. The experimental details of this reaction will not be discussed here, the reader being referred to the published literature. Only a short description will be given before discussing the stereoselectivity.

The simplest example of this synthesis is offered by the reaction of allyl chloride with acetylene and carbon monoxide in methanol, in the presence of tetracarbonylnickel (2):

$$CH_2=CHCH_2Cl + CH\equiv CH + CO + CH_3OH \xrightarrow{\ Ni\ }$$

$$\longrightarrow \begin{array}{c} CH \\ \diagup\diagup \quad \diagdown \\ CH_2 \qquad CH_2 \end{array} \begin{array}{c} H \qquad H \\ \diagdown \quad \diagup \\ C=C \\ \diagup \quad \diagdown \\ COOCH_3 \end{array} + HCl \qquad (2)$$

The reaction occurs at room temperature and atmospheric pressure.

From the synthetic point of view it should be noted that the catalytic intermediate which is responsible for the stereoselectivity has been recently prepared in situ [4n], simply by reducing nickel chloride in methanolic solution with a manganese-iron alloy containing about 80% of manganese, in the presence of thiourea and of a neutralising agent, such as magnesium oxide. It is only necessary to pass acetylene and carbon monoxide into the mixture and to add the allyl halide in order to obtain the formation of the 2-*cis* dienoic ester. This class of compounds thus becomes very easily accessible.

The intermediates in this synthesis are π-allylnickel complexes [4a, d, 6a, b, 7, 8]. Besides generating them in situ it has been possible to prepare nickel complexes of this type as dimers (I) [6a, b] or monomers containing additional ligands, such as phosphines or thioureas (II) [9], (III) [10], (IV) [5], (V), (VI) [11] (X = Cl, Br, I; R = alkyls or aryls; Tu = thiourea):

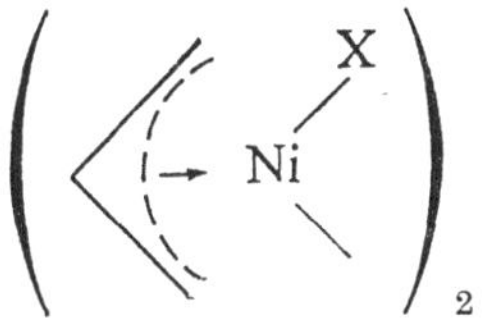

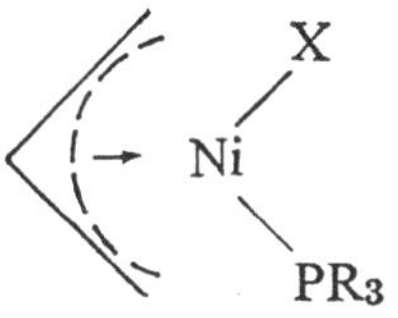

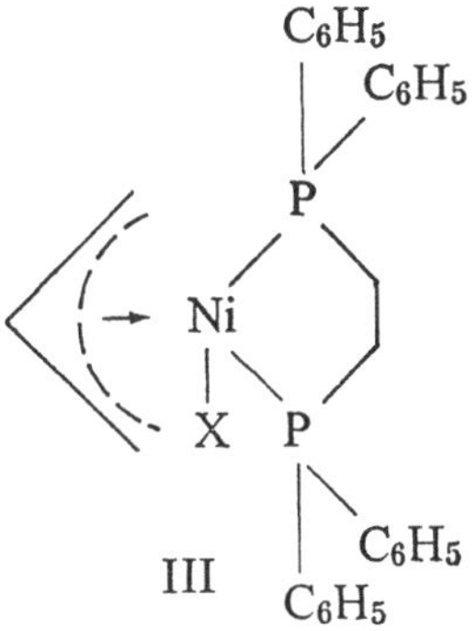

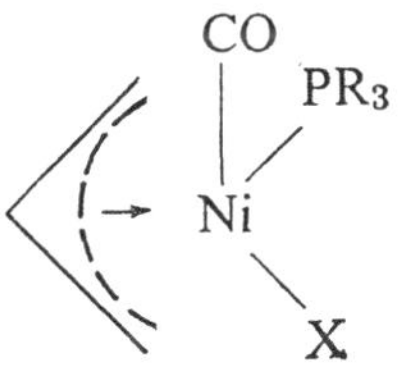

The structures of some of these complexes (III, V) have been investigated by X-ray methods. Only the structure of complex (V) (X = Cl; Tu = thiourea) is reported here, because it is a very effective catalyst of the carbonylation reaction with insertion of acetylene (*). In methanolic solution the complex has a remarkable ionic character which is also proved by the X-ray analysis, chlorine being at a distance of 3,45 Å from nickel. The other ligands are arranged in a planar square, whereas chlorine completes a sort of square pyramid [12].

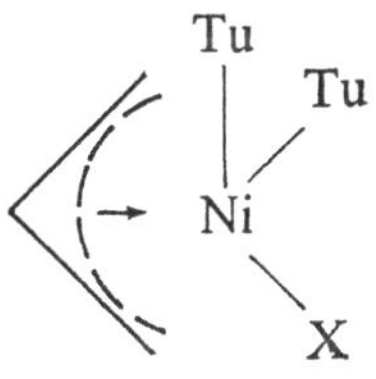

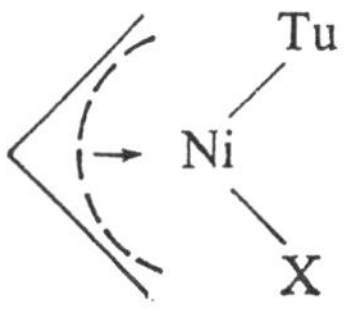

(*) Phosphines are not effective in the catalysis owing to their inability to promote the fast reoxidation of the zerovalent nickel, caused by the attack of the allyl halides at room temperature and atmospheric pressure.

The tendency to reach pentaco-ordination appears to be a most important factor in determining the reactivity of the allylic ligand and the stereoselective insertion of acetylene, as will be shown later.

For the purpose of studying the way of insertion of acetylene, however, it is more convenient to study the reaction on a complex which, under the usual

Structure of π-allylbis(thiourea)chloronickel

conditions (room temperature and atmospheric pressure), does not give a catalytic reaction but a stoichiometric one.

This is the case of type (IV) complexes [5] such as in the reaction here reported:

$$+CH\equiv CH+3CO+CH_3OH \longrightarrow \tag{3}$$

$$+Ni(CO)_3P(C_6H_5)_3+HX$$

If, before the reaction, a new molecule of triphenylphosphine id added, carbon monoxide is not evolved and the reaction does not proceed further.(*) This fact indicates that at least three places of the complex must be free in order that the reaction may occur. This is because triphenylphosphine must, of necessity, occupy the co-ordination site which should have been occupied by acetylene in the case of insertion. Some phosphines are known, from the work of Wilke and co-workers [13], to be mobilizing agents for the allyl group. Thus complex (VII)

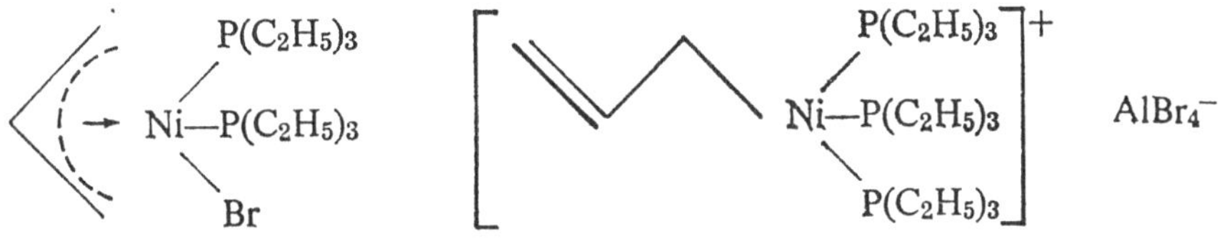

VII   VIII

has been shown to be in fast equilibrium between π and σ forms and complex VIII reaches a stable σ form. In our case, after the addition of the new molecule of triphenylphosphine, the complex contains two molecules of triphenylphosphine and one molecule of carbon monoxide, which is known to behave analogously to triphenylphosphine as ligand [14].

Consequently, the allyl group should have been displaced from the π to the σ form, thus liberating one of the two co-ordination sites it previously held. Acetylene, in the absence of added triphenylphosphine, must occupy this co-ordination site and thus become inserted *cis* to the σ allyl group. At this point the latter can migrate on to acetylene, probably as in the following scheme (4) (from which all non essential ligands have been omitted and the relative position of X has not been considered):

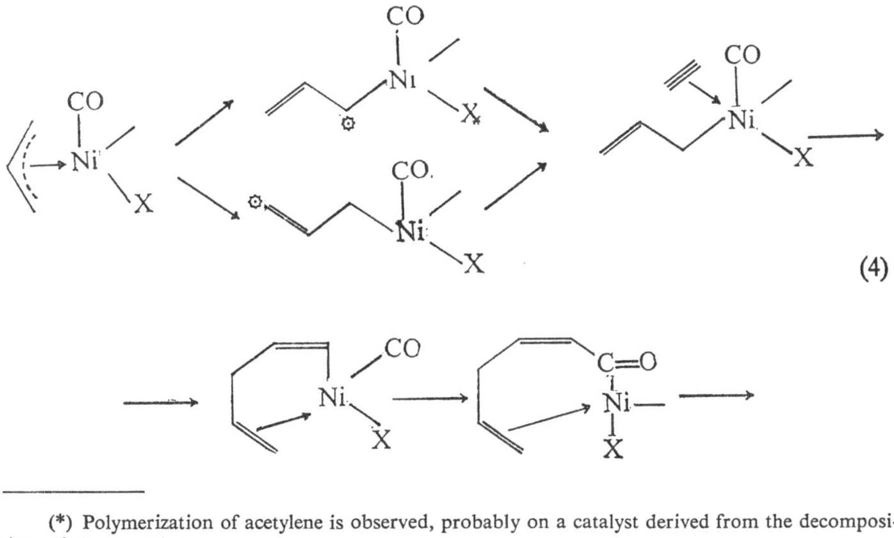

(4)

(*) Polymerization of acetylene is observed, probably on a catalyst derived from the decomposition of the complex.

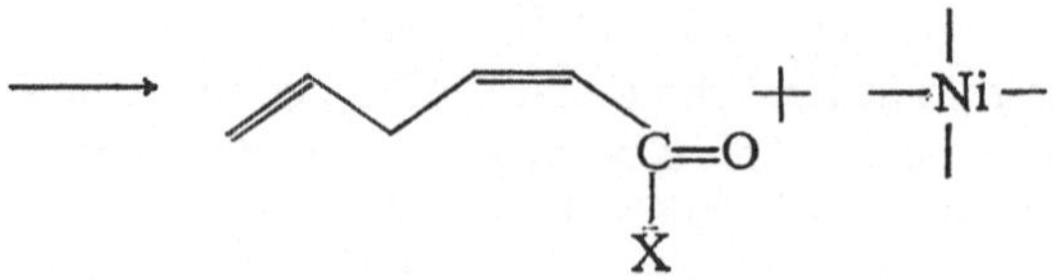

As shown in the scheme, the allyl group should migrate on to acetylene and the resulting pentadienyl group on to carbon monoxide. (For *cis* insertion see Ref. [15]). A nucleophilic attack of X on the hexadienoyl complex liberates hexadienoyl halide and zerovalent nickel.

This scheme provides an idea of the explanation of the stereoselectivity of acetylene insertion. The allyl group attacks acetylene from the *cis* position and this circumstance should favour the formation of a *cis* pentadienyl group. Co-ordination of the double bond of the original allyl group should, also, influence the attainment of this configuration. The pentadienyl group thus formed should attack the co-ordinated carbon monoxide from the *cis* position, as Calderazzo has already shown in the case of the methyl migration on manganese carbonyl [15a]. The resulting double bond is thus a *cis* one. The same conclusion is also reached if the acetylene insertion between allyl group and carbon monoxide is seen as an one step process. In the attempt to establish if a pentadienyl group is formed in a first stage, acetylene was reacted with π-allylnickel bromide in benzene solution, but only the acetylene polymer was obtained. Working in an excess of allyl bromide in a benzene-acetonitrile mixture, it was possible to isolate a new class of 1,4-diallylbutadienes with both double bonds of the *cis* configuration [16]:

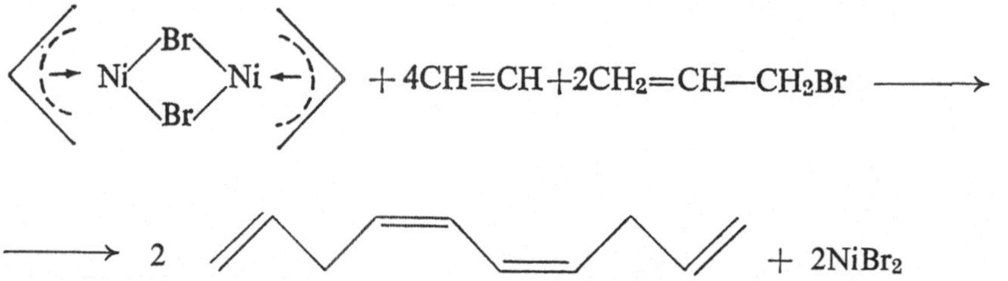

$$+ 4CH \equiv CH + 2CH_2 = CH - CH_2Br \longrightarrow \tag{5}$$

$$2 \quad \cdots \quad + 2NiBr_2$$

The products of this reaction, (5) which appears to be a general one, are very unstable and easily polymerize during their preparation. In most cases products were isolated only in low percentage yields, and only in the case of the cyclohexenyl group the *cis, cis*-dicyclohexenylbutadiene was obtained in a 25% yield:

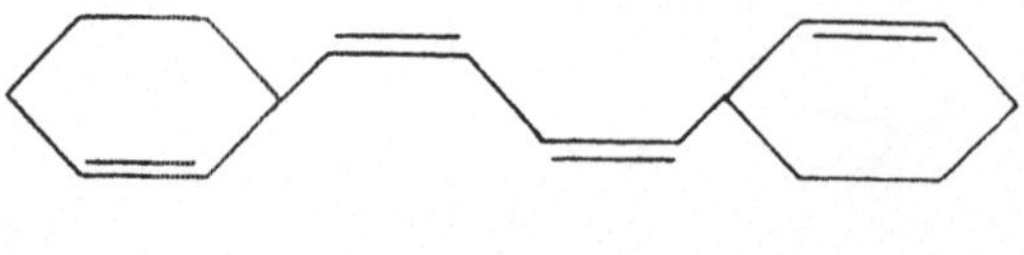

The *cis, cis* configuration is characterized by an ultraviolet absorption in the region of 237-240 mμ and by IR absorptions in the region between 700 and 750

cm^{-1}. The NMR spectrum reveals a highly symmetric structure with two groups of signals at 3.7 and 4.57 τ.

The synthesis of diallylbutadienes confirms that the formation of the pentadienylic structure on nickel is stereoselective and that it can migrate on to molecules other than carbon monoxide.

The way in which the second allyl group reacts, to interrupt this sort of polymerization of acetylene initiated by the allyl group, is not yet clear. A *cis, cis* insertion of two molecules of acetylene between two molecules of carbon monoxide from tetracarbonylnickel has been observed in the case of the formation of ε-lactones from acyl halides [17].

$$RCOCl + 2\ CH{\equiv}CH + Ni(CO)_4 + H_2O \longrightarrow$$

(6)

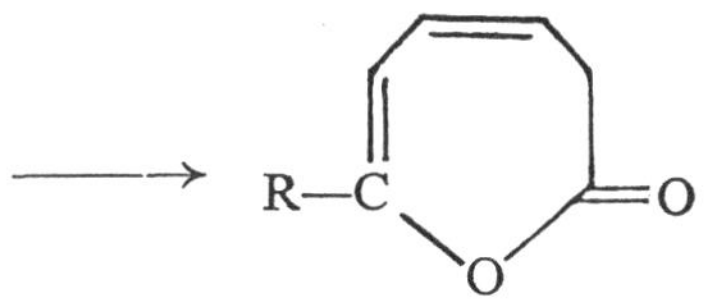

$$+ Ni(OH)Cl + 2CO$$

Analogy between the diallylbutadienes (5) and ε-lactones (6) syntheses would imply that the second allyl group should also become co-ordinated.

As far as the reactions of acetylene and carbon monoxide are concerned, it should be noted that the formation of a *cis* double bond has important consequences from several aspects. By working in an aprotic solvent in the case of acetylene and also in solvents containing water in the case of higher acetylenes, the reaction proceeds further than the insertion of carbon monoxide and gives rise to penta- and hexa-atomic cyclic structures. For example, *m*-cresol can be prepared in a very simple way from methallyl chloride, acetylene and carbon monoxide in a yield near to 80% [4] [18]. This is possible because of the *cis* structure of the non-cyclic intermediate formed on the nickel complex. The six-carbon cycle, however, is preferred only in the case of methallyl-type derivatives. In the other cases a five carbon cycle is favoured. A series of cyclopentenonic structures obtained from allyl halides, such as allyl chloride or bromide, crotyl chloride, methallyl chloride etc., acetylene or mono and disubstituted acetylenes, such as methylacetylene or phenylacetylene and carbon monoxide have already been published [4 *f, g, h, i*]. Only the following one is reported:

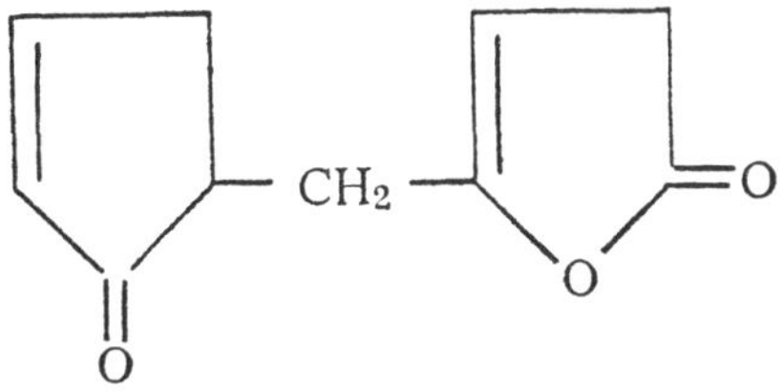

to point out that the cyclopentenonic group cannot be formed under the usual conditions of these syntheses (room temperature and atmospheric pressure) if the chain before cyclisation does not possess the *cis* configuration. The lactonic group is also derived from a *cis* structure formed by acetylene insertion, the intermediate being the lactonyl complex (IX):

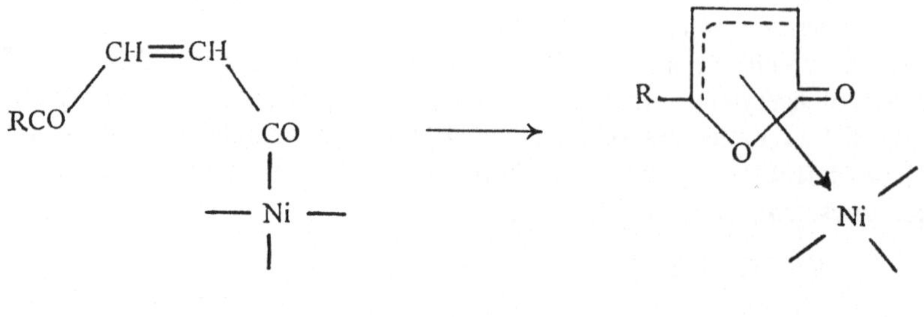

$$\tag{7}$$

IX

Thus, if *trans*-hexadienoyl chloride $CH_2 = CHCH_2CH = CHCOCl$ is reacted with acetylene and tetracarbonylnickel, the cyclopentenonic cycle is not formed, but the lactonic one is, and gives rise to the product here reported:

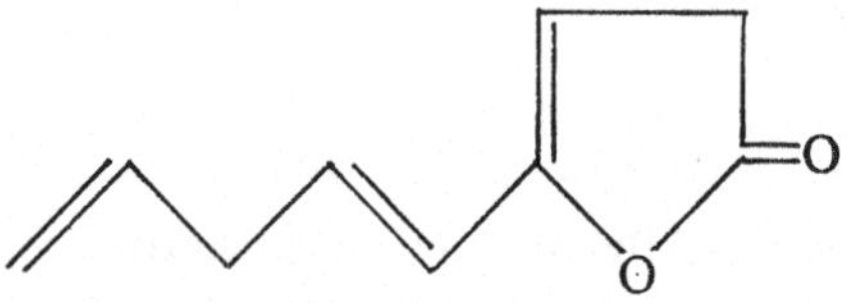

2.2. Conversion into a trans double bond

So far it has been shown that a triple bond can be converted on nickel into a *cis* double bond. It is also possible, however, to convert a triple bond into a *trans* double bond, under certain conditions that will be summarized in the following paragraphs:

a) Insertion of an acetylene molecule into a nickel complex, followed by protonation in place of carbonylation [4f]. An example is given below (8). Phenylacetylene is reacted in an aqueous acetone solution with allyl chloride and tetracarbonylnickel:

$$2C_6H_5C\equiv CH + CH_2=CHCH_2Cl + Ni(CO)_4 + HX \longrightarrow$$

$$\tag{8}$$

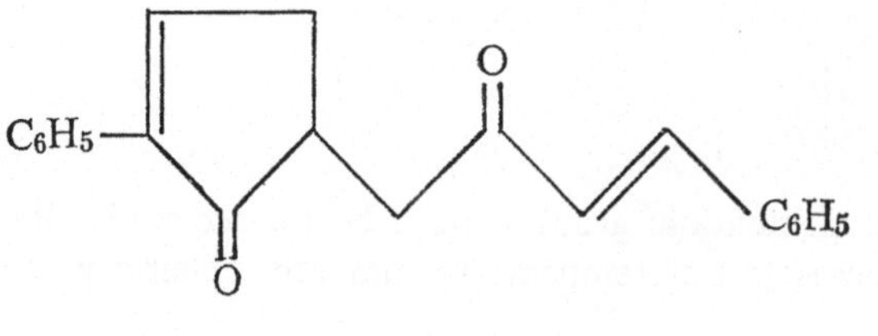

$$+ \; NiClX + 2CO$$

The reaction implies the step-by-step addition of the allyl group to phenylacetylene and of the resulting group to carbon monoxide, followed by cyclization and addition of a new molecule of carbon monoxide and of a new molecule of phenylacetylene. At this stage carbon monoxide is not inserted and a proton is taken up, instead, from water or acids HX.

It can be observed that a position *cis* to the growing chain RCOCH=CR—, which should have been occupied by carbon monoxide in the case of a carbonylation, is now taken by the attacking molecule of HX as follows:

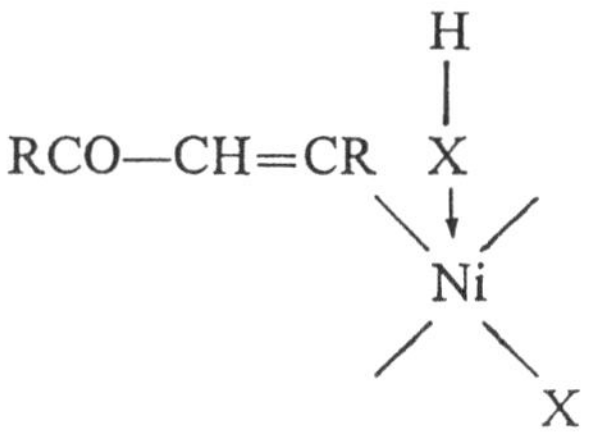

No *cis* insertion takes place and protonation of the carbon chain occurs instead. *Trans* double bonds appear to be formed in general.

b) Reaction of alkyl propiolates with allyl halides in protonic solvents: An example is given below (9). Methyl propiolate reacts with allyl halides on nickel giving rise mainly to methyl 2-*trans*-5-hexadienoate [45] and heavier products:

$$HC\equiv C-COOCH_3 + CH_2=CHCH_2Cl + Ni + HX \longrightarrow \tag{9}$$

The course of the reaction seems to be analogous to that of the preceeding one.

c) Insertion of an acetylene molecule into a nickel complex, followed by reaction with an allyl halide. An example is given below (10). Allyl chloride is reacted with acetylene and tetracarbonylnickel in an ethereal solvent [4 *i*]:

$$2CH_2=CHCH_2Cl + 2CH\equiv CH + 2CO + Ni \longrightarrow \tag{10}$$

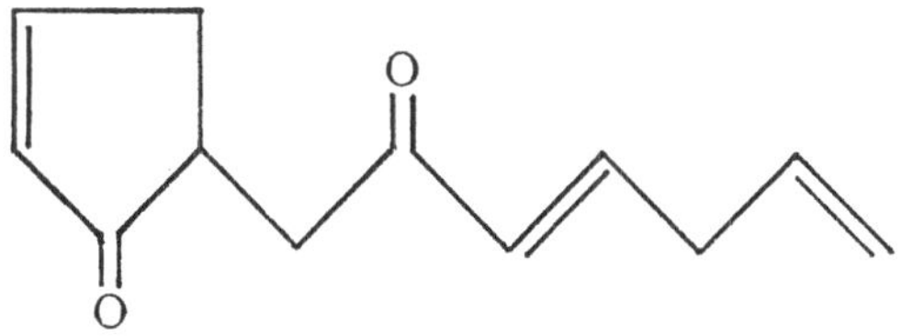

In this case the molecule of the allyl halide, which is going to split the complex, should be attacked by the growing chain on the complex according to a kind of nucleophilic substitution:

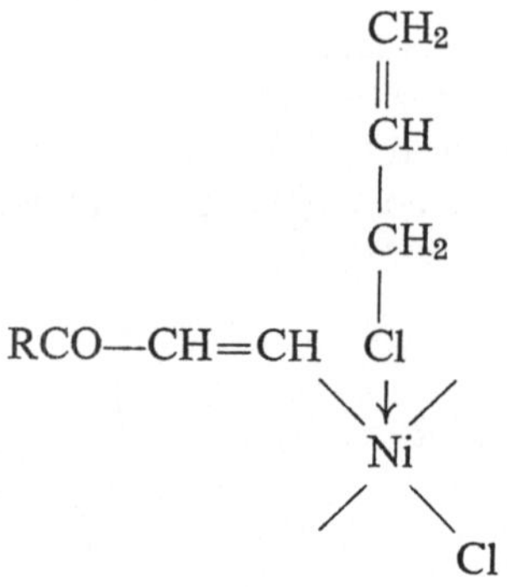

$$
\begin{array}{c}
CH_2 \\
\| \\
CH \\
| \\
CH_2 \\
|
\end{array}
$$

RCO—CH=CH Cl

Ni

Cl

It can be argued, tentatively, that every time a triple bond is inserted into a nickel complex between two co-ordinated groups, the resulting double bond is a *cis* one, whereas when, after a first insertion of the triple bond between a co-ordinated group and the metal, the complex is split by a protonic acid or by a haloderivative, the *trans* configuration of the resulting double bond is generally favoured.

This conclusion is in accordance with all the results obtained so far for the insertion of one molecule of acetylene. In the case of the insertion of two molecules of acetylene, (reaction (5)), the *cis, cis* configuration is preferred. It is possible that in this case diallylbutadienes originate from an insertion reaction between co-ordinated allyl groups and not from the attack of the allyl halide molecule as shown above.

3. CONVERSION OF THE VINYL DOUBLE BOND INTO A TRANS DOUBLE BOND

The conversion of a vinyl double bond into a *trans* double bond can be easily achieved by reacting an activated α-olefin with allyl halides in the presence of zero-valent nickel. An example is offered by the reaction of methyl acrylate with allyl chloride and nickel (as tetracarbonyl or as zerovalent nickel in general) at room temperature and atmospheric pressure. Working in an inert solvent two reactions occur, the first leading to elimination of hydrogen (11), the second to uptake of hydrogen (12) [19, 20]:

$$CH_2=CHCH_2Cl + CH_2=CHCOOCH_3 + Ni \longrightarrow$$

$$+ [NiHCl] \quad (11)$$

$$CH_2=CHCH_2Cl + CH_2=CHCOOCH_3 + Ni + H^+ \rightarrow \tag{12}$$

$$\rightarrow CH_2=CHCH_2CH_2CH_2COOCH_3 + NiCl^+$$

The reaction (11) leads stereoselectively to the formation of a *trans* double bond. The reaction should be considered a true insertion, analogous to those of acetylene, with the difference that in this case there is an elimination of hydrogen which produces the *trans* double bond as follows:

$$CH_2=CH-CH_2CH_2CH-COOR \rightarrow CH_2=CHCH_2CH=CHCOOR + [NiHCl].$$
$$\underset{\underset{Cl}{\overset{|}{\underset{|}{Ni}}}}{\overset{|}{|}}$$

The presence of a nickel hydride intermediate has not been proved, but it can be inferred from the formation of the hydrogenated by-product.

It should be noted that the stereoselectivity of this reaction depends on the solvent used and on the temperature. Using non polar solvents at a temperature not higher than room temperature, substantially all *trans* products are obtained. Aprotic dipolar solvents cause the formation of increasing amounts of the *cis* product and also of products resulting from the isomerisation of the double bonds, probably because they exert an influence on the dissociation of the σ-co-ordinated chain. The elimination of hydrogen thus loses its stereoselectivity.

The use of acrylonitrile in this type of reaction leads to the formation of a σ-co-ordinated complex (X) containing the precursor of the final *trans* nitrile (13) [19]. The reaction has been completely followed on co-ordinated ligands (14) [20]:

$$+ 2CH_2=CHCN \rightarrow$$

$$CH_2=CHCH_2CH_2CHCN \tag{13}$$
$$\underset{\underset{NCCHCH_2CH_2CH=CH_2}{\overset{|}{\underset{|}{Ni}}}}{} + NiBr_2$$
$$(X)$$

$$2 \underset{\underset{NCCH=CH_2}{\overset{\uparrow}{Ni}}}{\overset{\overset{CH_2=CHCN}{\downarrow}}{}} + 2CH_2=CHCH_2Br \rightarrow (X) + NiBr_2 + 2CH_2=CHCN \tag{14}$$

Complex (X) is polymeric. Its NMR spectrum is poorly resolved, nevertheless it reveals the high-field absorption of the proton on the nickel-bonded carbon. The vinyl groups behave spectroscopically as non co-ordinated double bonds. The complex is fairly stable and in order to get the *trans* nitrile it must be pyrolysed. It is possible that the elimination of hydrogen occurs through formation of a hydride as follows (arrows do not imply a concerted reaction):

$$CH_2=CHCH_2CH—CH—CN$$

$$\begin{array}{cc} | & | \\ H & Ni \\ & | \end{array}$$

During the pyrolysis, however, some *cis* nitrile is formed, together with the hydrogenated product (5-hexenenitrile) and products resulting from the isomerisation of the double bonds.

Complexes analogous to (X) have been obtained from various organic halides; the main products from pyrolysis were the doubly unsaturated *trans* nitriles [19]:

allyl halides	Ni-complex
allyl chloride	*bis*(1-cyanopent-4-enyl)nickel
crotyl chloride	*bis*(1-cyanohex-4-enyl)nickel
cinnamyl chloride	*bis*(1-cyano-5-phenylpent-4-enyl)nickel
benzyl bromide	*bis*(1-cyano-3-phenylpropyl)nickel
phenyl iodide	*bis*(1-cyano-2-phenylethyl)nickel

4. REACTIONS OF THE ALLYL DOUBLE BOND

In treating the reactions of the triple and vinyl double bonds, the behaviour of the allyl groups bearing substituents which could give rise to stereoisomers was not considered, for the sake of clarity. It can now be noted that under the conditions examined so far, as well as in others to be shown, the allyl group behaves stereoselectively, irrespective of its original configuration. In the following table some examples of reactions, in which the allyl group behaves stereoselectively, are collected:

Carbonylation [4a,b,l]

$$CH_3CH=CHCH_2Cl+CO+CH_3OH \xrightarrow{Ni} trans\ CH_3CH=CHCH_2COOCH_3 \quad (15)$$

Carbonylation with acetylene insertion [4a,b,l,m,n,5]

$$CH_3CH=CHCH_2Cl+CH\equiv CH+CO+CH_3OH \xrightarrow{Ni}$$

$$\longrightarrow 2\text{-}cis\text{-}5\text{-}trans\ CH_3CH=CHCH_2CH=CHCOOCH_3 \quad (16)$$

Substitutive hydrogenation [4c]

$$C_6H_5CH=CHCHCN+Ni+HCl \rightarrow trans \ C_6H_5CH=CHCH_2CN+NiCl_2 \quad (17)$$
$$\underset{Cl}{|}$$

Insertion of acrylonitrile [19]

$$CH_3CH=CHCH_2Cl+CH_2=CHCN \xrightarrow{Ni} trans, trans \ CH_3CH=CHCH_2CH=CHCN \quad (18)$$

Further information has been obtained by carrying out reactions on π-allylnickel complexes. In most cases the allyl halides bearing an alkyl substituent on a terminal carbon of the allylic system react with zerovalent nickel giving rise to complexes in which the allyl group is under the *syn* configuration:

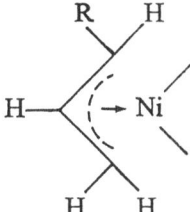

syn configuration

This configuration can be changed, however, by rotation of the allyl group in the σ form [21g]. It should be noted that the π-allyl group can be seen moving steadily from the π to σ form and viceversa. If a fast equilibrium exists between the two forms, so that the NMR instrument is not able to distinguish between the *anti* (*a* and *b*) and *syn* (*c* and *d*) protons:

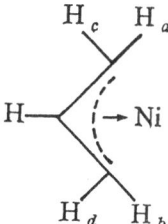

the system is called dynamic [13, 21]. On the other hand, however, the π-allyl group may not be completely static, but the $\pi \rightleftarrows \sigma$ equilibrium may be so slow that the *syn* and *anti* protons can be distinguished in the NMR spectrum.

Suppose now that an end substituted *syn* π-allyl group passes to the σ form and spends a sufficient time in this form to undergo rotation. On reverting to the π form a conversion from the *syn* to the *anti* configuration, and viceversa may occur. This conversion is dependent on the structure of the allyl group, on the ligands, the solvent and the temperature. It can be so slow that it can differentiate between *cis* and *trans* configurations in the transition state of reactions with other groups. The reaction is thus kinetically controlled. It should be noted, however, that the mobility of the allyl group may be greatly increased by the influence of the ligands present in the transition state. In this way a thermodynamic equilibrium may be reached. If stereoselectivity is attained, it can thus be derived from kinetically, as well as from thermodynamically, controlled reactions.

In the case of the reaction of a "dynamic" complex, a thermodynamic equilibrium may be reached, but also conditions exist under which a "directing group" on the allyl group or on the metal can induce the allyl group to adopt a preferred configuration. A striking example is offered by the reactions of complex (XI) (written in monomeric form):

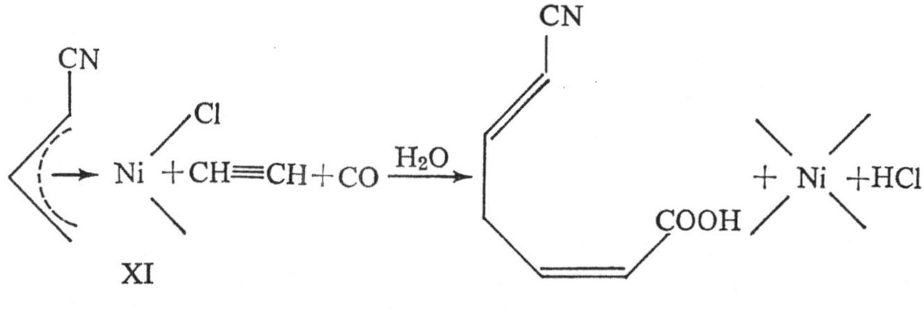

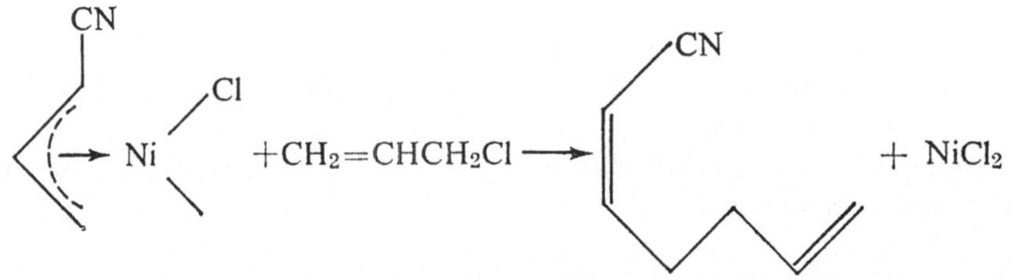

The carbonylation reaction in acetone gives mainly a *trans* nitrile, whereas the coupling reaction in benzene gives mainly a *cis* nitrile [4h, 22-23]. It should also be added that from both pure *cis* and *trans* 1-chloro-3-cyano-prop-2-ene and π-allylnickel halides almost the same result (87,5% *cis* at 0°) is obtained:

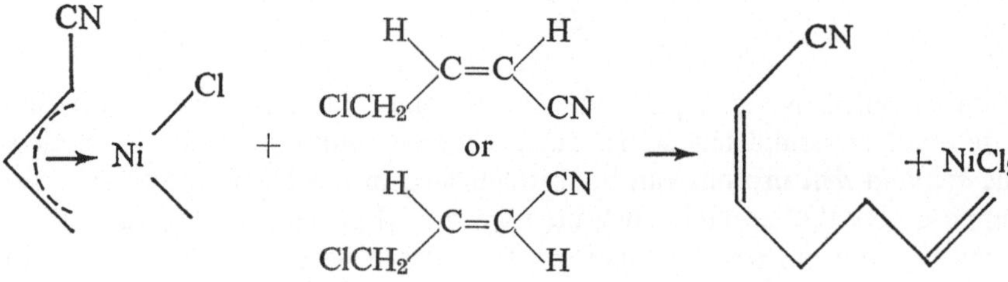

The formation of a *trans* nitrile in the carbonylation can be justified by the influence of the ligands of the previously discussed pentaco-ordinated intermediate. The behaviour of the cyanopropenyl group in the coupling reaction is striking if one considers the fact that, under the same conditions, another electron-withdrawing group, such as the carbomethoxy, in place of the cyano-group gives a mixture of *cis* and *trans* esters with predominance of the *trans* form. It is to be concluded that the cyano-group exerts an influence on the configuration and it probably does this by co-ordination [49]. The influence of directing groups, present in the co-ordinated allylic chain, on the configuration of the products is being investigated [23]. The literature offers at least one example of a complex where an alkyl-substituted allyl group has the *anti* configuration, probably as a consequence of the directing influence of a vinyl group, present in the same chain and co-ordinated to cobalt. This is the following complex (XIII) [25], whose structure has been determined by X-ray methods:

XIII

Care must be taken, however, in drawing general conclusions from these observations. Kinetic control of the formation of co-ordinated allyl groups on a metal can easily lead to the *anti* configuration. For instance the following complex (XIV) [26], obtained from butadiene, has the butenyl group (from butadiene and hydrogen) under the *anti* configuration:

XIV

Thus the presence of the co-ordinated vinyl in (XIII) does not necessarily mean that a directing effect is at work or that this effect would be operative in the transition state of any reaction of the co-ordinated allyl group. It can be said, however, that specific conditions exist for directing the reactions of the nickel-co-ordinated allyl groups towards a high degree of stereoselectivity. This can be obtained by choosing appropriate substituents on the allyl group, as well as appropriate ligands and solvents. The work of Wilke's group is relevant to this subject [13], [30], [48].

These considerations also help us to understand the very intriguing problem of the coupling reactions of allylic halides [27] in which one half behaves as a π-allylic one, whereas the other behaves differently [22, 28, 29]. The reaction is analogous to a nucleophilic substitution on the carbon of the second molecule of allyl halide. One possible scheme is the following, where L means one (or perhaps two) ligand that can also be allyl halide itself [22, 29d]:

$$(22)$$

$$+ \ NiX_2$$

Corey [29d] has demonstrated the occurrence of a replacement reaction which leads to the expulsion of the allyl group originally present in the complex:

$$(23)$$

In this way a new π-allylic structure is formed. If R is different from R' it may also react differently from a steric point of view.

This explains why in coupling reactions mixtures of stereoisomers are frequently found, as observed by Corey [29a,d]. By choosing the appropriate substituents and solvents, however, it is possible to obtain stereoselective reactions of the π-allyl group. This can be done using a large variety of structures bearing different types of substituents. For example, crotyl chloride couples under the action of tetracarbonylnickel, giving mainly the straight-chained *trans, trans*-2,6-octadiene and 25-30% of a branched *trans* isomer:

$$trans, trans \ CH_3CH=CHCH_2\text{---}CH_2CH=CHCH_3$$

$$2CH_3CH=CHCH_2Cl + Ni$$

$$trans \ CH_3CH=CHCH_2\text{---}CHCH=CH_2$$
$$\underset{CH_3}{|}$$

The π-co-ordinated moiety behaves stereoselectively if the reaction is carried out in an alcoholic solvent or in aprotic solvent of low polarity and at room temperature. Dipolar aprotic solvents give substantial amounts of the *cis* isomers. The branching of the second half derives, as already seen, from the attack of the π-allylic half on the internal allylic position of the second.

   In an analogous way it is possible to observe the stereoselective behaviour of several other π-allylic groups. It can be observed however, that the attainment of stereoselectivity becomes increasingly difficult as the structural effects of the *syn* or *anti* configurations approach each other. It is easy, for example, to obtain a *trans* double bond when the choice between a methyl group and a hydrogen atom determines the *syn* or the *anti* configuration. This can be seen by comparing the structures:

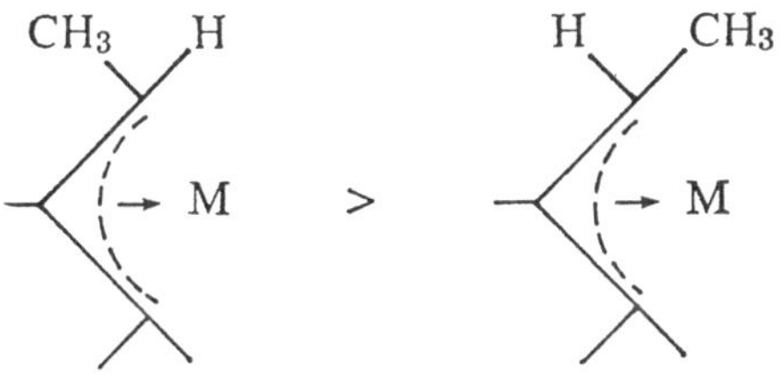

If however, the choice is between two similar groups such as CH_3 and RCH_2, it is more difficult to obtain a stereoselective reaction, unless R contains a group able to exert steric effects or to interact with the metal:

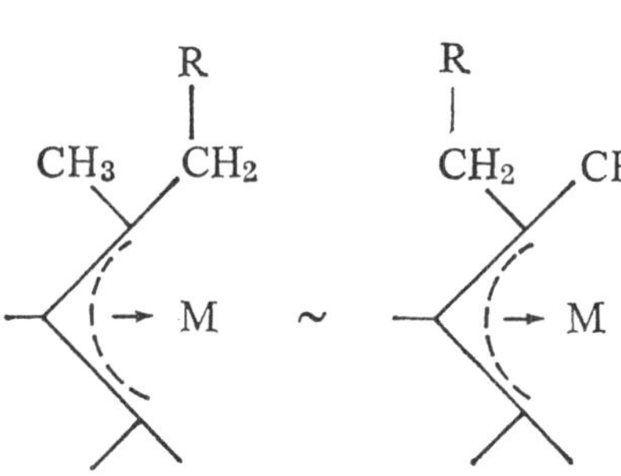

   The second molecule which acts on the π-allylic complex in the coupling reaction can be identical to the π-allylic one or different, as shown by Corey [29]. Isomerization of the second allyl group can take place easily. In the case of a mixture of stereoisomers being formed the stereoselectivity can be improved by minimizing the isomerization. The structure of the allyl group, the solvent and also the leaving group are important in this respect, for example allylic acetates give much less isomerization than allylic halides.

An example of a stereoselective coupling reaction recently achieved is the synthesis of terpenoid alcohols. It is worthwhile to recall the postulated biosynthetic pathway, which involves the reaction of dimethylallyl pyrophosphate with isopentenyl pyrophosphate (24):

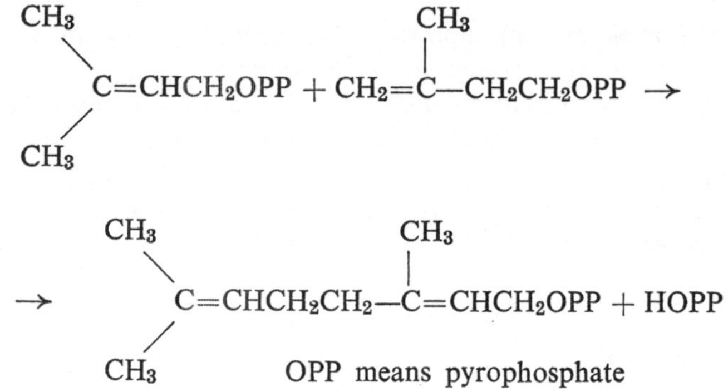

$$
\underset{\underset{\text{CH}_3}{}}{\overset{\text{CH}_3}{}}C=CHCH_2OPP + CH_2=\underset{|}{\overset{\text{CH}_3}{C}}-CH_2CH_2OPP \rightarrow
$$

(24)

$$
\rightarrow \quad \underset{\underset{\text{CH}_3}{}}{\overset{\text{CH}_3}{}}C=CHCH_2CH_2-\underset{|}{\overset{\text{CH}_3}{C}}=CHCH_2OPP + HOPP
$$

OPP means pyrophosphate

The biochemical synthesis gives stereoselectively either the *trans* alcohol (geraniol) or the *cis* one (nerol) [24].

The stereoselective synthesis on nickel is based on the reaction of isoprene hydrochloride with acetoxymethylbutenylnickel chloride (XV) [23]:

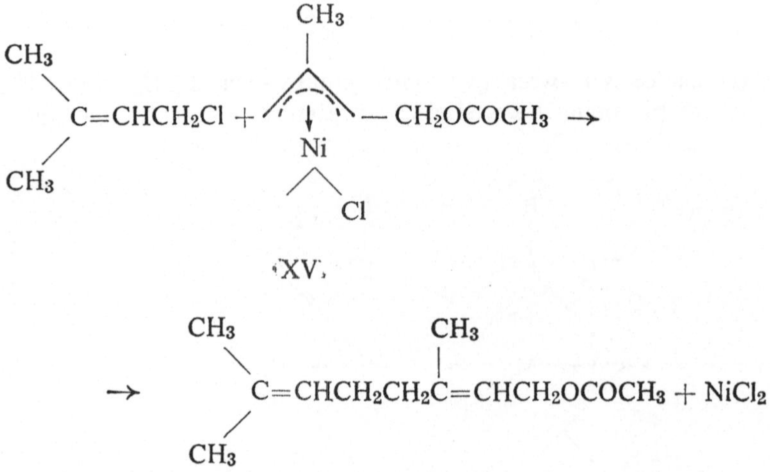

$$
\underset{\underset{\text{CH}_3}{}}{\overset{\text{CH}_3}{}}C=CHCH_2Cl + \text{(XV)} - CH_2OCOCH_3 \rightarrow
$$

$\cdot$XV$\cdot$

(25)

$$
\rightarrow \quad \underset{\underset{\text{CH}_3}{}}{\overset{\text{CH}_3}{}}C=CHCH_2CH_2\underset{|}{\overset{\text{CH}_3}{C}}=CHCH_2OCOCH_3 + NiCl_2
$$

The product thus obtained is 85% *trans*, 15% *cis*. Apart from the obvious differences from the biological pattern, this synthesis shows how a sterically stable allylic carbanion can effect an $S_N 2$ type reaction on an allylic carbon atom.

This method and analogous ones, based on the use of complex (XVI):

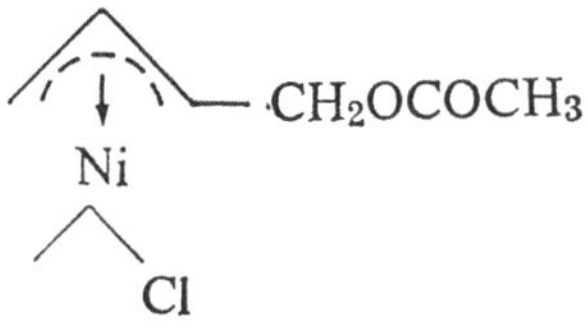

XVI

allows the step-by-step addition of units, having the isoprene or butadiene structure to allylic units. It is possible, indeed, to convert the acetoxy group into an halide and repeat the same reaction on complexes (XV) and (XVI). This method of synthesis thus allows a number of natural products and their analogous to be obtained.

Beside the reaction already cited, there are other reactions in which the allyl group behaves stereoselectively. The following is an example taken from the work of Wilke and co-workers [13]. Working on bis-π-crotylnickel (XVII) they obtained the stereoselective coupling of the two halves at a low temperature (-40°) to trans, trans-2,6-octadiene:

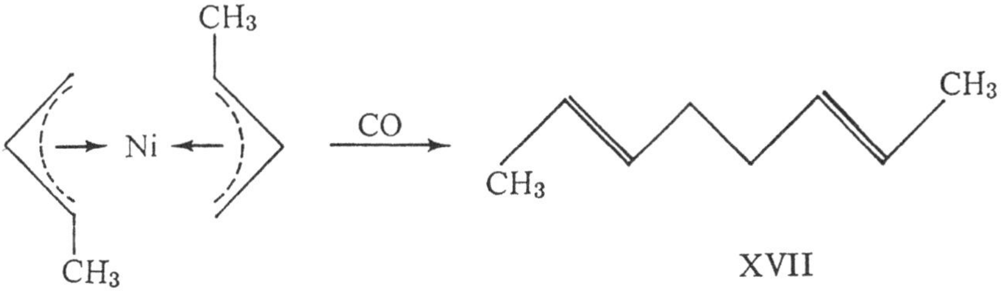

(26)

XVII

Increasing the temperature resulted in the loss of selectivity.

Although this reaction is not the same as the substitution reaction [29d] effected by an allyl group on an allyl halide, the important result is that stereoselectivity can be achieved under proper conditions.

Another example of a stereoselective synthesis on a nickel complex is given by the work of Wilke and co-workers. They found that bis (cyclooctadiene) nickel reacts with acetone and butadiene to give a π-allyl complex (XVIII) which is able to insert a new molecule of acetone [13c]. By hydrolysis a trans hexendiol is obtained:

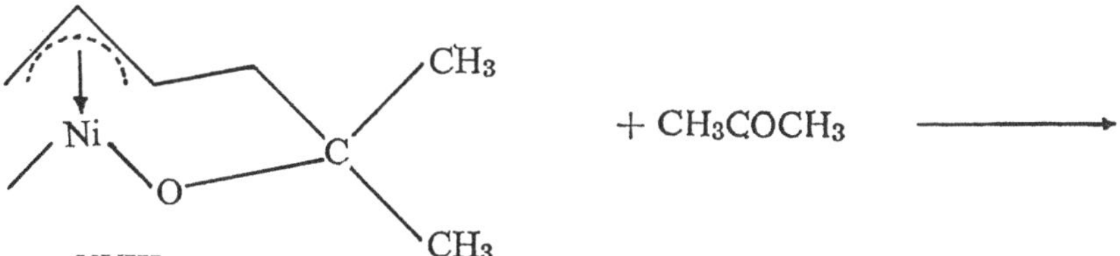

XVIII $+ CH_3COCH_3$

$$\xrightarrow{\quad} \quad \xrightarrow[-\ Ni(OH)_2]{2H_2O}$$

(27)

Coupling reactions have been utilized for the formation of cycloolefins [13, 29, 30]. Diallylnickel and analogous π-allylic compounds have been employed. Details of these reactions will not be discussed here [48], although it should be pointed out that the stereoselective formation of cyclic olefins possessing *cis, cis* configuration as cycloocta-1,5-diene, *cis, trans* configuration as cyclodeca-1,5-diene, *trans, trans, trans* configuration as 1-methylcyclododeca-1,5,9-triene, or *cis, cis, trans* configuration as 4,5-dimethylcyclodeca-1,4,7-triene has been achieved via complex (XIX) (L = ligand, e. g. trialkyl or aryl phosphines or phosphites; isoprene in the third case):

Corey [29*b*] obtained 1,6-dimethylcyclododeca-1,5,9-triene as a 2:1 mixture of *trans, trans, trans* and *cis, trans, trans* isomers by reacting solely *trans* 1,12-dibromo-3,10-dimethyldodeca-2,6,10-triene or a mixture with 20% *cis, trans, trans* isomer at 78 °C in 1,2-dimethoxyethane:

$$\text{BrCH}_2\text{CH}=\text{CCH}_2\text{CH}_2\text{CH}=\text{CHCH}_2\text{CH}_2\text{C}=\text{CHCH}_2\text{Br} \quad \xrightarrow{\text{Ni(CO)}_4}$$

with CH₃ groups:

$$\overset{|}{\underset{\text{CH}_3}{}} \qquad \overset{|}{\underset{\text{CH}_3}{}}$$

(29)

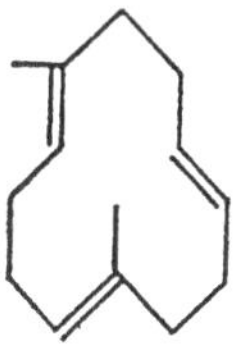

trans, trans, trans and *cis, trans, trans*

The dibromides of structure $\text{BrCH}_2\text{CH}=\text{CH(CH}_2)_n\text{CH}=\text{CHCH}_2\text{Br}$, when reacted with tetracarbonylnickel in dimethylformamide at 50°C gave mainly *trans, trans* cyclic olefins independent of the initial configuration [29*c*]:

$n = 6 \qquad \longrightarrow$

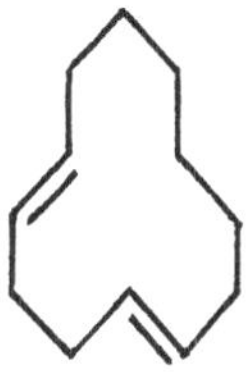

$n = 8 \qquad \longrightarrow$

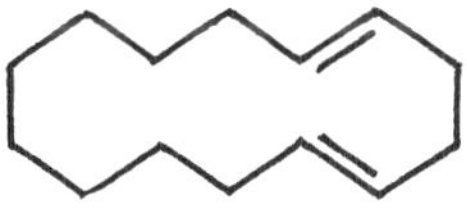

(30)

$n = 12 \qquad \longrightarrow$

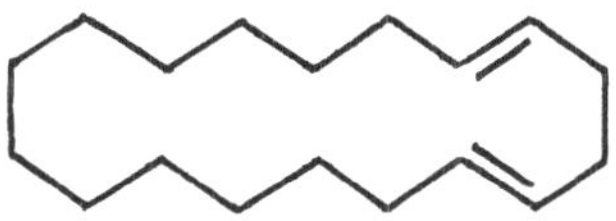

A recent development of the stereoselective reactions using nickel is based on the conversion of a cyclic structure into a straight-chained one, containing two double bonds [31]. The starting material is the class of vinyllactones of the type:

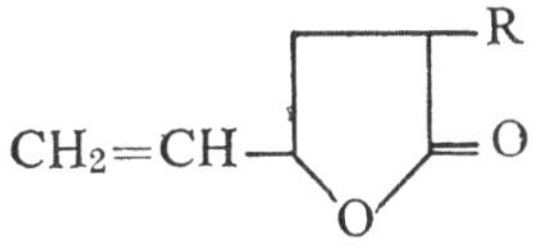

$$\text{CH}_2=\text{CH}-$$

where R can be H or an alkyl group.

These vinyllactones are very convenient starting materials. The simplest member can be easily prepared from butadiene monoxide and malonic ester and the general class can be easily obtained by the reaction of a radical (deriving, for instance, from the decomposition of peroxides) with acrylic acid and butadiene, according to a general synthesis recently found by Minisci and co-workers [32]. The lactones can be converted into dimers, mainly straightchained and having the *trans* configuration, simply by refluxing them in tetrahydrofuran with tetra-carbonylnickel:

$$(31)$$

$$\longrightarrow \ -OOCCH_2CH_2CH=CHCH_2CH_2CH=CHCH_2CH_2COO- + Ni^{++}$$

The product can apparently be formed only by opening the lactonic ring through the intermediacy of a complex of type (XX).

If the reaction with tetracarbonylnickel is carried out in the presence of acetylene and methanol insertion of acetylene and carbon monoxide occurs with the formation of the methyl ester of nona-2-*cis*-5-*trans*-dien-1,9-dioic acid:

$$(32)$$

The reaction takes place at 40-50° and atmospheric pressure in a mixture of tetra-hydrofuran and methanol. A lower temperature may be used in the presence of a catalytic amount of hydrochloric acid. Water can also be used, with the formation of the corresponding dibasic acid.

The insertion of acetylene gives the expected *cis* double bond. The conversion of the allylic-type vinyl double bond into a *trans* double bond provides another example of stereoselectivity deriving from co-ordination. It should be noted that nickel succeeds in opening the lactonic ring through a type of "alkylic scission" that is not common in normal ionic reactions [47]. This leads to the formation of the complex structure that has been shown above (XX). A definite complex has not been isolated from the reaction in methanol. By refluxing the vinyllactone in tetrahydrofuran with tetracarbonylnickel a red solid is obtained, whose decomposition with acids gives the coupling products shown above and vinyllactone. The solid has not been examined by NMR due to its insolubility in suitable solvents. Its probable formula is a polymeric form of (XX). It should be added that, in accordance with the reactions of the π-allyl groups observed so far on nickel, a straight-chained group is obtained from the reaction of the π-allyl group derived from the vinyllactone. The behaviour may be completely different in other media and with other metals. For instance, the co-ordinated crotyl group on palladium gives straight-chained products in the carbonylation reaction, but it reacts with the internal carbon of the allylic group in the recently described insertion of butadiene and carbon monoxide [33].

4.1. Summary of observations on stereoselectivity

The results presented in the preceeding pages can be summarized as follows:

a) A triple bond can be converted into a *cis* double bond by means of an insertion reaction between two ligands co-ordinated to nickel;

b) A triple bond can be converted into a *trans* double bond by splitting a nickel complex with acids or haloderivatives after the insertion of acetylene between an organic residue and nickel;

c) An activated vinylic double bond can be converted into a *trans* double bond by reaction with organic halides on nickel;

d) An allylic-type double bond can be converted by nickel into a *cis* or *trans* double bond, provided that suitable allylic structures, ligands and solvents are chosen.

5. STEREOSELECTIVE POLYMERIZATION

Nickel complexes have been employed in the polymerization reactions to obtain stereoselective polymerizations of butadiene [13a, b, 40-42].

Bis-(π-allyl) nickel chloride in benzene polymerizes butadiene to give 89% of the *cis*-1,4-polymer [40, 41, 42]. On changing to the bromide and then to the iodide increasing amounts of *trans* polybutadiene are obtained. This difference is no longer observed, the *cis* polymer being obtained, if the products of the reaction of π-allylnickel bromide with Lewis acids are employed.

A high selectivity in the 1,4-*cis* polymer (99%) has been recently obtained using π-allyl-Ni-OCOCF₃ as catalyst [46]. In all cases addition of compounds able to occupy a coordination site of π-allyl-Ni-complexes such as water, alcohol, tetrahydrofurane, $P(OC_6H_5)_3$ etc. results in the formation of the *trans* polymer [40, 46].

Also [methyl maleate] nickel-bromide is able to polymerise butadiene in aqueous-alcoholic solution with the formation of the *trans* 1,4-polymer [43].

Allyl groups having been found in the polymers obtained from π-allylnickel halides it has been proposed [40] that the polymerization reaction is analogous to an insertion reaction of butadiene into an allyl-Ni bond.

The formation of *cis* or *trans* polymers has been attributed to different modes of coordination of the butadiene molecule to π-allylnickel halides having two or one vacant coordination sites, respectively, which allow the coordination of the butadiene molecule with two or one double bonds. The ionic complexes obtained from π-allylnickel halides and Lewis acids or by splitting of the weak chloro-bridge of dimeric π-allylnickel chlorides by butadiene are likely to provide the additional site required for the double coordination [40].

In relation to the stereochemistry of the coordinated groups, Frolov and coworkers [44] have recently tried to ascertain whether bis-(π-crotyl) nickel chloride, which catalyses the formation of *cis* 1,4-polymer from butadiene, evolves *trans* or *cis* but-2-ene by hydrogenolysis. They established that 97-98% *trans* but-2-ene was evolved. Thus the coordinated crotyl gives a *trans* product, whereas its reaction with butadiene (which is expected to proceed through an analogous π-allylnickel complex) gives a *cis* polymer. This suggests that a directing action (possibly by the polymer chain [44]) is effective in the transition state.

Independently of the original configuration of the allyl group and of the monomer, what seems clear in any case is that certain coordination sites must be vacant in order to allow appropriate ligands (including the monomer and the polymeric chain) to control the stereoselectivity. These coordination sites may also be sterically different from each other. For example the recent finding that π-allylnickel trifluoracetate gives a 50% *cis*, 50% *trans* 1,4-polymer from butadiene in the presence of trifluoroacetic acid or chlorobenzenes has been interpreted as an alternate insertion of *cis* and *trans* units which is controlled by different coordination sites of the complex [46].

No discussion of the problem of the origin of stereospecificity in polymerisation reactions will be given here, but it has been deemed worthwhile to propose it to the reader.

6. UTILIZATION OF THE STEREOISOMERS

The products that can be obtained through stereoselective syntheses on nickel are suitable for other reactions whose description is beyond the scope of this review. The reader is referred to the literature on this subject [4, 34, 35, 36, 38]. There are some reactions, however, that require a specific configuration of the double bond. Two examples will be given.

A double bond can be reacted with sulphonium ylids to prepare cyclopropane derivatives. Corey [37] reacted diphenylsulphonium isopropylide with methyl 5-methylsorbate, a *trans* isomer obtained by stereoselective isomerization of the corresponding *cis*-5-methylhexa-2,5-dienoate (from the carbonylation reaction 4[c]). The methyl ester of chrysanthemic acid, the *trans* isomer, was obtained stereoselectively:

$$CH_3$$
$$\diagdown$$
$$C=CH-CH=CH-COOCH_3 \; + \; \underset{CH_3\;\;\;\;C_6H_5}{\overset{CH_3\;\;\;\;C_6H_5}{C\cdots\overset{+}{S}}} \; \longrightarrow$$
$$CH_3$$

$$\longrightarrow \; \underset{CH_3}{\overset{CH_3}{C=CH-C}}\underset{\underset{CH_3\;\;CH_3}{C}}{\overset{H\;\;\;\;\;H}{C-COOCH_3}} + (C_6H_5)_2S$$

The other example refers to the formation of α, β-unsaturated δ-lactones from 2,5-dienoic acids or esters by means of 60-95 % sulphuric acid at room temperature [39]:

$$R-CH=CH \overset{OR'}{\diagdown}=O \; \longrightarrow \; R-CH_2 \overset{O}{\diagup}=O \; + R'OH$$

This reaction is very simple and gives very high yields but requires the *cis* configuration of the dienoic acids or esters employed. The latter are easily prepared through the syntheses already described.

As to the *cis* or *trans* polymers of dienes, their utilization as elastomers is well known and offers very interesting prospects for the future.

7. REFERENCES

[1] E. L. ELIEL, *Stereochemistry of Carbon Compounds*, p. 436, (1962), Mc Graw Hill, (New York).

[2] W. REPPE, "Annalen", *582*, 1, (1952).

[3] E. R. JONES, T. Y. SHEN and M. C. WHITING, "J. Chem. Soc.", 48, (1951).

[4] a) G. P. CHIUSOLI, "Chimica e Industria" (Milan), *41*, 503, 506, 513, 762, (1959) "Gazz. Chim. Ital.", *89*, 1332, (1959).
  b) G. P. CHIUSOLI and S. MERZONI, "Chimica e Industria" (Milan), *43*, 259, (1961).
  c) G. P. CHIUSOLI, G. B. BOTTACCIO and A. CAMERONI, "Chimica e Industria" (Milan), *44*, 131, (1962).
  d) G. P. CHIUSOLI and S. MERZONI, "Chimica e Industria" (Milan), *45*, 6, (1963).
  e) G. P. CHIUSOLI, S. MERZONI and G. MONDELLI, "Chimica e Industria" (Milan), *46*, 743, (1964).
  f) G. P. CHIUSOLI and G. BOTTACCIO, "Chimica e Industria" (Milan), *47*, 165, (1965).
  g) G. P. CHIUSOLI, G. BOTTACCIO and C. VENTURELLO, "Chimica e Industria" (Milan), *48*, 107, (1966).
  h) L. CASSAR and G. P. CHIUSOLI, "Chimica e Industria" (Milan), *48*, 323, (1966).
  i) L. CASSAR, G. P. CHIUSOLI and M. FOÀ, "Chimica e Industria" (Milan), *50*, 515, (1968).
  l) G. P. CHIUSOLI, S. MERZONI and G. MONDELLI, "Tetrahedron Lett.", 2777, (1964).
  m) G. P. CHIUSOLI and L. CASSAR, "Angew. Chem. Int. Ed.", *6*, 124, (1967).
  n) G. P. CHIUSOLI, M. DUBINI, M. FERRARIS, F. GUERRIERI, S. MERZONI and G. MONDELLI, "J. Chem. Soc. C", 2889, (1968).

[5] F. GUERRIERI and G. P. CHIUSOLI, "Chem. Comm.", 781, (1967); id., "J. Organometal. Chem.", *15*, 209, (1968).

[6] a) E. O. FISCHER and G. BÜRGER, "Z. Naturforschg.", *16b*, 77, (1961);
  b) id, ibid. *17b*, 484, (1962).

[7] G. P. CHIUSOLI and S. MERZONI, "Z. Naturforschg.", *17b*, 850, (1962).

[8] R. F. HECK, "J. Amer. Chem. Soc.", *85*, 2013, (1963).

[9] R. F. HECK, J. C. W. CHIEN and D. S. BRESLOW, "Chem. and Ind." (London), 986, (1961).

[10] M. L. CHURCHILL and T. A. O'BRIEN, "Chem. Comm.", 246, (1968).

[11] F. GUERRIERI, "Chem. Comm.", 983, (1968).

[12] A. SIRIGU, "J. Chem. Soc. D", 256, (1969).

[13] a) G. WILKE et al., "Angew. Chem. Int. Ed.", *2*, 105, (1963).
  b) G. WILKE, B. BOGDANOVIC, P. HARDT, P. HEIMBACH, W. KEIM, M. KRÖNER, W. OBERKIRCH, K. TANAKA, E. STEINRÜCKE, D. WALTER and H. ZIMMERMANN, "Angew. Chem. Int. Ed.", *5*, 171, (1966).
  c) G. WILKE, *Plenary Lectures presented at the X International Conference on Coordination Chemistry, Tokyo, 1967*, p. 179, (1968), Butterworths, (London).

[14] L. S. MERIWETHER and M. L. FIENE, "J. Amer. Chem. Soc." *81*, 4200, (1959).

[15] a) F. CALDERAZZO and K. NOACK, "Coord. Chem. Rev.", *1*, 118, (1966), and references cited therein.

b) R. F. HECK, "Advan. Chem. Ser.", *49*, 181, (1965).

c) P. COSSEE, "Rec. Tr. Ch. Pays Bas", *85*, 1151, (1966).

[16] F. GUERRIERI and G. P. CHIUSOLI, "J. Organometal. Chem.", *19*, 453 (1969).

[17] M. FOÀ, L. CASSAR and M. TACCHI VENTURI, "Tetrahedron Lett.", 285, (1967).

[18] L. CASSAR, G. P. CHIUSOLI and M. FOÀ, unpublished results.

[19] a) M. DUBINI, F. MONTINO and G. P. CHIUSOLI, "Chimica e Industria" (Milan), *47*, 839, (1965), and unpublished results.

b) M. DUBINI and F. MONTINO, "J. Organometal. Chem.", *6*, 188, (1966).

[20] G. P. CHIUSOLI, "Chimica e Industria" (Milan), *43*, 365, (1961).

[21] See for literature on π-allyls and on π-σ interconversion:

a) W. R. MAC LELLAN, H. H. HOCHER, H. N. CRIPPS, E. L. MUTTERTIES and B. W. HOWK, "J. Amer. Chem. Soc.", *83*, 1601, (1961).

b) M. L. H. GREEN and P. L. G. NAGY, *Advances in Organometallic Chemistry*, editors F. G. A. Stone and R. West, p. 325, (1964), Academic Press, (New York).

c) E. O. FISCHER and H. WERNER, *Metal π-Complexes*, (1966), Elsevier, (Amsterdam).

d) C. W. BIRD, *Transition Metal Intermediates in Organic Syntheses*, (1967), Logos Press, (London).

e) F. A. COTTON, J. W. FALLER and A. MUSCO, "Inorg. Chem.", *6*, 179, (1967).

f) J. POWELL and B. L. SHAW, "J. Chem. Soc. A", 1839, (1967).

g) P. CORRADINI, G. MAGLIO, A. MUSCO and G. PAIARO, "Chem. Comm.", 618, (1966).

h) K. VRIEZE, C. MAC LEAN, P. COSSEE and C. W. HILBERS, "Rec. Tr. Ch. Pays Bas", *85*, 1077, (1966); id., "J. Organometal. Chem.", *6*, 672, (1966).

i) R. F. HECK, *Organic Syntheses via Metal Carbonyls*, editors I. WENDER and P. PINO, Vol. I, p. 373, (1968), Wiley, (New York).

l) J. K. BECCONSALL and S. I. O'BRIEN, "Chem. Comm.", 302, (1966).

m) G. L. STATTON and K. C. RAMEY, "J. Amer. Chem. Soc.", *88*, 1327, (1966).

n) J. POWELL, S. D. ROBINSON and B. L. SHAW, "Chem. Comm.", 78, (1965).

o) J. C. W. CHIEN and H. C. DEHM, "Chem. and Ind." (London), 745, (1961).

p) H. C. VOLGER and K. VRIEZE, "J. Organometal. Chem.", *9*, 297, 527, (1967); id., ibid., *13*, 479, 495, (1968).

q) K. VRIEZE, P. COSSEE, A. P. PRAAT and C. W. HILBERS, "J. Organometal. Chem.", *11*, 353, (1968).

r) K. VRIEZE, A. P. PRAAT and P. COSSEE, "J. Organometal. Chem.", *12*, 533, (1968).

s) J. A. BERTRAND, H. B. JONASSEN and D. W. MOORE, "Inorg. Chem.", *2*, 601, (1963).

t) A. J. DEEMING and B. L. SHAW, "J. Chem. Soc. A", 1562, (1969).

[22] G. P. CHIUSOLI and G. COMETTI, "Chimica e Industria" (Milan), *45*, 401, (1963).

[23] F. GUERRIERI and G. P. CHIUSOLI, "Chimica e Industria" (Milan), *51*, 1252, (1969), and unpublished results.

[24] J. W. CORNFORTH, R. H. CORNFORTH, C. DONNINGER and G. POPJAK, "Proc. Roy. Soc. B", *163*, 519, (1966).

[25] a) G. NATTA, U. GIANNINI, P. PINO and A. CASSATA, "Chimica e Industria" (Milan), *47*, 524, (1965).

104

b) G. ALLEGRA, F. LO GIUDICE, G. NATTA, U. GIANNINI, G. FAGHERAZZI and P. PINO, "Chem. Comm.", 1263, (1967).

[26] L. PORRI, G. VITULLI, M. ZOCCHI and G. ALLEGRA, "J. Chem. Soc. D", 276, (1969).

[27] *a*) I. G. FARBENIND. A. G., Belgian Pat. 448884, (1943).
b) I. D. WEBB and G. T. BORCHERDT, "J. Amer. Chem. Soc.", *73*, 2654, (1951).

[28] M. DUBINI, G. P. CHIUSOLI and F. MONTINO, "Tetrahedron Lett.", 1591, (1963).

[29] *a*) E. J. COREY and M. F. SEMMELHACK, "Tetrahedron Lett.", 6237, (1966); id., "J. Amer. Chem. Soc.", *89*, 2755, (1967).
b) E. J. COREY and E. HAMANAKA, "J. Amer. Chem. Soc.", *86*, 1641, (1964).
c) E. J. COREY and E. K. W. WAT, "J. Amer. Chem. Soc.", *89*, 2757, (1967).
d) E. J. COREY, M. F. SEMMELHACK and L. S. HEGEDUS, "J. Amer. Chem. Soc.", *90*, 2416, (1968).

[30] *a*) P. HEIMBACH and W. BRENNER, "Angew. Chem.", *79* 814, (1967).
b) P. HEIMBACH, W. BRENNER, H. HEY and G. WILKE, *Proc. XIth International Conference on Coordination Chemistry, Haifa and Jerusalem, 1968*, p. 10, (1968) Elsevier, (Amsterdam).

[31] G. P. CHIUSOLI, M. FERRARIS and S. MERZONI, *paper presented at the XIIth International Conference on Coordination Chemistry, Sydney, 20-27 August 1969.*

[32] F. MINISCI, R. GALLI, M. CECERE, U. MALATESTA and I. CARONNA, "Tetrahedron Lett.", 5609, (1968).

[33] D. MEDEMA, R. VON HELDEN and C. F. KOHLL, "Inorganica Chimica Acta", *3*, 255, (1969).

[34] *a*) G. P. CHIUSOLI and S. MERZONI, "Chimica e Industria" (Milan), *43*, 255, (1961).
b) G. P. CHIUSOLI, G. AGNÈS, C. A. CESELLI and S. MERZONI, "Chimica e Indudustria" (Milan), *46*, 21, 1964.

[35] *a*) G. P. CHIUSOLI, M. FERRARIS, F. GUERRIERI, S. MERZONI and G. MONDELLI, It. Pat. 760135.
b) M. DUBINI, M. FERRARIS, F. MONTINO and G. P. CHIUSOLI, French Pat. 1570823, (1967).

[36] *a*) G. P. CHIUSOLI, U.S.P. Pat. 2978446, (1961).
b) G. P. CHIUSOLI and S. MERZONI, It. Pat. 630365, (1959).

[37] E. J. COREY and M. JAUTELAT, "J. Amer. Chem. Soc.", *89*, 3912, (1967).

[38] *a*) G. P. CHIUSOLI and G. AGNÈS, "Chimica e Industria" (Milan), *46*, 25, (1964).
b) G. AGNÈS and G. P. CHIUSOLI, "Chimica e Industria" (Milan), *49*, 465, (1967).

[39] G. AGNÈS and G. P. CHIUSOLI, "Chimica e Industria" (Milan), *50*, 194, (1968).

[40] *a*) L. PORRI, G. NATTA and M. GALLAZZI, "Chimica e Industria" (Milan), *44*, 426, (1964); id., ibid., "J. Pol. Science C", 2525", (1967).
b) G. NATTA and L. PORRI, "High Polymers", *23*, (Pt2), 597, (1968).

[41] R. D. BABITSKII, B. A. DOLGOPLOSK, V. A. KORMER, M. I. LOBACH, E. I. TINYAKOVA and V. A. YAKOVLEV, "Dokl. Akad. Nauk SSSR", *161*, 583, (1965).

[42] M. I. LOBACH, B. D. BABITSKII and V. A. KORMER, "Russian Chemical Reviews", *36*, 477, (1967).

[43] M. DUBINI and F. MONTINO, "Chem. Comm.", 749, 1966; ibid., "Chimica e Industria" (Milan), *49*, 1283, (1967).

[44] V. M. FROLOV, A. V. VOLKOV and O. P. PARENAGO, "Dokl. Akad. Nauk SSSR", *177*, 1359, (1967).

[45] G. P. CHIUSOLI, C. VENTURELLO and S. MERZONI, unpublished results.

[46] PH. TEYSSIE, F. DAWANS, J. P. DURAND and J. C. MARECHAL, *Colloques Internationaux du Centre National de la Recherche Scientifique*, n. 191, *La Nature et les Proprietés des Liaisons de Coordination, Paris*, October (1969).

[47] see also N. L. BAULD, "Tetrahedron Lett.", 859, (1962).

[48] After completion of this chapter the following papers appeared, describing stereoselective cyclo-oligomerizations of butadiene alone or with olefins or acetylenes:
   a) B. BOGDANOVIC, P. HEIMBACH, M. KRÖNER and G. WILKE with E. G. HOFFMANN and J. BRANDT, "Annalen", *727*, 143, (1969).
   b) W. BRENNER, P. HEIMBACH, H. HEY, E. W. MÜLLER and G. WILKE, ibid., *727*, 171, (1969).
   c) P. HEIMBACH and G. WILKE, ibid. *727*, 183, (1969).
   d) W. BRENNER, P. HEIMBACH and G. WILKE, ibid. *727*, 194, (1969).
   e) P. HEIMBACH and R. TRAUNMÜLLER, ibid., *727*, 208, (1969).

[49] Work in progress by GUERRIERI and CHIUSOLI has recently shown that also the stereochemistry of this coupling reaction can be completely reversed in favour of the *trans* isomer, by working in the presence of appropriate ligands in coordinating solvents.

Chapter 3

Dimerization and Co-dimerization of Olefinic Compounds by Co-ordination Catalysis

G. LEFEBVRE and Y. CHAUVIN

Institut Français du Pétrole, 92. Rueil-Malmaison, France

108

1. GENERAL INTRODUCTION

1.1. Scope

Low molecular weight olefins and dienes, produced in large quantities and at low cost by petrol refining processes, such as steam cracking, may give rise by dimerization and co-dimerization to valuable intermediates. These olefinic and diolefinic dimers may find industrial applications in very important processes such as polymerization, oxo-synthesis, alkylation and so on.

On the other hand, co-ordination complex catalysts afford a new very attractive mean of achieving olefin dimerization with the high selectivity already attained in polymerization by Ziegler-Natta catalysts.

These are the reasons for the extensive work done on this subject during the last few years and for the great number of patents and publications covering this field.

However, until now, few reviews [1] [2] have been published on this subject. The scope of this review is to give an up to date picture of olefin dimerization by co-ordination catalysts. It is not intended to be exhaustive. It will deal with olefin and vinylic compounds; linear and cyclic dimerization of dienes, which has been the subject of recent reviews [2] [3] [4] [5] will not be discussed here. However dienes will be taken into account when they enter into co-dimerization with olefinic compounds.

1.2. Basic concepts of dimerization

Products resulting from the polymerization and oligomerization of mono- and diolefins may be defined by means of four essential features in their mode of formation:

1) The degree of polymerization or number of monomer units forming the hydrocarbon chain. It depends on the frequency of chain transfer reactions with respect to the propagation reaction.

2) The mode of linking of the monomer units which, in simple cases (mono-olefins and conjugated dienes) can be head-to-head, tail-to-tail or head-to-tail, i. e.

$$
\begin{array}{ccc}
-C-C-C-C- & -C-C-C-C- & -C-C-C-C- \\
\;\;\;|\qquad\;\;\;| & \;\;|\;\;\;| & \;\;|\qquad\;\;\;| \\
\;\;\;R\qquad R & \;\;R\;\;R & \;\;R\qquad R
\end{array}
$$

In the case of conjugated dienes linking can involve two or four carbon atoms

$$
\begin{array}{cc}
-C-C=C-C- & \quad -C-C- \\
& \qquad\quad| \\
& \qquad\quad C \\
& \qquad\quad\| \\
& \qquad\quad C
\end{array}
$$

3) The mode of opening of the olefinic bond. If an asymmetric carbon atom is formed, it may have a regular or random conformation (concept of tacticity). If a new double bond forms, this will introduce geometric isomerism into the chain (*cis* or *trans*).

4) The position of the double bond(s).

A co-polymer is to be defined on the basis of the above characteristics and, in addition, by the recurrence frequency and distribution of each of the monomers in the chain.

The same is true for dimerization and co-dimerization products but in this case each factor responsible for each type of molecular order may contribute to the formation of a defined chemical entity which may be isolated and characterized by analysis. This will then introduce the concept of selectivity either for each of the above factors or for each of the products. What could be achieved only in a statistical manner for macromolecules can now be done "individually" for dimers or lower oligomers. The different features leading to a defined structure will be discussed hereafter:

a) *The selectivity for the degree of polymerization* is the important aspect of dimerization. It the case of dimerization and "linear" co-dimerization, which can be considered as the first steps in a polymerization, the main question is: why does the growing chain, once initiated, stop? What are the conditions contributing to the instability of the growing species after the addition of a second olefin molecule?

This important problem, i. e. the predominance of chain transfer reactions with respect to the propagation reaction, is linked to the conditions of shifting of hydrogen atoms in the hydrocarbon compounds. Although this can vary with the nature of the activating agents, it is frequently observed that the activation energy of hydrogen displacements is higher than the simple electronic transfer. Thus, an increase in temperature contributes to the lowering of the molecular weight of polymers (often to a point leading quite selectively to the dimer) or favours a linear oligomerization rather than a cyclic oligomerization.

Another obvious condition for obtaining selectivity for the degree of polymerization is that the product formed must have a sufficiently low affinity for the catalyst. This is not always the case, especially when the product shows little branching, e. g. butene formed by dimerization of ethylene. It will then be sufficiently reactive to dimerize and co-dimerize with the reactant. An indirect method can be used to preserve selectivity, i. e. using a high concentration of the reactant and limiting the conversion.

b) *The mode of linking of the monomer molecules* in dimerization depends on two fundamentally different steps, i. e. the initiation step and the propagation step.

In a polymer, the contribution of the initiation step to the microstructure can be neglected, whereas in a dimer it is essential, thus making the study of the factors influencing linking more complex.

c) *The ability of the catalyst to mobilize the hydrogen atoms* is, also, responsible for the position of the double bond in the dimer. Either this position is defined in the initiation or transfer step, or it results from a subsequent isomerization of the formed olefin. Dimerization catalysts can in fact catalyze prototropic displacements to a greater or lesser extent.

It is therefore not easy, by the nature of the active species, to separate the dimerization and isomerization. The determination of the intimate mechanism of this catalysis will require a simultaneous study of the nature of the products of dimerization and the conditions causing the migration of the double bond in the olefins. It should be noted, in this respect, that a much wider variety of catalytic systems is known that are capable of catalyzing the isomerization of olefins. Actually, the growing step (or "insertion" step or "*cis*-rearrangement") is frequently the limiting step, sometimes for simple steric reasons and sometimes because the active centre is incapable of accommodating two olefin molecules simultaneously. Therefore, one condition for obtaining selectivity as far as the position of the double bond is concerned is that the products, once formed, should not be isomerized (or only a little). When isomerization occurs, it can be kept artificially at a minimum by increasing the concentration of the reactants in the medium and thus minimizing conversion.

d) The possibility of asymmetric synthesis in a catalytic dimerization has never been reported.

1.3. Survey of ionic dimerization catalysis

Olefin dimerizations and co-dimerizations through co-ordination are reactions whose study is relatively recent and which have been investigated to a much lesser extent than polymerizations. Their mechanism is often incompletely known and frequently only consists of hypotheses.

On the other hand, several ionic processes are known (and, at least at first sight, they are simpler than co-ordination processes) which selectively provide dimers whose nature, resulting from a preferential mode of linking of the monomer units, can be interpreted on the basis of considerations of olefin polarizability and charge stability alone.

Here it would be usefull to recall, as briefly as possible, several salient facts and some principles of base and acid catalysis in dimerization and isomerization before examining the more complex case of co-ordination catalysis.

1.3.1. BASE CATALYSIS

Among the numerous dimers potentially available by anionic catalysis [7], two propylene dimers are produced on an industrial scale, i. e. 2-methyl-1-pentene (for isoprene synthesis) and 4-methyl-1-pentene (for stereospecific polymerization).

The specific "obtainment" of α-olefins and the high selectivity frequently observed in these reactions can be related to the relative carbanion stability.

It is not our purpose to discuss the concepts of thermodynamic and kinetic stability of carbanions [6] but, as a first simplification, carbanions will be classified according to the following order of "stability" and "reactivity" with respect to the proton or olefin addition:

$$
\begin{array}{c}
R—CH_2—CH—\overset{\ominus}{CH}—CH_2 \\[4pt]
R—CH=CH^{\ominus} \\[4pt]
R—CH_2{}^{\ominus} \\[4pt]
R_1—\overset{\ominus}{CH}—R_2 \\[4pt]
R_1—\overset{\ominus}{\underset{|}{C}}—R_2 \\
R_3
\end{array}
$$

increasing "reactivity"   increasing "stability"

Carbanions are more stable and hence less subject to rearrangement as they are less substituted and characterized by a greater degree of resonance. This means that the product obtained from an anionic polymerization or dimerization, which by nature contains fewer double bonds and more substitutions than the monomer, will, in general, be less reactive than this monomer with respect to carbanions and will be less subject to rearrangements, thus fulfilling one of the conditions required for achieving selectivity.

On the other hand, the anionic reactivity of olefins, whose addition to carbanions occurs in accordance with Markownikoff's rule, follows approximately the following order:

butadiene > ethylene > propylene > 1-butene > 2-butene > isobutene

However, a carbanion cannot be considered as a free entity. The metallic counter-ion and its environment exert a considerable influence on its reactivity in general and, more particularly, on its ability to provide for the mobility of hydrogen atoms. Thus metals forming highly polar bonds with the organic groups (e. g. sodium, potassium) have a stronger tendency to enhance proton transfers (metallation). Those forming bonds of a covalent nature (e. g. boron, aluminium) facilitate hydride ion elimination. These reactions are the basis of anionic isomerization, dimerization and alkylation selectivities, and consequently, will be discussed here in some detail.

1) *Proton transfer*

This occurs more easily as the carbanion which is to be formed is more stable (however kinetic limitations may have an effect).

Thus, in the isotopic exchange of hydrogen catalyzed by t. BuOK with deuterated ammonia, five of the hydrogens of propene exchange rapidly compared to the sixth hydrogen or the hydrogens of ethylene (allylic carbanions being more stable than vinyl carbanions because of the allylic resonance):

$$CH_2 = CH - CH_2^{\ominus} \leftrightarrow {}^{\ominus}CH_2 - CH = CH_2 \ .$$

Likewise, in the isomerization of methylpentenes, some kinetic limitation occurs, i. e. starting from 4-methyl-1-pentene or 4-methyl-2-pentene, the prevailing allylic structure is:

$$\begin{array}{c} C \\ | \\ C-C-C-C-C \\ \underline{} \\ \ominus \end{array}$$

Starting from 2-methyl-2-pentene or 2-methyl-1-pentene, the prevailing allylic structure is:

$$\begin{array}{c} C \\ | \\ C-C-C-C-C \\ \underline{} \\ \ominus \end{array}$$

Such behaviour is due to the difficulty of forming secondary and tertiary carbanions and explains the low rate of establishment of the equilibrium:

$$\begin{array}{ccc} C & & C \\ | & & | \\ C-C=C-C-C & \rightleftarrows & C-C-C=C-C \end{array}$$

In the case of olefins which may undergo poly-addition to a carbanion, it is sometimes possible to use a temperature for which the ratio of *rate of addition to rate of transfer* has a value such that selective dimerization is observed [7]. In fact, by putting propylene into contact with an organopotassium compound at 150-200 ºC in an aliphatic hydrocarbon solvent, the rather selective formation of 4-methyl-1-pentene is obtained. This fact can be explained by the following reasons:

a) The most acidic hydrogen in the medium is the allylic hydrogen of the propylene, so that exchange occurs:

$$R-CH_2^{\ominus}K^{\oplus} + CH_3-CH=CH_2 \rightarrow R-CH_3 + \overline{CH_2-CH-CH_2}^{\ominus} \ K^{\oplus}$$

b) Addition of a molecule of propylene occurs on to the allylic anion and according to the Markownikoff's rule:

$$\overline{CH_2-CH-CH_2}^{\ominus} \ K^{\oplus} + CH_3-CH=CH_2 \rightarrow K^{\oplus \ominus}CH_2-\overset{\overset{\textstyle CH_3}{|}}{CH}-CH_2-CH=CH_2$$

c) In the medium, the most acidic proton is still the allylic proton in the propylene and a transfer takes place due to the instability of the active species with respect to the reactant:

$$CH_3-CH=CH_2 + K^{\oplus\ominus}CH_2-\overset{\overset{\displaystyle CH_3}{|}}{CH}-CH_2-CH=CH_2 \rightarrow$$

$$\rightarrow CH_3-\overset{\overset{\displaystyle CH_3}{|}}{CH}-CH_2-CH=CH_2 + \overset{\ominus}{CH_2}\text{------}CH\text{------}CH_2 K^{\oplus}$$

The by-products formed in this dimerization may be attributed to:

α) an anti-Markownikoff addition (n-hexenes);

β) an allylic isomerization by metallation of the secondary carbon in the 4-methyl-1-pentene (4-methyl-2-pentene);

γ) a possible vinylic metallation of propylene.

The dimerization of propylene by alkali metals is, in fact, a special case of telomerization (or alkylation) in which the telogene is the olefin itself.

But other hydrocarbons may be alkylated by the olefin, e. g. toluene, iso-butene, etc.

2) *Hydride addition and elimination*

β-elimination occurs more easily as this carbon atom is more highly substituted. For instance, when stoichiometric quantities of both an olefin with an internal double bond and diborane ($B^{\oplus}H^{\ominus}$) are reacted, a primary alkyl-boron is obtained, the unsubstituted carbanion being the more stable one:

$$R-CH=CH-CH_3 + \overset{\diagdown}{\underset{\diagup}{B}}{}^{\oplus}H^{\ominus} \rightarrow R-CH_2-\overset{\overset{\displaystyle \diagdown B{}^{\oplus}\diagup}{}}{\underset{\ominus}{CH}}-CH_3 \rightarrow$$

$$R-CH_2-CH_2-CH_2^{\ominus}\overset{\diagdown}{\underset{\diagup}{B}}{}^{\oplus} \rightarrow R-CH_2-CH=CH_2 + \overset{\diagdown}{\underset{\diagup}{B}}{}^{\oplus}H^{\ominus}$$

Aluminium can achieve the same reaction but, in addition, can cause the insertion of an α-olefin. The second addition gives rise to a particularly mobile hydrogen:

$$CH_3-CH=CH_2 + \overset{\diagdown}{\underset{\diagup}{Al}}{}^{\oplus}H^{\ominus} \longrightarrow \overset{\diagdown}{\underset{\diagup}{Al}}{}^{\oplus\ominus}CH_2-CH_2-CH_3 \overset{C_3H_6}{\longrightarrow}$$

$$\overset{\diagdown}{\underset{\diagup}{Al}}{}^{\oplus\ominus}CH_2-\overset{\overset{\displaystyle CH_3}{|}}{\underset{\underset{\displaystyle H}{|}}{C}}-CH_2-CH_2-CH_3 \rightarrow \overset{\diagdown}{\underset{\diagup}{Al}}{}^{\oplus}H^{\ominus} + CH_2=\overset{\overset{\displaystyle }{}}{\underset{\underset{\displaystyle CH_3}{|}}{C}}-CH_2-CH_2-CH_3$$

At 200 °C and 200 atmospheres, 2-methyl-1-pentene selectivity exceeds 99%. Very little poly-addition and isomerization occur (through anti-Markownikoff hydride addition).

In the particular case of ethylene, addition of a second molecule does not provide conditions for sufficient mobility of the secondary hydrogen at the moderate reaction temperature employed:

$$\diagdown Al - C_2H_5 + nC_2H_4 \rightarrow \diagdown Al - (CH_2 - CH_2)_n - C_2H_5$$

The procedure is then carried out in two stages, i. e. the growth reaction is effected in the neighbourhood of 110 °C and the olefin is then displaced by raising the organoaluminium compound to 300 °C for fractions of a second.

1.3.2. ACID CATALYSIS

Despite their poor selectivity, cationic processes are extensively used for the industrial production of olefin oligomers, which are more often mixtures than pure dimers. Among them are: isobutene dimers (2,2,4-trimethylpentenes), propylene trimers and tetramers, propylene-butene dimers (isoheptenes) and oligomers of various olefins. The development of acid catalysis is essentially due to its low cost.

The poor selectivity is related to the particular carbonium ion stability. This will be now discussed briefly.

The first step in cationic polymerization, dimerization or isomerization is usually the addition of a proton to an olefin to form a carbonium ion. The tendency of carbonium ions to lose a proton, capture hydrocarbon hydrogens or add olefins follows approximately the following order:

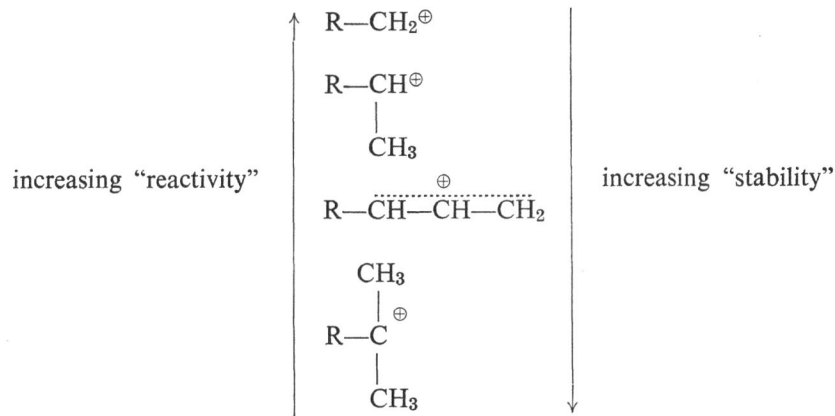

The tendency of olefins to add to a carbonium ion follows approximately the order:

$$CH_2{=}CH_2 < CH_2{=}CH{-}CH_3 < CH_2{=}CH{-}CH{=}CH_2 < CH_2{=}C\underset{CH_3}{\overset{CH_3}{<}}$$

This addition occurs in accordance with Markownikoff's rule.

Therefore, as opposed to the characteristics of carbanions, the stability of the carbonium ions increases with increasing substitution, allylic delocalization only contributing to a small extent to stabilization. We will now describe the details of these reactions.

1) *Proton addition and elimination*

Addition follows Markownikoff's rule, but the mode of elimination is a function of the environment of the carbonium ion. The main olefin obtained by proton elimination is:

a) either the one in which the double bond carries the largest possible number of alkyl groups (Saytzev's rule):

$$CH_3{-}\overset{\overset{\displaystyle CH_3}{|}}{CH}{-}CH_2{-}\overset{\oplus}{CH}{-}CH_3$$

$$CH_3{-}\overset{\overset{\displaystyle CH_3}{|}}{CH}{-}CH_2{-}CH{=}CH_2 + H^{\oplus} \qquad 20\%$$

$$CH_3{-}\overset{\overset{\displaystyle CH_3}{|}}{CH}{-}CH{=}CH{-}CH_3 + H^{\oplus} \qquad 80\%$$

b) or the one in which the double bond carries the least number of alkyl groups (Hoffman's rule) in the case of highly branched hydrocarbons:

$$CH_3{-}\underset{\underset{\displaystyle CH_3}{|}}{\overset{\overset{\displaystyle CH_3}{|}}{C}}{-}CH_2{-}\overset{\oplus}{C}{-}CH_3$$

$$CH_3{-}\underset{\underset{\displaystyle CH_3}{|}}{\overset{\overset{\displaystyle CH_3}{|}}{C}}{-}CH_2{-}\underset{\underset{\displaystyle CH_3}{|}}{C}{=}CH_2 + H^{\oplus} \qquad 80\%$$

$$CH_3{-}\underset{\underset{\displaystyle CH_3}{|}}{C}{-}CH{=}\underset{\underset{\displaystyle CH_3}{|}}{C}{-}CH_3 + H^{\oplus} \qquad 20\%$$

The proton thus eliminated can either add again to another olefin molecule, or recombine with the counter-ion (destruction of the active centre).

Combinations of addition and elimination reactions lead to the displacement of the double bond:

$$CH_3 - CH_2 - CH = CH_2 + H^{\oplus} \rightarrow CH_3 - CH_2 - \overset{\oplus}{C}H - CH_3 \rightarrow$$
$$CH_3 - CH = CH - CH_3 + H^{\oplus}$$

2) *Capture of a hydride ion*

Capture of a hydrogen by the carbonium ion occurs from the most highly substituted carbon atoms and more rarely from the allylic positions. Therefore, the polymer or dimer, which is usually more highly branched than the monomer, is very reactive towards the carbonium ion. The following reactions can take place:

a) Branching:

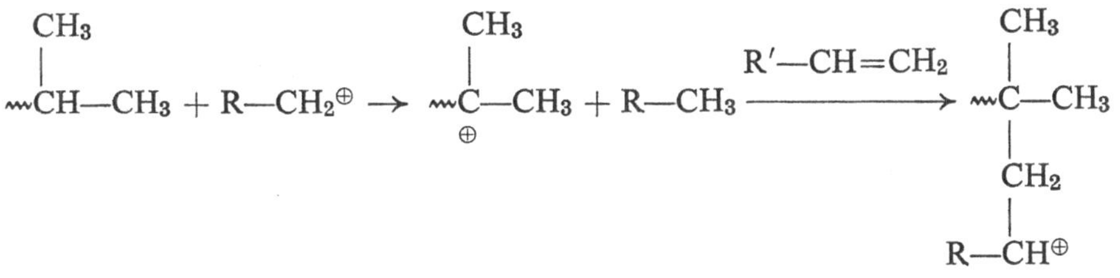

$$\underset{CH_3}{\overset{CH_3}{\overset{|}{\sim CH-CH_3}}} + R-CH_2^{\oplus} \rightarrow \underset{\oplus}{\overset{CH_3}{\overset{|}{\sim C-CH_3}}} + R-CH_3 \xrightarrow{R'-CH=CH_2} \sim C-CH_3$$

with resulting structure:

$$\begin{array}{c} CH_3 \\ | \\ \sim C-CH_3 \\ | \\ CH_2 \\ | \\ R-CH^{\oplus} \end{array}$$

b) Insertion with isomerization [8] when hydrogen abstraction is more rapid than insertion and occurs from the carbon located in α or β of the carbonium ion:

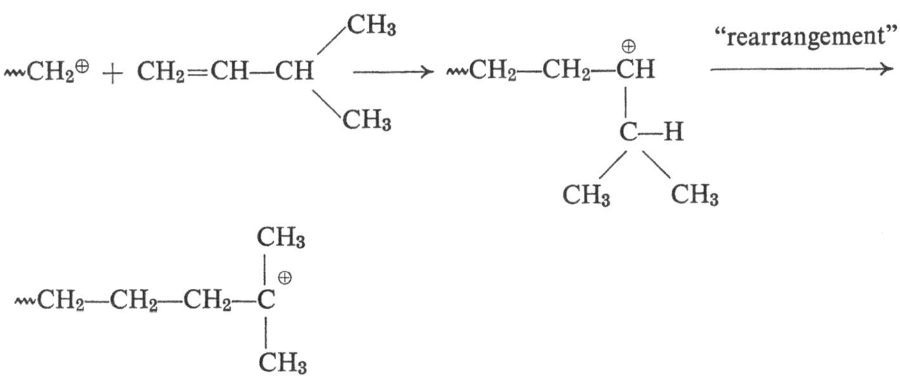

$$\sim CH_2^{\oplus} + CH_2=CH-CH \overset{CH_3}{\underset{CH_3}{<}} \longrightarrow \sim CH_2-CH_2-\overset{\oplus}{CH} \xrightarrow{\text{"rearrangement"}}$$

with

$$\begin{array}{c} C-H \\ / \quad \backslash \\ CH_3 \quad CH_3 \end{array}$$

$$\sim CH_2-CH_2-CH_2-\overset{CH_3}{\underset{|}{\overset{|}{C}}}\overset{\oplus}{} \\ CH_3$$

3) *Other reactions*

In addition to hydrogen displacements, carbonium ions can also induce fragmentations and 1,2-nucleophilic rearrangements of the following type:

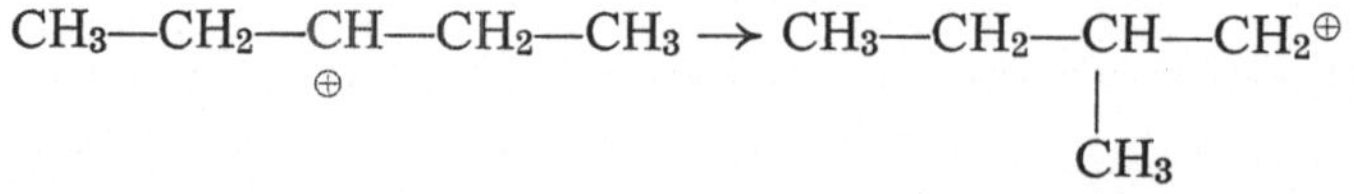

$$CH_3-CH_2-\underset{\oplus}{CH}-CH_2-CH_3 \rightarrow CH_3-CH_2-CH-CH_2{\oplus}$$
$$\overset{|}{\underset{CH_3}{}}$$

To conclude, the large number of reactions involving the carbonium ion are responsible for the low selectivities frequently observed in cationic polymerizations and dimerizations. These transfer occurs all the more readily, compared with those characterizing a carbanion, as the carbonium ion is unstable towards the reactants.

There are, nevertheless, special cases in which there is no easy transfer. Such is the case, for example, of olefins from which quaternary carbons are formed by polyaddition and for which only proton additions and eliminations are observed (e. g. isobutene):

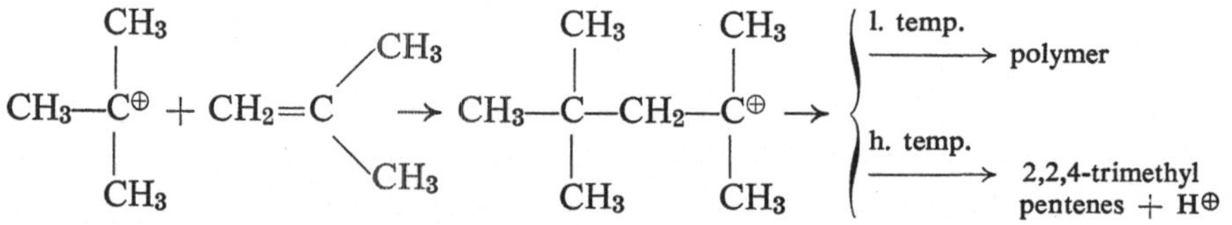

(Incidentally, the effect of temperature on chain transfer should be noted, i. e. production of dimer at high temperature, and production of elastomer at low temperature).

As in anionic catalysis, secondary reactions of the product can be minimized by working at high monomer concentrations and limiting the conversion, or else by eliminating the dimer as it is formed, for example by insolubilization.

1.4. Survey of co-ordination catalysis related to dimerization

1.4.1. INTRODUCTION

In the preceeding part it has been seen that the ionic processes of dimerization may be represented as limited ("degenerated") polymerizations, i. e. they involve the following three steps: *initiation* (i. e. formation of an activated group), "*insertion*" of a monomer molecule into the activated species, and *transfer* (deactivation of the chain). The type of hydrogen migration and the steric course of "*insertion*" are, at first sight, relatively easy to classify and interpret on the basis of the nature and stability of the charged species and the degree of ionic character of the bonds. Selective dimerization of olefins can be achieved in the same way as their polymerization, by the use of particular conditions (temperature, concentration, solvent...). The nature of dimers is well visualized in terms of the carbanions and carbonium ions stability and on the basis of general knowledge of reactivity of the these species.

The same is not true for co-ordination catalysis dimerization. Factors other than charges must be taken into account, i. e. interactions between olefins and transition metals are of various types (σ and/or π, σ or π allylic) and, in fact, little is understood about the factors which are important in determining their reactivity. On the other hand, products of a co-ordination dimerization are often of miscellaneous structures and minor changes in the catalytic systems result in major changes in the products. Obviously all the products cannot be as easily explained as in the case of conventional ionic mechanisms.

If a "degenerated" polymerization type mechanism is acting in dimerization, selective hydrogen transfer at the dimer stage cannot be only ascribed to the difference in the stabilities of σ metal-carbon bonds. So, in dimers as various as n-hexenes, dimethylbutenes and 2-methylpentenes simultaneously formed in propylene dimerization, intermediates are not of the same type (primary, secondary or tertiary σ bonded carbon atoms).

In addition, many products of dimerization cannot be immediately visualized as originating in a "degenerated" polymerization, e. g. 1,5-hexadiene and 1,3-hexadiene, found among the products of the co-dimerization of butadiene with ethylene, are not easily explained as resulting from, for example, a π-allylic mechanism. Another likely type of mechanism must be proposed.

Lastly, hydrogenated dimers obtained either under hydrogen pressure or by a non-catalytic reductive process may arise from a different pathway.

1.4.2. FORMAL MECHANISMS OF DIMERIZATION

Three types of mechanisms will be tentatively discussed although such a classification is not entirely satisfactory in the present state of knowledge.

1.4.2.1. *"Degenerated" polymerization*

By analogy with non-transition metals and on account of the low electropositive nature of the transition-elements, the dimerization co-ordination catalysis is often related to anionic catalysis ("insertion" step being preceeded, then followed, by co-ordination of the olefins). Metal-carbon and metal-hydrogen bonds which may form will be of a marked covalent nature, and their behaviour will often be closer to that of organo-aluminium compounds than to that of organo-alkaline metal compounds. In addition we have that:

The reactivity order of olefins is sometimes the same as observed in base catalysis.

The dimer "selectivity" is high, meaning that the reactivity of the products is less than that of the reagents.

No case of aliphatic alkylation is known and cases of aromatic alkylation are not very frequent.

No skeletal rearrangements of an acid type are known.

The mechanism invoked to interpret hydrogen transfers is therefore described in terms of either metallation of the vinylic and allylic carbon atom or addition and elimination of metal hydride (with formation of a π-allylic complex in the case of conjugated diolefins). Actually it is the latter mechanism which is most often suggested, even though the examples for which metal hydrides can be characterized in the reaction medium are relatively rare and the preformed hydrido-complexes sometimes prove to be inactive. Nevertheless, it has been frequently observed that it is useful and sometimes necessary for catalytic activity to introduce into the reaction medium, molecular hydrogen, protons or various compounds, and these are assumed to be the hydride ion sources.

As has been pointed out, the selective hydrogen tranfers at the stage of the dimer cannot be ascribed to the difference in the stability of σ-metal-carbon bonds owing to their diverse types. Then the selectivity may arise from the difference in π-bonding ability between monomer and dimer (co-ordination steps). It can be assumed that very fast equilibria are readily established between hydride and alkyl species (the "formal insertion" of a monomer molecule into the metal-carbon bond being the irreversible rate-determining chain-growth reaction),

$$M\text{—}H \ + \ CH_2\text{=}CHR \ \underset{}{\overset{\text{fast}}{\rightleftarrows}} \ M\text{—}CH_2\text{—}CH_2R$$

$$M\text{—}CH_2\text{—}CH_2R \ + \ CH_2\text{=}CHR \ \rightarrow \ M\text{—}CH_2\text{—}\overset{\overset{\displaystyle R}{|}}{C}H\text{—}CH_2\text{—}CH_2R$$

$$M\text{—}CH_2\text{—}\overset{\overset{\displaystyle R}{|}}{C}H\text{—}CH_2\text{—}CH_2R \ \overset{\text{fast}}{\rightleftarrows} \ M\text{—}H \ + \ CH_2\text{=}\overset{\overset{\displaystyle R}{|}}{C}\text{—}CH_2\text{—}CH_2R$$

Taking into account the limiting co-ordination steps

$$\overset{\overset{\displaystyle R}{|}}{\underset{\underset{\displaystyle M}{\downarrow}}{C}}H_2\text{=}C\text{—}CH_2CH_2R \ + \ CH_2\text{=}CHR \ \rightleftarrows \ CH_2\text{=}CHR \ + \ \overset{\overset{\displaystyle R}{|}}{\underset{\underset{\displaystyle M}{\downarrow}}{C}}H_2\text{=}C\text{—}CH_2CH_2R$$

the equilibrium can be shifted far to the right (cf. dimerization of ethylene by rhodium complexes).

A difficulty arises as far as the steric course of "insertion" is concerned. The factors which are important in determining the products are not yet well understood. Metal-hydrogen and metal-carbon bonds react with olefins on one hand like hydride ions and carbanions (the less substituted carbon being linked to the metal), and on the other hand like protons and carbonium ions (the more substituted carbon being linked to the metal), that is:

$$M—H + CH_2=CHR \quad \overset{a}{\nearrow} \quad \overset{\searrow}{b} \quad
\begin{matrix} M—CH_2—CH_2R \\ \\ R \\ | \\ M—CH—CH_3 \end{matrix}$$

$$M—R' + CH_2=CHR \quad \overset{a}{\nearrow} \quad \overset{\searrow}{b} \quad
\begin{matrix} M—CH_2—CH\overset{\diagup R}{\diagdown_{R'}} \\ \\ R \\ | \\ M—CH—CH_2R' \end{matrix}$$

factors influencing the a/b ratio are, up to now, not fully understood.

1.4.2.2. Concerted coupling

Concerted coupling is a well documented reaction in cyclic oligomerizations of dienes and acetylenes (the most extensively studied process using nickel). These reactions proceed by the stepwise addition of monomers to the metal, followed by the formation of the carbon-carbon bonds in a multi-centred bond process.

Open chain oligomers may originate from the same type of mechanism, cyclization being precluded by particular conditions for hydrogen shifts (high temperature, large steric size of the metal atom, presence of acidic compounds like alcohols, phenol and so on).

A similar mechanism may occur in the case of mono-olefins, and it has been formally represented as an activitation of a hydrogen-carbon bond, followed by coupling, e. g.

$$\begin{matrix} R—CH=CH..H \\ CH=CH_2 \\ | \\ R \\ \text{"activation"} \end{matrix} \quad \rightarrow \quad \begin{matrix} R—CH=CH{-}CH—CH_3 \\ | \\ R \\ \text{"coupling"} \end{matrix}$$

The term "dienylation" has been suggested by Wittenberg [4] when the hydrogen is moving from a diene molecule.

More recently Dall'Asta et al. [9] reformulated this mechanism in a more precise way, the hydrogen abstraction being represented as an oxidative addition on the metal followed by a reductive elimination and then coupling of the activated elements on a co-ordinated monomer. This mechanism has been suggested for the co-dimerization of butadiene with ethylene; the general scheme is as follows:

In the case of two co-ordinated double bonds:

$$
\begin{array}{ccc}
\underset{\text{``activation''}}{
\begin{array}{l}
\quad\ \ CH_2 \\
CH \quad\ \ CH{=}CH_2 \\
|\quad\ M \\
CH \qquad H \\
\quad\ \ CH_2
\end{array}}
& \rightarrow\ M^{(0)}\ + &
\underset{\text{``coupling''}}{
\begin{array}{l}
CH_2{-}CH{=}CH_2 \\
CH \\
\parallel \\
CH \\
\quad CH_3
\end{array}}
\end{array}
$$

In the case of one co-ordinated double bond:

$$
\begin{array}{ccc}
\underset{\text{``activation''}}{
\begin{array}{l}
\qquad\quad CH_2 \\
CH{-}CH \quad CH{=}CH_2 \\
\quad\qquad M \\
CH_2 \qquad\ H
\end{array}}
& \rightarrow\ M^{(0)}\ + &
\underset{\text{``coupling''}}{
\begin{array}{l}
\quad CH_2{-}CH{=}CH_2 \\
CH{-}CH_2 \\
CH_2
\end{array}}
\end{array}
$$

In the case of a "dienylation":

$$
\begin{array}{ccc}
\underset{\text{``activation''}}{
\begin{array}{l}
CH_2 \quad\ CH{=}CH{-}CH{=}CH_2 \\
\parallel\ \rightarrow M \\
CH_2 \qquad H
\end{array}}
& \rightarrow\ M^{(0)}\ + &
\underset{\text{``coupling''}}{CH_2{=}CH{-}CH{=}CH{-}CH_2{-}CH_3}
\end{array}
$$

In such a mechanism the nature of the products is very dependent on the number of free co-ordination positions of the metal and therefore on the excess of monomers, as in the case of cyclo-oligomerization of dienes. A similar mechanism may be formulated in the homo-dimerization of α-olefins.

1.4.2.3. *Reductive dimerization*

In some cases dimerization proceeds only under hydrogen pressure, or in the presence of various hydrogen donors or reducing agents. The dimerization products are formed, in part or exclusively, of dihydrodimers.

The mechanisms of these reactions are by no means well understood and are probably of very different types. Among the hypotheses that may be formulated, the following have been suggested:

a) hydrogen may act as a chain transfer agent, as in the case of conventional polymerization mechanisms, by the means of a metal-carbon bond hydrogenolysis (when β-elimination is not favoured) i. e.:

$$\underset{\overset{|}{M-CH_2-CH-R'}}{R} + H_2 \rightarrow M-H + \underset{\overset{|}{CH_3-CH-R'}}{R}$$

b) hydrogen may generate or re-generate a hydride, after coupling of two alkyl groups:

$$2M\underset{R'}{\overset{R}{<}} \rightarrow M-M + 2R-R'$$

$$M-M + H_2 \rightarrow 2M-H$$

The insertion steps of olefins into metal-hydrogen or metal-carbon bonds are of the same type as depicted in the preceeding mechanisms.

The oxidizing coupling reactions will not be discussed in this review owing to the entirely different course of the reaction.

1.4.3. Models of the intermediate species

Here we shall give some reactions by which metal-hydrogen and metal-carbon bonds are likely to be formed and inter-converted in dimerization catalysis.

1) *Metal hydride complexes*

a) Directly from hydrogen by addition, cleavage or reduction reactions:

$$[Co(Ph_2PCH_2CH_2PPh_2)]^+ + H_2 \rightarrow [Co(H)_2(Ph_2PCH_2CH_2PPh_2)]^+$$

$$[(C_5H_5)Mo(CO)_3]_2 + H_2 \rightarrow 2(C_5H_5)Mo(CO)_3H$$

$$PdCl_2(PPh_3)_2 + H_2 \rightleftarrows PdHCl(PPh_3)_2 + HCl$$

b) By oxidative addition of acidic compounds:

$$Ni^{(o)}(PR_3)_2 + CH_3-\overset{\overset{\displaystyle O}{\|}}{C}-OH \rightarrow (PR_3)_2Ni\overset{\displaystyle H}{\underset{\displaystyle O-\underset{\underset{\displaystyle O}{\|}}{C}-CH_3}{<}}$$

c) By reduction by means of a great variety of reducing agents:

$$NiCl_2(PR_3)_2 \xrightarrow{\ NaBH_4\ } (PR_3)_2Ni\overset{\displaystyle H}{\underset{\displaystyle Cl}{<}}$$

$$Fe(CO)_5 + 4OH^- \rightarrow [Fe(CO)_4]^{--} + CO_3^{--} + 2H_2O \xrightarrow{\ H^+\ } H_2Fe(CO)_4$$

d) By hydrogen transfer from a co-ordinated ligand:

$$(CH_2{=}CH_2)Fe(Ph_2PCH_2CH_2PPh_2)_2 \rightleftarrows$$

$$\rightleftarrows CH_2{=}CH_2 + HFe-C_6H_4-\underset{\underset{\displaystyle Ph}{|}}{P}CH_2CH_2PPh_2(Ph_2PCH_2CH_2PPh_2)$$

2) *Metal alkyl complexes*

a) From organo-aluminium compounds or Grignard reagents:

$$FeCl_3 + a,a'\text{-dipyridyl} \xrightarrow{\ \cdot AlEt_3\ }$$

b) By oxidative addition of alkyl halides:

$$Pd(PPh_3)_2 + CH_3I \rightarrow (PPh_3)_2Pd(I)CH_3$$

c) By oxidative addition of "radicals" (COD is 1,5 cyclooctadiene):

$$Ni(COD)_2 + Ph_3C\!-\!CPh_3 \rightarrow Ph_3C\!-\!Ni\!-\!CPh_3 + 2COD$$

d) By concerted addition-dimerization of a diolefin:

$$Ni^{(0)} + 2C_4H_6 + 2L \longrightarrow$$

3) Metal-hydride, metal-alkyl and π-complexes interconversion

The interconversions are the essential steps of dimerization acting by a "degenerated" polymerization mechanism.

These reactions have been mainly discussed by M.L.H. Green [10] in terms of metal-β-interaction. The "β-effect" explains the interconversion $\sigma \rightleftarrows \pi$ (by hydride abstraction) or σ-allyl $\rightleftarrows \pi$ (by protonation), and some examples are here reported:

$$(C_5H_5)(CO)_2Fe\!-\!CH_2\!-\!CH_3 \underset{NaBH_4}{\overset{Ph_3C^+ClO_4^-}{\rightleftarrows}} \left[(C_5H_5)(CO)_2Fe \leftarrow \| \begin{array}{c} CH_2 \\ CH_2 \end{array} \right]^+$$

$$(C_5H_5)(CO)_2Fe\!-\!CH_2\!-\!CH\!=\!CH_2 \overset{HCl}{\longrightarrow} \left[(C_5H_5)(CO)_2Fe \leftarrow \| \begin{array}{c} CH_2 \\ CH \\ | \\ CH_3 \end{array} \right]^+$$

The hydride abstraction and the protonation reactions may proceed via the formation of a β-carbonium ion which is stabilized by the transfer of some positive charge to the metal (the transition metal acting as "an electron reservoir"):

$$M\!-\!\overset{|}{\underset{|}{C}}\!-\!\overset{|}{\underset{|}{C}}\!- \rightarrow M\!-\!\overset{|}{\underset{|}{C}}\!-\!\overset{|}{\underset{|}{C}}\!-\!{\oplus} \rightarrow M^{\oplus} \leftarrow \| \begin{array}{c} C \\ C \end{array}$$

An iridium derivative containing both σ-bonded and π-bonded organic moieties has been obtained by the reaction of acrylonitrile on $IrH(CO)(PPh_3)_3$:

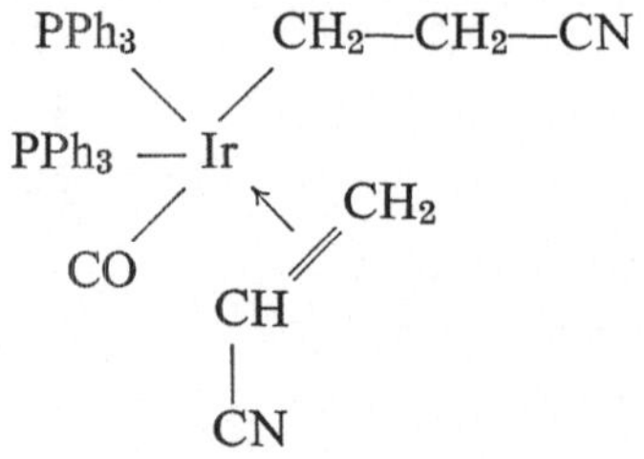

$$
\begin{array}{c}
PPh_3 \qquad CH_2\!-\!CH_2\!-\!CN \\
\diagdown \quad \diagup \\
PPh_3 \!-\! Ir \\
\diagup \quad \diagdown \; CH_2 \\
CO \qquad \diagup\!\!\diagup \\
\;\;\; CH \\
| \\
CN
\end{array}
$$

This complex can be considered as one of the closest models of the intermediate species involved in a dimerization reaction.

4) *The steric course of olefin insertion*

As has been pointed out, the factors determining the steric course of insertion are not well understood. The direction of addition may depend on the nature of the olefins, the complex involved and the temperature. For example:

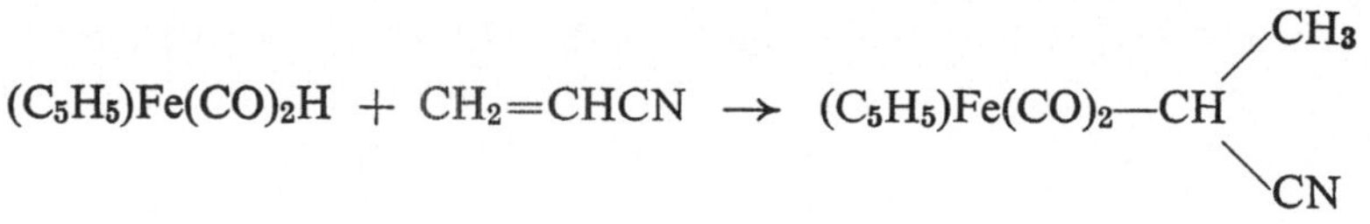

$$
(C_5H_5)Fe(CO)_2H \;+\; CH_2\!=\!CHCN \;\rightarrow\; (C_5H_5)Fe(CO)_2\!-\!CH \begin{array}{c} \diagup CH_3 \\ \diagdown CN \end{array}
$$

and:

$$
(CO)_3Fe \begin{array}{c} \diagup H \\ \diagdown CH_2\!-\!CH_2\!-\!CN \end{array} + CH_2\!=\!CHCN \;\rightarrow\; (CO)_3Fe \begin{array}{c} \diagup CH_2\!-\!CH_2CN \\ \diagdown CH_2\!-\!CH_2CN \end{array}
$$

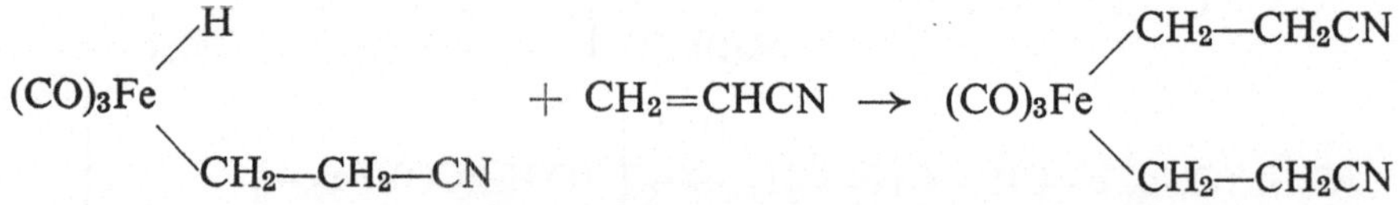

Steric factors have been recently proposed as very important.

5) *Coupling reactions*

Coupling reactions proceed via alkyl displacement and homolytic cleavage, and are exemplified by the following reactions:

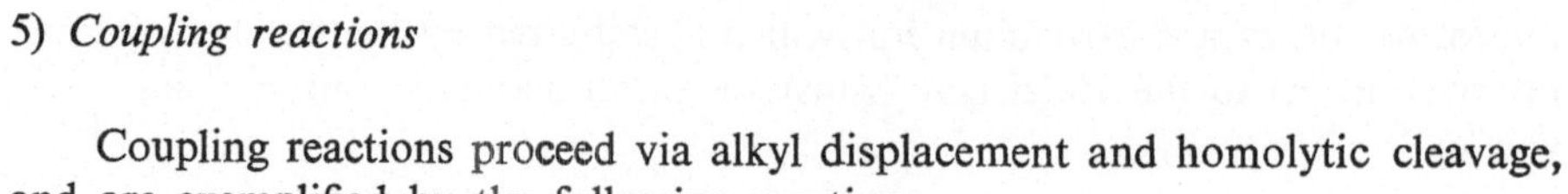

$$
\left\langle \!\! \begin{array}{c} \rightarrow Ni \leftarrow \end{array} \!\! \right\rangle + 4PR_3 \quad \longrightarrow \quad \diagup\!\!\diagup\bigwedge\!\!\bigvee\!\!\diagup\!\!\diagup \quad + Ni(PR_3)_4
$$

and:

$$CH_3-\hspace{-0.5em}\underset{\underset{DMF}{\diagdown}}{\overset{\overset{Br}{\diagup}}{\diagup}}\hspace{-0.5em}Ni \quad + RI \; \rightarrow \; R-CH_2-\overset{\overset{CH_3}{|}}{C}=CH_2 + NiBrI + DMF$$

1.4.4. Requirements for catalytic activity in dimerization

Actually it is difficult to define these requirements exactly. Consequently only some remarks and parallels will be tentatively drawn.

a) Apart from those based on titanium, the most commonly used catalysts for dimerization and co-dimerization of olefins have been, up to now, based on the *VIIIth group elements* and within them, the elements of the first transition series (Fe, Co, Ni) and of the second transition series (Ru, Rh, Pd) being particularly active. On the other hand, it is well documented that although high molecular weight polymers can be obtained from conjugated dienes with mono-metallic or bi-metallic complexes of the VIIIth group elements, these cannot be achieved starting from α-olefins. It has been tentatively suggested that, in the case of these elements, the π-allylic bond is an essential feature of the stabilization of the growing chain. Such a stabilization is destroyed by "insertion" (or coupling) of the double bond of a mono-ene into a π-allylic or σ-vinyl bond. Consequently, reaction with mono-olefins results in hydrogen migration, i. e. isomerization, at most dimerization when "insertion" or "coupling" occurs.

b) As commonly assumed in the case of other types of olefin activation, it seems necessary for the metal to be of a sufficiently "soft" character. So, the elements of the first transition series (Fe, Co, Ni) are generally active when bonded with less than two "hard" anions. When starting from higher oxidation states the necessary "reduction" can be effected by a variety of alkyl derivatives of aluminium or magnesium as well as by other reducing agents. The reducing power of some derivatives may be enhanced in the presence of a Lewis acid. The situation is less clear for the elements of the second transition series which are of more "soft" character even in their upper oxidation states. Reduction, if necessary, may be achieved also by means of a great variety of organic compounds, such as hydroxylic solvents or the monomer itself.

c) The *addition of "soft" ligands* like phosphines, may provide better selectivity, especially in co-dimerizations when two monomeres are of very different co-ordinating power, e.g. butadiene and ethylene. Phosphines may also influence the mode of linkage of the monomer molecules.

d) The *introduction of Lewis acids* in a given catalytic system sometimes results in either an increase of the reaction rate or even a profound modification of the course of the reaction. This effect is by no means completely understood and is most often specific for each metal and each type of monomer. Many examples of this effect are known in the case of conjugated dienes, i. e. enhancement of the polymerization rate, transformation of an oligomerization catalyst into a polymerization catalysts. This will be discussed later in the case of mono-enes.

Different explanations have been tentatively suggested to account for the "Lewis acids effect":

release of a free co-ordination position in the transition metal by, for example, dissociation of a dimeric complex or competition for anionic and neutral ligand.

increase of the positive charge carried by the metal.

release of a proton.

e) *Solvents* often play an important role; either they promote the catalytic activity (e.g. owing to their reducing power) or they enhance the dimerization rate. The nature of the solvent is sometimes critical, i. e. being too "soft" it competes with the olefin for co-ordination, and being too "hard" it can neutralize the needed Lewis acidity. Nevertheless, in several cases the dimerization rate increases with the dielectric constant of the medium. This fact is usually an indication of the ionicity of the complex.

However, as will be seen later, it is often difficult to discuss the definite nature of the environment of the catalytic species obtained in the various "recipes".

2. TRANSITION METALS AS CATALYSTS IN DIMERIZATION OF OLEFINS

2.1. Introduction

Among the various possibilities of presenting the numerous cases of olefin dimerization we have chosen the classification based on the nature of the transition metal used. Although such a plan is somewhat inconsistent, it avoids a certain number of repetitions. As has been seen and will appear later, the hypotetical nature of most of the mechanisms excludes a classification based on them. In fact, every metal has its own characteristics of reactivity and selectivity which, up to now, have not been related to its electronic structure.

It can be asserted that Nickel is "the" metal of olefin dimerization, because it is the only metal which can control the mode of linking of olefins (G. Wilke and al.); its specific activity is high and it is one of the less expensive of the transition elements. These are the reasons for which many papers and patents have been devoted to Nickel catalysed dimerization.

On the other hand, up to now most of the fundamental knowledge of dimerization mechanisms has come from studies made on rhodium (Cramer).

Finally, ruthenium, owing to its specific activity toward acrylic derivatives (Rhone-Poulenc's workers), and cobalt and iron (Japanese workers), due to the possibilities they offer of co-activating dienes and olefins, have been the subjects of important studies directed towards industrial applications.

The different metals will be considered in the order of the periods of the Periodical table. However, when it is necessary for the clarity of the subject some inversions will be introduced.

At the end of this review an appendix is given classifing the dimerization with respect to individual olefins.

We have attempted to collect the greatest number of references but it has not always been possible to give a detailed picture of each of them (particulary in the case of patents). Finally, owing to the increase of work on the subject some recent references may have been omitted.

2.2. Titanium

Titanium compounds, and especially halides, usually associated with Aluminium compounds, are commonly used to catalyse the formation of high molecular weight polymers from ethylene, α-olefins and dienes. Most of the mechanisms and theories concerning the propagation and stereoregulation of chain polymerization have been developed on this element, but relatively little work has been done on the conditions under which hydrogen atoms are activated by titanium. The same titanium catalytic systems used for polymerization give dimers and oligomers from ethylene and α-olefins by minor changes in the constituents of the catalytic mixture or in the experimental conditions. A mixture of titanium (or zirconium) esters and trialkylaluminium produces polyethylene or n-butenes from ethylene [11] [12]. Natta described [13] the conditions favouring one or other of these products, i. e. the polymer with a molar ratio $\frac{AlR_3}{Ti(OR)_4} > 20$ and the dimer with a ratio $\frac{AlR_3}{Ti(OR)_4} < 10$ [14b].
In the latter case, dimerization selectivity is high (more than 90%), most of the other products being hexenes and octenes. The major part of the dimers formed is 1-butene with a small amount of 2-butenes. This means that the system is not a very isomerizing one, and it behaves like triethylaluminium alone, except for the fact that the experimental conditions are much less severe. Using the same catalytic system, the dimerization of propylene and the codimerization of ethylene with propylene and ethylene with butene have been achieved [14]. Variations of this system resulting in improved selectivity have been proposed, such as tetraaryltitanate or trialkyltitanate with trialkylaluminium [15] or methylcyclopentadienyl titanium esters associated with trialkylaluminium [16]. More recently, it has been pointed out that trialkylaluminium may be replaced by N, N-dialkylaminoalanes, e.g. $H_2AlN(CH_3)_2$ [17].

Other titanium derivatives such as cyclopentadienyl halides, which in the presence of aluminium compounds give polyethylene, lead under certain conditions to the formation of dimers. For example, cyclopentadienyl titanium trichloride associated with amalgams of alkali metals [18] [19] or with Grignard reagents [20] mainly gives 1-butene.

Lastly, the $TiCl_4$—AlR_2Cl system, which is a typical Ziegler catalyst for the linear polymerization of ethylene between $+ 20 ^oC$ and $+ 100 ^oC$ in a hydrocarbon medium, when used from $- 100 ^oC$ to $- 50 ^oC$ in a halogenated hydrocarbon medium promotes the oligomerization of ethylene and α-olefins into branched chain olefins. The nature of the active centre in the above catalytic system has been discussed by Bestian et al. [21] and a mechanism has been proposed.

The active species formed by mixing $TiCl_4$ and $(CH_3)_2AlCl$ is the complex CH_3TiCl_3/CH_3AlCl_2 which, in the presence of an olefin, is stable only at a low temperature. In fact, if into the system the olefin is introduced at room temperature the resulting organotitanium compound $R\ TiCl_3$ is, unlike the corresponding methyl derivative, unstable, and gives $TiCl_3$ by homolytic decomposition. This latter is a well-known polymerization catalyst. On the basis of an interesting solvent effect observed in the catalysis, the following ionic solvated structure has been ascribed to the active species:

$$\left[\begin{array}{c} Cl \\ | \\ Cl-Ti \\ | \\ CH_3 \end{array} \right]^{+} \left[\begin{array}{c} Cl \\ | \\ Cl-Al-CH_3 \\ | \\ Cl \end{array} \right]^{-}$$

the cationic part of which co-ordinates the olefin as follows:

$$\left[\begin{array}{c} Cl \\ | \quad\quad CH_2 \\ Cl-Ti \leftarrow \| \\ | \quad\quad CH_2 \\ CH_3 \end{array} \right]^{+}$$

The titanium-carbon bond plays the major role, i.e. it reacts on the co-ordinated olefin, but aluminium is a powerful activator of this reaction; in fact the insertion of 1-pentene into the titanium-carbon bond of the ionic complex is 300 times faster than in the non ionic CH_3TiCl_3 species.

The dimerization of ethylene by the ionic complex is highly dependent on the solvent (halogenated hydrocarbons > aromatic hydrocarbons > aliphatic hydrocarbons) and occurs with poor selectivity (see Table 1). On the other hand, the co-dimerization of ethylene with α-olefins proceeds at a rate comparable to the homo-dimerization of ethylene, and results in the "ethylation" of the

Table 1

PRODUCTS OF OLIGOMERIZATION OF ETHYLENE IN CH_2Cl_2 AT $-$ 70 °C [21]. CATALYST: CH_3TiCl_3—CH_3AlCl_2 (45 mM); 400 ml OF CH_2Cl_2; 100 g OF C_2H_4/H. REACTION TIME: 6 HOURS.

Fraction	Yield based on ethylene conversion	Composition of fraction	
C_4	14%	100%	1-butene
C_6	39%	15%	1-hexene
		7%	2 and 3-hexene
		78%	2-ethyl-1-butene
C_8	21%	7.5%	1-octene
		7.5%	2,3 and 4-octene
		82%	2-ethyl-1-hexene
		3%	4-ethyl-1-hexene
C_{10}	7%		n-decenes
			2-ethyl-1-octene
$> C_{10}$	19%	Chiefly oligomers formed from 2-ethyl 1-alkenes	

β-carbon of the olefin; consequently the poor selectivity in ethylene dimerization is ascribed to the β-ethylation of the linear α-olefins primarily formed. Under such conditions, the isomerization activity of this system is very low and most

Table 2

PRODUCTS OF ISOMERIZATION AND OLIGOMERIZATION OF α-OLEFINS IN CH_2Cl_2 AT $-$ 70 °C [21]. CATALYST: CH_3TiCl_3—CH_3AlCl_2 (15 mM); 330 ml OF α-OLEFINS. REACTION TIME: 15 HOURS.

Starting olefin	Conversion to isomers	Conversion to oligomers	Dimers (% of oligomers)	Branched dimers (% of dimers)
1-pentene	16	45	67	2 (n-propyl) 1-heptene 90%
1-hexene	14	49	65	2 (n-butyl) 1-octene 90%
1-heptene	17	51	63	—

of the olefins formed are α-olefins (the small quantity of olefins with an internal double bond are the primary products of dimerization).

The rate of homo-dimerization of α-olefins is much lower, the dimerization selectivity being comparatively higher; the dimers have mainly branched structure (see Table 2). 2-olefins (with 90% "*cis*" isomers) are the main products of α-olefin isomerization. The disubstituted α-olefins are not isomerized.

The scheme proposed for this catalysis is of the same type as for "Ziegler-Natta" polymerization (co-ordinated anionic mechanism). In the case of ethylene, the scheme has the following main characteristics (see scheme 1):

a) Chain growth is accomplished by transfer of the carbanion to the complexed ethylene.

b) Chain transfer reactions are the result of hydride transfer from the α-carbon atom of the alkyl-group to the olefin.

c) The σ-ethyltitanium-α-olefin complex "B" is transformed by three processes:

displacement of the α-olefin from the π-complex by ethylene (formation of 1-butene),

addition of the ethyl anion on the α-carbon atom of the olefin according to Markownikoff's rule. The hydride transfer of the 2-ethylalkyl group is accompanied to a slight extent by a growth step. This results in the formation of 4-ethyl-1-hexene,

addition of the ethyl anion into the α-carbon of the olefin (anti-Markownikoff addition). The alkyl-group so formed is particularly readily displaced by ethylene and yields the straight-chain olefins with internal double bond.

The quantitative examination of the processes involved shows that the chain transfer reactions proceed at a higher rate than the chain propagation. After the hydride transfer, the α-olefin remains attached to the titanium with a relatively strong bond, and the rate of its displacement by ethylene is slow compared to that of transfer of the ethyl anion to the α-and β-carbon atoms of the complexed α-olefin (displacement/transfer ratio: 1/3). In the absence of CH_3AlCl_2, CH_3TiCl_3 has low oligomerizing activity and only produces linear ethylene oligomers: thus aluminium activates the hydride transfer reaction more than the growth reaction.

The processes occuring on oligomerization of α-olefins are less complicated. Here, however, the isomerization must be taken into consideration. A constant transfer of hydride to olefin takes place in the starting complex and the process becomes irreversible when the hydride transfer proceeds via the anti-Markownikoff route to the α-carbon atom of the olefin and subsequent release of

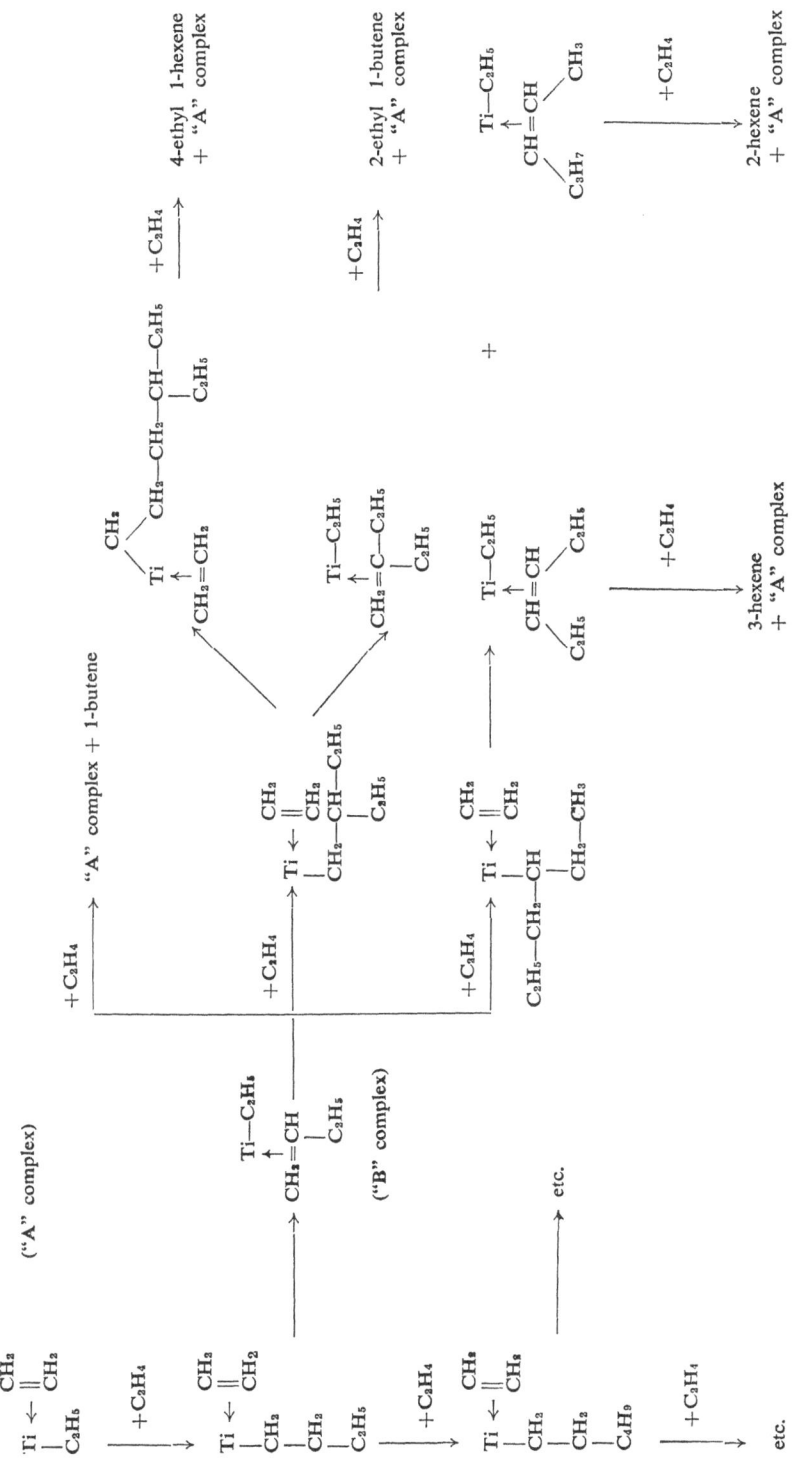

Scheme 1: oligomerization of ethylene by Titanium [21].

hydride from the CH_2 group adjacent to the titanium-carbon bond:

$$
\begin{array}{lll}
& CH_2 & \\
Ti \leftarrow \| & & \\
| \quad CH-CH_2R & & \\
CH_2 & & \longrightarrow \ CH_2{=}CH \\
| & & \\
CH_2 & & \\
| & & \\
CH_2R & &
\end{array}
$$

The double bond migration does not proceed further than the "2" position because the 2-olefins are then rapidly displaced by α-olefins.

A similar catalytic system, obtained by mixing $TiCl_4$—$Al_2Et_3Cl_3$ and propylene oxide [22], oligomerizes at 20 °C α-olefins (e.g. 1-octene), giving oils with an interesting viscosity index. Contrary to the previous system, the dimer fraction consists mainly of linear olefins with the double bond in the "*trans*" form, resulting from a head-to-head reaction.

2.3. Nickel

2.3.1. GENERAL FEATURES OF NICKEL CATALYSIS

It is worth remembering that catalysis entered the field of organometallic compounds by way of nickel which, for the first time, enabled ethylene to be catalytically inserted into a carbon-magnesium bond [23]. On the other and, the "nickel effect" observed during the reaction of ethylene with triethylaluminium [24] was the origin of the development of the so-called "Ziegler-Natta catalysis". The importance of nickel in co-ordination catalysis is not merely historical. It has played and still plays an essential, fundamental and practical role in the research of olefin and diolefin transformations on account of the diversity of the catalysed reaction types and the requirements of each with respect to the environment and the form in which the metal must be presented. Before going into a detailed description, we shall place the dimerization catalysis of olefins among various other types of unsaturated hydrocarbon transformations catalyzed by nickel. It would also be appropriate to say here that a great deal of the knowledge we have in nickel catalysis is due to the work of G. Wilke, P. Heimbach, B. Bogdanovic et al. [3] [5] [25].

1) *Cyclo-oligomerization of diolefins; Co-trimerization of butadiene with ethylene*

These reactions are catalyzed by "bare" nickel, a complexed metal atom from which the neutral or anionic ligands can be displaced by substances taking part in the catalytic reaction [3]. The "bare" nickel may may be prepared, for example, by "reducing" a nickel salt (Ni^{++}) (after exchanging "hard" anions with "soft" anions) with a suitable organometallic compound in the presence of electron donors if necessary. A typical source of "bare" nickel is bis-allyl nickel: in this case, cyclo-oligomerization of butadiene involves the following steps: *butadiene displaces the two allyl groups* and the bare nickel tends to fill up its electron system as much as possible in order to attain the next higher inert gas configuration, by sharing in the π-electron systems of the conjugated double bonds of the butadiene molecules. The open chain thus formed is attached to the nickel by π-or σ-allyl bonds, depending on the nature and concentration of ligands. Then, by carbon-carbon linking, four cyclic oligomers may be obtained as follows,

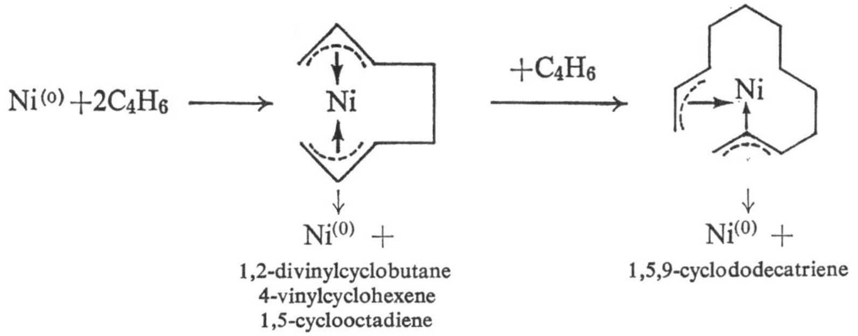

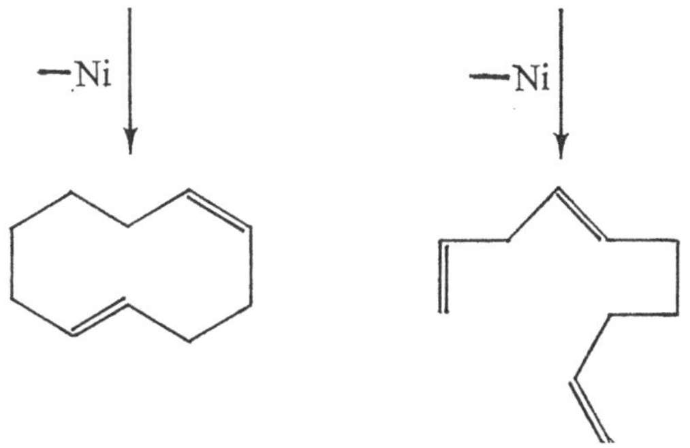

$$Ni^{(0)} + 2C_4H_6 \longrightarrow$$

$$+C_4H_6 \longrightarrow$$

$$Ni^{(0)} +$$

1,2-divinylcyclobutane
4-vinylcyclohexene
1,5-cyclooctadiene

$$Ni^{(0)} +$$

1,5,9-cyclododecatriene

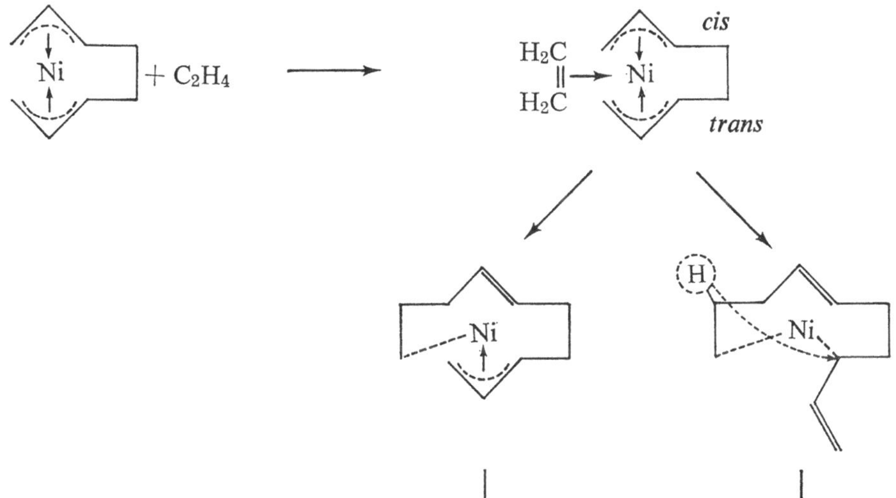

$$Ni \quad + C_2H_4 \longrightarrow$$

$$H_2C \atop \| \atop H_2C \longrightarrow Ni$$

cis

trans

On the same type of complex, a mixture of ethylene and butadiene affords a linear co-trimer (1-*trans*-4,9-decatriene) and a cyclic co-trimer (*cis, trans*-1,5 cyclodecadiene) forming a ten-membered chain bonded to nickel by a σ-bond and a π-or σ-allylic bond:

−Ni

−Ni

The σ-allylic bond reacts with a hydrogen transferred from the β-position of the σ nickel-carbon bond. At high temperature (110 °C) the hydrogen transfer is favoured and more than 65% n-decatriene is formed. At lower temperature (20 °C) the yield of cyclodecadiene is 80% [5] [26].

2) *Linear polymerization of dienes*

If nickel is bonded with one "hard" anion (e.g. bis π-allylnickel halide) butadiene gives a high molecular weight polymer, the microstructure of which depends on the bonded anion ([27] and references therein). Increasing the electronegativity of the anion (e.g. by adding Lewis acid with the formation of a complex anion) the polymerization rate and the *cis*-1,4 content of polybutadiene increase. The species then has a marked ionic character and may be:

$$\left[\left\langle\!\!\!\left\langle\!\!-\text{NiS} \right.\right.\right]^{+} \left[\text{AlX}_4 \right]^{-} \quad or \quad \left[\text{Al}_2\text{X}_7 \right]^{-}$$

(with $S = $ solvent)

The insertion of butadiene takes place on the allyl moiety which is incorporated in the polymer chain.

3) *Dimerization of olefins*

Despite the fact that "bare" nickel may co-activate ethylene with two butadiene molecules the activation of olefins can apparently only be done with great difficulty using nickel alone, even if it is associated with a "hard" anion. However, in the presence of a Lewis acid such compounds as π-allyl nickel halide are extremely active catalysts for the dimerization of olefins [3b] [28]. The Lewis acid, which in the case of linear polymerization of dienes was merely an accessory means of activation, is a major component of the catalyst in this case, probably because any simple anion is weak enough to compensate for the limited coordinating power of mono-olefins. The catalyst activity is further increased by phosphines. The probable activation scheme is as follows:

In contrast to their poor solubility in hydrocarbons, active species are readily soluble in polar solvents such as halohydrocarbons.

Complex (a) reacts with two moles of carbon monoxide thus resulting in an inactive complex. Complex (b) reacts with only one mole of carbon monoxide. The two active species should, accordingly, have two and one free co-ordination positions respectively.

Even at low temperature, species (a) and (b) dimerize ethylene into a mixture of n-butenes (the catalyst has a high isomerising activity) and they dimerize propylene into a mixture of n-hexenes, 2-methylpentenes and 2,3-dimethyl-butenes. The relative proportions of different products in the latter mixture depend to a large extent on the nature of the co-ordinated phosphine.

The mechanism will be discussed later. However, it should be noted that the *allyl group does not participate in the reactions* [25]. It is neither displaced (as in cyclo-oligomerization) nor inserted (as in polymerization) but found on the nickel at the end of the reaction. Its role is probably solely that of a stabilizer of a certain electronic state.

4) *Co-dimerization of dienes with olefins*

In the preceding cases it was possible to determine the environment in which nickel catalyzes a given transformation. The same is not yet possible for the co-dimerization reaction of dienes with olefins. This reaction has been only catalyzed by a Ziegler type mixture (i. e. $NiCl_2$—$2PR_3$—AlR_2Cl—halogenated hydrocarbons) from which it was not possible to isolate the catalytic species [29]. By means of this system, ethylene and butadiene give mixture of co-dimers:

$$CH_2=CH_2 + CH_2=CH-CH=CH_2 \rightarrow CH_3-CH=CH-CH_2-CH=CH_2 +$$

$$+ CH_2=CH-\overset{\overset{\displaystyle CH_3}{\displaystyle |}}{CH}-CH=CH_2$$

and various by products resulting from the isomerization and the rearrangement of the co-dimers.

As a result, the reaction follows an entirely different pathway from the one observed in the presence of "bare" nickel (formation of linear and cyclic co-trimers). This difference may probably be related to the presence of Lewis acid (chloroalkylaluminium) in the medium. The Lewis acid activates the hydride transfer reaction more than the growth reaction.

Under otherwise indentical conditions, but in the absence of butadiene, ethylene is readily dimerized to a mixture of n-butenes, and in the absence of a trivalent phosphorus moiety, *trans*-1,4-polybutadiene is formed without co-dimerization [29a]. Hence, in this catalytic system it can be assumed that the nickel environment must be quite similar to the one favouring both the dimerization of olefins and the linear polymerization of butadiene, i. e.

$$\left[\left\langle\!\!\!\!\!\!\!\!\!\!\;\rightarrow Ni \leftarrow L\right]^{+} \left[AlR_{4-x}\;X_{x}\right]^{-}\right.$$

with $2 \leqslant x \leqslant 4$

The indispensable role of phosphine in obtaining co-dimers is to favour the co-ordination of ethylene over that of butadiene by blocking one of the co-ordination positions in such a way that an allyl group and one mono-olefin double bond can be present simultaneously. It can be supposed that the allyl group effectively participates in the reaction in this case.

5) *Isomerization of the double bond*

Many forms of nickel, with the exception of "bare" nickel, promote the double bond shift [30]. Contrary to polymerization and oligomerization, isomerization is an "easy" reaction which does not require the presence of a specific environment. Consequently, in most of the cases, the isomerization might be an intramolecular reaction. Nevertheless, it is difficult to determine whether a common element exists in all these systems. Several of them require a hydrogen source in one or other form, e.g. a metal hydride (or a metal alkyl affording a hydride), protonic or molecular hydrogen. Many authors link this observation to the probable formation of a nickel hydride, and hence attribute the isomerization to a hydride addition-elimination mechanism.

6) *Miscellaneous reactions*

Numerous other reactions of olefins are catalyzed by nickel complexes, e.g. the insertion of ethylene into the metal-carbon bond of arylmagnesium halide [23] [31], the skeletal rearrangement of 1,4-dienes [32], the "nickel effect" in the reaction of ethylene with triethylaluminium [24].

For all these reactions the environment of the metal atom has not been discussed.

2.3.2. DESCRIPTION OF VARIOUS NICKEL SYSTEMS FOR OLEFIN DIMERIZATION

In the preceding part, we have attempted to place the catalysis of olefin dimerization among various other olefin transformations. However, the approach was highly simplified because there are a great variety of ways of obtaining active catalytic systems without always being able to connect the systems logically with each other. "Recipes" become all the more varied as the reaction proves to be of potential industrial interest. Actually, it can be said that nickel in all its forms may dimerize ethylene and propylene, provided that it is associated with a suitable Lewis acid and eventually with a selected reducing agent.

It is beyond the scope of this review (and it is of little interest) to draw up an exhaustive list. We shall give a few examples selected on accound of their ease of use or, on the contrary, because they are the most extreme ones with regards to the form in which the nickel is used (Table 3).

The point common to all these "recipes" is the presence of a Lewis acid which is not necessarily a reducing agent when the nickel is bonded to less than two "hard" anions ("bare" nickel, π-allylnickel halide, Ni(I) etc.) and which is a reducer when the bivalent nickel is bonded to two "hard" anions (NiCl$_2$, Ni (acac)$_2$ etc.). It may be pointed out that AlEt$_2$(OEt) is a weak Lewis acid, and in fact the system Ni(acac)$_2$/AlEt$_2$(OEt) is comparatively less active. On the other hand, in the system NiCl$_2$/PR$_3$/AlCl$_3$ [36] [47] the phosphine can eventually play the role of a reducing agent.

Now the question may be, what is the active form of the nickel after this mixing? Bogdanovic and Wilke [28c] used the interaction of Ni(acac)$_2$ with Al$_2$Et$_3$Cl$_3$ in the presence of cyclo-octene and then treatment with an aqueous ammonia solution, to isolate a π-allyl of the cyclo-octene dimer, i. e.

which supports a π-allyl intermediate, common for all the systems. Nevertheless, the catalytic species formed "in situ" by the reaction of Ni(II) with Al$_2$R$_{6-x}$X$_x$ are generally more active. Many attempts have been made to isolate intermediates [39e] [39g] [44e] [49a].

It has been possible to dimerize ethylene by nickel in the absence of a Lewis acid. By heating nickelocene to 200 °C in the presence of ethylene, butenes are formed [62]. Through the homolytic decomposition of nickelocene, excited nickel atoms are probably produced which catalyse the dimerization. This would seem to bring us near heterogeneous catalysis. Similar results are obtained in the presence of π-(cyclopentenyl) π-(cyclopentadienyl) nickel [63]. The activity of these systems nonetheless remains weak (1g of nickelocene affords 5 ml of butenes in 18 hours at 200 °C).

On the other hand styrene may be dimerized in diphenylbutene by means of π-allyl nickel iodide without Lewis acid [70].

Table 3

NICKEL-BASE CATALYSTS FOR DIMERIZATION OF MONO-OLEFINS

Nickel compound	Reducing agent or Lewis acid	Additional ligands	References
Ni acetylacetonate	$Al_2R_{6-x}X_x$	—	[30, 34]
Ni acetylacetonate	$Al_2R_{6-x}X_x$	$P(Ph_3)_3$	[33]
Ni acetylacetonate	$Al_2R_{6-x}X_x$	PR_3+COD	[35]
Ni acetylacetonate	$AlCl_3$	PR_3+NEt_3	[36]
Ni acetylacetonate	AlR_3 or $AlR_2(OR)$	—	[37]
Ni fluoroacetylacetonate	$AlR_2(OR)$	—	[38]
Ni oleate	$Al_2R_{6-x}X_x$	—	[39]
Ni oleate	$Al_2R_{6-x}X_x$	PR_3	[39f]
Ni naphtenate	$Al_2R_{6-x}X_x$	$PRCl_2$	[40]
Ni di-isopropylsalicylate	$Al_2R_{6-x}X_x$	—	[41]
Ni di-isopropylsalycylate	$AlEt_2F$	—	[42]
Ni alkylbenzenesulfonate	$Al_2R_{6-x}X_x$	—	[43]
$NiX_2 \cdot$ tetramethylcyclobutadiene	$Al_2R_{6-x}X_x$	PR_3	[44a,b,c,d,e,f]
$NiX_2 \cdot 2PR_3$	$Al_2R_{6-x}X_x$	—	[44g, 45, 46, 47, 48, 49]
Ni $(RCOO)_2 \cdot 2PR_3$	$Al_2R_{6-x}X_x$	—	[44j, 44h]
$NiX_2[OP(NR_2)_3]_2$	$AlR_6 {}_{x}X_x$	—	[50]
$NiX_2 \cdot 2PR_3$	$AlCl_3$	PR_3+NEt_3	[36, 47]
$[PR_4]^+[NiX_3PR_3]^-$	$Al_2R_{6-x}X_x$	—	[51]
$[Ni\ (dimethylsulfoxide)_6]^{++}[NiCl_4]^{--}$	$Al_2R_{6-x}X_x$	—	[52]
$Ni(NO)ClPR_3$	$Al_2R_{6-x}X_x$	—	[53, 44g]
$(\pi\text{-allyl}NiX)_2$	$AlCl_3$	—	[3b, 25, 28, 55]
$(\pi\text{-allyl}NiX)_2$	Various Levis acids	—	[39e, 39f, 54]
$NiX(PR_3)_3$	$AlCl_3$		[56, 57]
$Ni(PR_3)_4$	$AlCl_3$		[57, 58]
$Ni(CO)_2(PR_3)_2$	Lewis acid		[44i, 56, 59, 60]
$Ni(acrolein)_2$	Lewis acid		[61]
Nickel Raney	AlR_2Cl		[64a]

X = halogen $x = 2, 3$ or 4 $COD = 1,5\ C_8H_{12}$

2.3.3. THE REACTIVITY OF OLEFINS

The dimerizing activity of such catalysts is limited to simple olefins and, sometimes in the presence of phosphine, it is limited to the co-dimerization of dienes with olefins. Vinylic compounds such as acrylates are unreactive. This is probably due to the "required" presence in the catalytic system of a Lewis acid of the "hard" type, which is incompatible with functional groups (which are "hard" bases). On the other hand "bare" nickel catalyses the co-trimerization of two molecules of butadiene with one molecule of acrylate in the same way as it does in the case of ethylene [69].

According to their reactivity, olefins are classified in the following order: ethylene > propylene > n-butenes > cyclopentene > isobutene (which is unreactive, but may be cationically polymerized). It is also possible to carry out the co-dimerization of these olefins.

Dimerization catalysts usually have high isomerizing activity. Thus, butenes and higher molecular weight straight chain olefins are isomerized up to their thermodynamic equilibrium before any dimerization occurs. In fact, whichever butene isomer is used, an equilibrium n-butenes mixture, formed by rapid migration of the double bond, takes part in dimerization (at 20°C, 3.5% 1-butene, 23.5% *cis*-2-butene, 73% *trans*-2-butene), and the reactivity of the butenes cannot be attributed either to 1-butene alone or to the mixture of 1-butene and 2-butenes. It is no longer true in the case of n-pentenes. The product of the hydrogenation of the ethylene-pentene co-dimerization has been shown to be constituted of three different isomers: n-heptane, originating from the reaction of carbon 1 of 1-pentene, triethylmethane, formed by the reaction of carbon 3 of 2-pentene, and 3-methyl hexane, from the linking of carbon 2 of 1-or 2-pentene [64a]. Thus the catalytic system $Ni(acac)_2/AlEt_2Cl$ affords at 0 °C the following isomer distribution.

$$(1.8\%)\ \overset{1}{CH_2}\!\!=\!\!\overset{2}{CH}\!-\!\overset{3}{CH_2}\!-\!CH_2\!-\!CH_3 \rightleftarrows \overset{1}{CH_3}\!-\!\overset{2}{CH}\!=\!\overset{3}{CH}\!-\!CH_2\!-\!CH_3\ (98.2\%)$$

$$+\ C_2H_4 \qquad\qquad +\ C_2H_4$$

C—C—C—C—C—C—C   C—C—C—C—C—C   C—C—C—C—C
                          | |
                          C   C
                                               |
                                               C
30.5%   39%   30.5%

Taking into account the concentrations, the carbon 1 of 1-pentene is thus 54 times more reactive than the carbon 3 of 2-pentene.

The dimers formed can react in turn on the monomer to give trimers. This reaction is particularly important in the case of ethylene which produces isohexenes in addition to n-butenes [44c] [44h].

142

2.3.4. SOME ASPECTS OF THE ISOMERIZATION REACTION

Isomerization was observed in all cases of dimerization but has been specially studied by some authors [30] [39b] [39e] [64b]. Generally speaking the straight chain olefins are readily brought to thermodynamic equilibrium, apparently without any great differences between successive migration steps along the chain. On the other hand, the presence of a branching in the vicinity of the double bond decreases its reactivity, and the passage of the double bond from or towards a tertiary carbon atom is comparatively very slow. Hence, during the isomerization of 4-methyl-1-pentene, equilibrium is readily established between the 1,2,3-carbons,

$$
\begin{array}{c}
\text{C} \\
| \\
\text{C--C--C--C--C} \\
\text{5\quad 4\quad 3\quad 2\quad 1}
\end{array}
$$

and is attained more slowly between all the isomers. This is shown concretely in the triangular diagram (fig. 1) on which are plotted 4-methyl-1-pentene, 4-methyl-2-pentene, and 2-methyl-2-pentene+2-methyl-1-pentene, and which represents the evolution of the composition of 4-methyl-1-pentene going to the thermodynamic equilibrium. In the early stages of the reaction, the *cis*-2-isomer is preferentially formed from the terminal olefin.

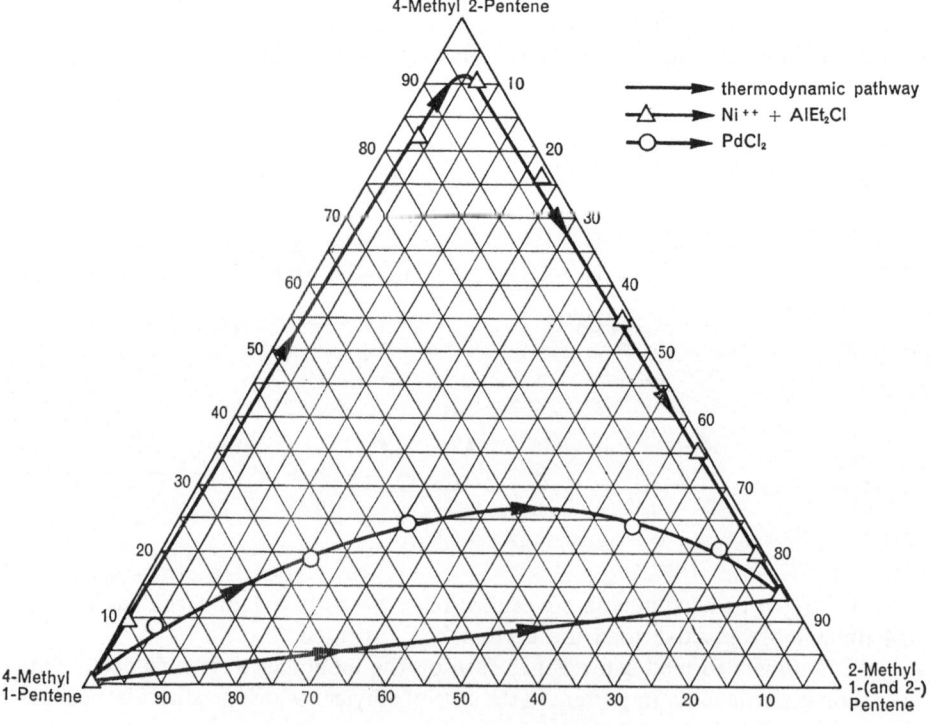

Figure 1. Isomerization of 4-methyl-1-pentene.

The isomerizing behaviour of nickel based catalysts is quite similar to that of t.BuOK.

2.3.5. FACTORS INFLUENCING THE COURSE OF THE DIMERIZATION REACTION

The principal parameters acting on the nature of the dimers formed are here examined.

2.3.5.1. *Conversion*

The analysis of the final reaction products of dimerization only imperfectly reflects the nature of the primary products, double bond migration occurring under dimerization conditions. Working in "batch" (i. e. when all the monomer is introduced at the beginning of the reaction) provides evidence for the isomerization of the products. Gas chromatographic analysis performed throughout the reaction, reveals that the steric course of the coupling remains constant, although the positions of the double bonds vary [45]. In some cases it is possible to determine with relative accuracy the initial composition of the dimers.

In figure 2 the percentage of 4-methyl-1-pentene, formed by dimerizing propylene with various nickel complexes, is plotted versus the conversion.

2.3.5.2. *Nature of the co-ordinated phosphine*

The introduction of phosphines into the catalytic system results in the rate being enhanced [28] [33]. However, the most important modification concerns the distribution of the isomers produced, which depends largely on the organic group in the phosphorus ligand. The ratio of the rate of isomerization to the rate of dimerization is also dependent on the phosphine. This "phosphine effect" was revealed by G. Wilke and B. Bogdanovic in the case of π-$C_3H_5NiX.PR_3$ activated by aluminium chloride [3b] [28]. The same effect is also operative in various Ziegler "recipes" in which phosphine is allowed to co-ordinate with nickel prior to dimerization [35a] [39e] [44] [45] [47] [48] [51] [60].

The influence of some phosphines on the distribution of the C_6–isomers obtained by propylene dimerization is reported in Table 4. The 2,3-dimethylbutene content approximately parallels the basicity of the phosphines or the Taft constants of the organic groups linked to phosphorus. Nevertheless, when three different organic groups are linked to the same phosphorus atom ($PR^1R^2R^3$), the dimethyl-butene content does not follow the sum of the Taft constant values for R^1, R^2 and R^3 groups. However, a coefficient proportional to the amount of dimethyl-butene formed in the presence of each phosphine PR_3^1, PR_3^2 and PR_3^3 can be attributed to each group. There is also an additive relation between these coefficients [45a].

The analysis of the reaction course is all the more complicated as within each structure the isomer distribution also depends on the phosphine. For example

144

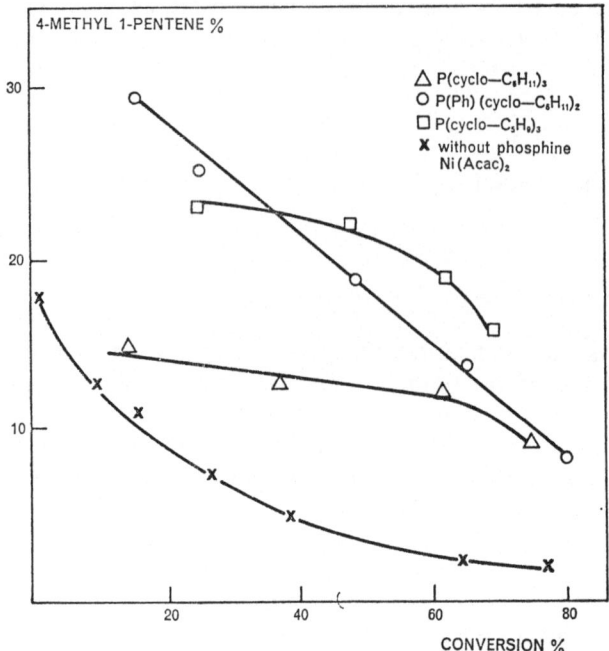

Figure 2. Dimerization of propylene. Percent of 4-methyl-1-pentene v.s. conversion
Catalytic system: NiCl₂ · 2PR₃, AlEt₂Cl

within the methylpentene structure, the main primary products obtained in the presence of less "basic" phosphines are 4-methyl-2-pentene and 4-methyl-1-pentene. On the contrary, in the presence of strongly "basic" phosphines these are 4-methyl-1-pentene and 2-methyl-1-pentene. Furthermore, the isomerizing activity [28c] is in the following order:

$$PMe_3 > PPh_3 > PEt_3 > \text{without phosphine} > P(iPr)_3$$

However, the isomerizing ability depends on the olefin, i. e. dimethylbutenes are more easily isomerized by catalytic systems containing highly "basic" phosphines, and methylpentenes are more easily isomerized by systems without phosphine or having weakly "basic" phosphines.

A similar "phosphine effect" is also observed, but to a smaller extent, in various co-dimerization reactions (see for example, Table 5 and 6) [64a].

Phosphites do not exert a similar effect but hexamethylphosphoroustriamide favours the linear coupling [50].

At present, the nature of the "phosphine effect" is by no completely understood. However, it is clear that the basicity is not the only important factor. Recent work done by Wilke, Bogdanovic and coworkers [161] has shown that steric effects are the most important mainly in the insertion reaction into the Ni—C σ bond.

Ligands having other heteroatoms than phosphorus, such as amines, pyridines, ethers, thioethers, sulfoxides [52], arsines, stibines etc., do not seem to influence the steric course of the reaction. However they often increase the selectivity in the dimer content.

Table 4

INFUENCE OF PHOSPHINES ON THE PROPYLENE DIMERIZATION. CATALYST: π-C_3H_5 NiX · PR_3, AlX_3, —20 °C, [28b]

PR_3	n-hexenes	2-methylpentenes	2,3-dimethylbutenes	Isomerized products
		C | C—C—C-|-C—C—C	C C | | C—C-|-C—C	
no phosphine	21.3 %	73.2 %	5.5 %	29.5 %
P(phenyl)$_3$	21.6	73.9	4.5	58.1
P(methyl)$_3$	9.9	80.3	9.8	75.8
P(n-butyl)$_3$	7.1	69.6	23.3	42.7
P(cyclohexyl)$_3$	3.3	37.9	58.8	4.9
P(i-propyl)$_3$	1.8	30.3	67.9	3.7

Table 5

INFLUENCE OF PHOSPHINES ON THE PROPYLENE-BUTENE. CO-DIMERIZATION: CATALYST, $NiCl_2$ · $2PR_3$, $AlEt_2Cl$, WITHOUT SOLVENT, 0 °C [64a].

	n-hexenes	2-methylhexenes	3-methylhexenes	2,3-dimethyl-pentenes
		C | C-C-C-C-|-C-C	C | C-C-C-|-C-C-C	C C | | C-C-C-|-C-C
	C-C-C-C-|-C-C-C			
no phosphine (*)	16 %	15 %	43.5 %	25.3 %
P(phenyl)$_3$	20	31	36	13
P(n-butyl)$_3$	13	27	39	21
P(cyclohexyl)$_3$	11	13	43	33

(*) Ni(acac)$_2$

2.3.5.3. *Nature of the Lewis acidity*

The most commonly used Lewis acids are alkylaluminium halides (especially chlorides) and aluminium halides when the nickel is bonded with less than two "hard" anions, or when the "recipe" contains a reducing agent.

Table 6

INFLUENCE OF PHOSPHINES ON THE PENTENE-ETHYLENE CO-DIMERIZATION. CATALYST: NiCl$_2$ · 2PR$_3$, AlEt$_2$Cl, WITHOUT SOLVENT, 0 °C [64a].

	n-heptenes	3-methylhexenes	3-ethylpentenes
	C—C—C—C—C-¦-C—C	C—C—C—C-¦-C—C (with C branch)	C—C—C-¦-C—C (with C—C branch)
no phosphine(*)	30.5 %	39 %	30.5 %
P(phenyl)$_3$	30	45	25
P(n-butyl)$_3$	16.5	57	26.5
P(cyclohexyl)$_3$	18	72.5	9.5

(*) Ni(acac)$_2$

There are no marked differences between the distributions of the dimers when, for example, AlEt$_2$Cl, Al$_2$Et$_3$Cl$_3$, AlEtCl$_2$ or AlCl$_3$ are used with the same nickel complex. However, the activity of the catalyst is dependent on the Lewis acid. For example, in a chlorobenzene medium, the rate of dimerization increases in the following order:

$$AlEt_2Cl < AlEtCl_2 < Al_2Et_3Cl_3$$

In contrast, when the Ziegler-type catalyst comprises a "weaker" Lewis acid, the content of straight chain dimers increases. So the catalyst, obtained by mixing a halide-free alkylaluminium (e.g. AlEt$_3$ or Et$_2$AlOR) with chelates of divalent nickel without any phosphine, converts propylene into dimers containing more than 80% of n-hexenes [37] [38]. In the same way, using AlEt$_2$F (which is a "weaker" acid than AlEt$_2$Cl), the content of n-hexenes is about 50%, to be compared to the 25% n-hexene content which is obtained in the presence of AlEt$_2$Cl [42]. Analogously, starting from 1-pentene, AlEt$_2$F + Ni(II) gives more than 40% of unbranched decenes. However, the activity of such catalysts is rather low and depends on the organic part of the chelate.

The π-allyl nickel halides may be activated by numerous other Lewis acids e.g. TiCl$_4$, WOCl$_4$, MOCl$_5$ etc. [39f] [39e] [54]. The same is true for the various Ni(0) and Ni(I) complexes such as Ni(CO)$_2$(PR$_3$)$_2$ [44i] [56] [59] [60], Ni(PR$_3$)$_4$ [57], NiCl(PR$_3$)$_3$ [56] [57] etc. The isomer distribution is also dependent on the acid used [60] but this effect cannot be discussed easily at the present time. Using

$Ni(CO)_2(PR_3)_2$ the co-catalytic activity decreases as follows [64c]:

$$NbF_5 > SbF_5 > TaF_5 > BF_3Et_2O > AlCl_3 > InBr_3$$

As mentioned earlier, the need for an acid co-catalyst is not always well understood. The most surprising fact is that even in the presence of a hundred fold excess of acid, no "free" acid remains in the medium, i. e. its Friedel-Craft behaviour may be suppressed by trace amounts of nickel and no alkylation or oligomerization of "cationic type" can be observed (see for example, ref. [49a]).

2.3.5.4. Nature of the solvent

Owing to the presence of a "hard" Lewis acidity, "hard" solvents have a detrimental effect on the reaction. In fact, it seems that dimerization can be effected only in hydrocarbon or halo-hydrocarbon media, or even without a solvent.

Broadly speaking, the rate of the reaction is all the more rapid as the dielectric constant is higher. For instance, the dimerization of propylene by means of $NiCl_2$ (tetramethylcyclobutadiene)—$AlEtCl_2$—$0.5PBu_3$ complex is faster in chlorobenzene than in benzene [44b]:

$$\frac{rate_{(chlorobenzene)}}{rate_{(benzene)}} \sim 5$$

However, this is not always true and can depend on the nature of the Lewis acid.

On the other hand, the ratio of the isomerization rate to the dimerization rate increases with the dielectric constant. Accordingly, when unisomerized dimers are wanted, the dimerization is better conducted without a solvent.

2.3.5.5. Reaction temperature

The dimerization reaction is commonly conducted between $-20\ ^\circ C$ to $+40\ ^\circ C$, as at higher temperatures de-activation of the catalyst occurs. Temperature has little effect on the products. If the reaction is carried in the absence of phosphine, the dimethylbutene content slightly increases with increasing temperature. In contrast, it seems that the "phosphine effect" decreases at the same time, i. e. in the presence of strongly "basic" phosphine, the amount of dimethylbutenes is not so high at $40\ ^\circ C$ as at $0\ ^\circ C$ [33c]. This may probably be ascribed to the dissociation of the phosphine ligand from the catalytic complex. Nevertheless, the situation is not very clear and the opposite effect has been reported [44b]. Once more, this can depend on the catalytic system.

2.3.6. KINETICS OF OLEFIN DIMERIZATION

Despite the instability of the catalytic system even at low temperature (simultaneous occurrence of an induction period and de-activation), a kinetic study has been carried out in the case of $NiCl_2 \cdot$ tetramethylcyclobutadiene, $AlEtCl_2$, $P(nBu)_3$ for the dimerization of propylene [44b], the dimerization of ethylene [44c] and the co-dimerization of ethylene with propylene [44d]. The reaction is second order with respect to monomers, and first order with respect to nickel, but much more complex with respect to aluminium compounds. The overall experimental activation energy is given as lying between 7 and 9 kcal./mol. The rate constant of ethylene dimerization is about forty times greater than for propylene.

Some other catalytic systems are much more stable even up to 40 °C, for instance $NiCl_2 \cdot 2PR_3$, $AlEt_2Cl$ or $Ni(acac)_2$, $AlEt_2Cl$. With a slightly modified system a study of possible technical importance has been achieved, i. e. the co-dimerization of n-butenes with propylene [64a]. The following experimental values have been obtained for the second order rate constants (at 10 °C):

$$k_{propylene} = 20 \ (mol/1)^{-1} \, h^{-1}$$

$$k_{propylene\text{-}butene} = 0.6 \ (mol/1)^{-1} \, h^{-1}$$

$$k_{butene} = 0.012 \ (mol/1)^{-1} \, h^{-1}$$

The order with respect to the nickel complex has not been reported; however, beyond a concentration of 0.5 to 2 m.M/1 the dimerization rate does not increase further [49a].

2.3.7. CO-OLIGOMERIZATION OF DIENES WITH OLEFINS

As has been pointed out earlier, two cases are to be considered, depending on the nature of the catalyst.

a) The first one, in which the catalyst is based on "bare" nickel (or nickel (0)) or derived by reaction of a nickel salt of the "hard" type with a reducing agent which cannot give a Lewis acidity, for instance, nickel acetylacetonate, formate or dimethylglyoximate with diethylaluminium ethoxide, butyllithium, diethylzinc and so on. The use of such catalysts results in butadiene and other conjugated dienes co-trimerisation with one molecule of ethylene into cyclic and linear oligomers depending on the temperature [5] [26] [65] [66] [67] [68]. Styrene and butadiene [67] [159] give 1-phenyl-1,4,9-decatriene and 4-phenyl-1,7-cyclodecadiene. Methylacrylate and butadiene [69] give a good yield of methyl-2,5,10-undecatrienoate:

$$CH_2{=}CH{-}CH_2{-}CH_2{-}CH_2{-}CH{=}CH{-}CH_2{-}CH{=}CH{-}COOCH_3$$

However, the co-trimerization of a non activated α-olefin such as propylene, has not been reported.

   b) The second one, in which the catalyst is obtained by mixing various nickel complexes such as Ni(II) or Ni(0) with an alkylaluminium halide, and a phosphine or phosphite [29a]. Such a mixture leads to the formation of 1 : 1 adducts from conjugated dienes and ethylene or propylene. A typical catalyst is $NiCl_2 \cdot 2PR_3$—Al R_2Cl. With such a system butadiene and ethylene give a mixture of 1,4-hexadiene and 3-methyl-1,4-pentadiene. Hexamethyl phosphoramide may be used in place of phosphine or phosphite [50].

   Isoprene and ethylene give 4-methyl-1,4-hexadiene and 2,3-dimethyl-1,4-pentadiene. The products result from the attaching of ethylene to positions 1 and 3, excluding position 4 (cf. iron and cobalt catalysts).

   Piperylene and ethylene give a single product, 3-methyl-1,4-hexadiene. Thus, ethylene bonds exclusively at position 4. 2-chloro-1,3-butadiene, 1,3-cyclohexadiene, 1,3-cyclooctadiene also form 1 : 1 dimers with ethylene. Yields of co-dimers are drastically reduced by substituting ethylene with propylene, 1-hexene and other substituted mono-olefins.

   At high diene conversions, the species catalyses a number of well-known side reactions, resulting in lower selectivities of co-dimers. Among these reactions are:

   the dimerization of the mono-olefin itself,

   the isomerization of the 1,4-diene to the 2,4- or 1,3-isomer,

   the formation of various oligomers originating from the addition of olefins to 2,4- or 1,3-diene formed by isomerization,

   the skeletal rearrangements of 1,4-dienes [32]. These rearrangements are of two types. The first one is responsible for the formation of 1,4-hexadiene from 3-methyl-1,4-pentadiene which is ascribable to the "reversible" fixation of ethylene at positions 1 and 2 of the diene, as follows

$$\overset{\overset{\displaystyle CH_3}{\displaystyle |}}{CH_2=CH-CH-CH=CH_2} \; \rightleftarrows \; [CH_2=CH-CH=CH_2 + CH_2=CH_2] \rightleftarrows$$

$$\rightleftarrows \; CH_2=CH-CH_2-CH=CH-CH_3$$

Exchange of two olefins is also possible:

$$CH_2=CH-CH_2-CH=CH-CH_3+CH_3-CH=CH_2 \rightleftarrows$$

$$\rightleftarrows \; \overset{\overset{\displaystyle CH_3}{\displaystyle |}}{CH_2=C-CH_2-CH=CH-CH_3} + CH_2=CH_2$$

Such a reversibility is also found in cobalt catalysts.

The second type of rearrangement is responsible for the formation of the unexpected *trans*-2-methyl-1,3-pentadiene from 1,4-hexadiene. This rearrangement is less easily visualized. However, experimental findings are consistent with the intervention of a cyclopropane ring in the same way as depicted in "homoallylic rearrangements" of Grignard reagents, the catalytic reaction involving a hypothetical nickel hydride:

$$
\begin{array}{l}
\text{CH}=\text{CH}-\text{CH}=\text{CH}-\text{CH}_3 \\
\text{CH}_2 \quad \text{CH}_2 \quad \text{CH}_2
\end{array}
\xrightarrow{\text{NiH}}
\left[
\begin{array}{l}
\text{Ni} \\
| \\
\text{CH}=\text{CH}-\text{CH}_3 \\
\text{CH}_3 \quad \text{CH}_2 \quad \text{CH}
\end{array}
\right]
\longrightarrow
$$

$$
\longrightarrow
\left[
\begin{array}{l}
\text{CH}—\text{CH}-\text{CH}_3 \\
\text{CH}_3 \quad \text{CH}_2 \quad \text{CH} \\
\qquad\qquad\qquad | \\
\qquad\qquad\qquad \text{Ni}
\end{array}
\right]
\xrightarrow{-\text{NiH}}
\begin{array}{l}
\text{CH}_3 \\
| \\
\text{C}=\text{CH}-\text{CH}_3 \\
\text{CH}_2 \quad \text{CH}
\end{array}
$$

2.3.8. MECHANISM OF THE DIMERIZATION REACTION

At present the mechanism of olefin dimerization by nickel complexes is not well understood. The question is all the more complicated as apparently the same species catalyses numerous types of reactions, i. e. dimerization, isomerization, co-dimerization, molecular rearrangements, and even hydrogenation. Many possibilities have been discussed, often only based on speculations. Of course, it is not improbable that several types of mechanisms may be operative for different reactions on the same catalyst.

The most frequently envisaged dimerization mechanism is of the "degenerated" polymerization type, involving a nickel hydride addition-elimination [39e] [39g] [44a] [44b]. Up to now this has explained also the other reactions quite well. In fact unlike iron and cobalt based catalysts, all the products of the co-dimerization of dienes with olefins by nickel catalysts are consistent with this mechanism. The reaction is visualized as a 2,1 addition of Ni—H to the less substituted double bond of a conjugated diene, and then "insertion" of an olefin into the metal-carbon bonded complex [29]. The reaction cannot be achieved without phosphine type ligands; the lack of phosphine leads to formation of polydienes. Clearly phosphine blocks the simultaneous co-ordination of two diene molecules. Likewise, molecular rearrangement of 1,4-dienes requires the presence of phosphine; on the other hand, without phosphine migration of the double is the sole reaction which occurs.

Recently many nickel hydride complexes have been characterised and their reactions will provide a better approach to the intimate mechanism.

2.4. Iron and cobalt

The catalytic behaviour of iron and cobalt in dimerization exhibits sufficient similarity to justify their study in a single chapter. However, this behaviour is greatly different from that of nickel.

In fact, if nickel is the more effective and more studied metal for the catalytic cyclo-oligomerization of conjugated dienes and for the dimerization of olefins, iron and cobalt are particularly active for the linear dimerization of dienes and their co-dimerization with olefins.

2.4.1. OLEFIN DIMERIZATION

Cobalt and iron have been far less extensively studied than nickel. However, numerous papers related to dimerization using nickel complexes also claim the use of cobalt or iron [35, 39a,e,f, 44h,i, 74]. In the case of cobalt the Ziegler type systems similar to those used with nickel are described, but their activity seems limited to ethylene.

The systems Co(III) (or Co(II)) and a reducing organometallic compound have been specially proposed [71] [72] [73], i. e. the system Co(III) acetylacetonate and triethylaluminium convert ethylene at 30 °C into n-butenes with a selectivity of 99.5% [71]. The n-butenes are, in these conditions, a mixture of 95% 2-butenes and 5% 1-butenes. The molar ratio AlR_3/Co must be between 2 and 5, as beyond that the activity decreases. On the other hand the addition of triphenylphosphine decreases the rate of the reaction.

Olefins higher than ethylene are not dimerized by two component cobalt catalysts such as $CoCl_2 + AlEt_2Cl$ as long as the cobalt is not contaminated with nickel. However, in the case of propylene, the use of methylene chloride seems to greatly increase the catalytic activity [74].

In addition to these systems, already described in the case of nickel, a well defined complex, nitrogentris(triphenylphosphine)cobalt hydride, $HN_2Co(PPh_3)_3$, hase been used to dimerize ethylene or propylene [75]. Here, dimerization of ethylene takes place at room temperature without the use of a Lewis acid. For instance, working at 25-30 °C with a m-xylene solution containing 0.63 mmol. of the above cobalt hydride, ethylene is absorbed at the initial rate of ca. 100 ml/hr., but the rate of ethylene absorption decreases with time, presumably due to partial decomposition of the catalyst. This decomposition is less marked at 0 °C. Propylene dimerization carried out in the same manner gives 2-methyl-1-pentene as the main product. The addition of 3 moles of tris(n-butyl)phosphines per mole of cobalt complex considerably reduces the dimerization rate. A mechanism hase been proposed in which the olefin is first "inserted" between the Co—H bond, and then a second mole of olefin is "inserted" between the Co-alkyl bond. Diplacement of the dimer by the olefin regenerates the cyclic process ("degenerated polymerization").

2.4.2. Olefin-diene co-dimerization

Four different dienes may be obtained in the ethylene-butadiene codimerization, namely: 1-*cis* 4-hexadiene, formed by a formal migration of hydrogen from ethylene to butadiene via a 1-4 type addition, 1-*trans* 3-hexadiene resulting from a migration of hydrogen from ethylene to butadiene and 3-methyl-1,4-pentadiene coming (like the first) from a migration of hydrogen from ethylene to butadiene but via a 1-2 type of addition. In addition, 2,4-hexadiene may result from a subsequent isomerization reaction, and 1,5-hexadiene has been obtained in valuable yields by certain workers [81].

In fact, iron and cobalt complexes are highly effective for a selective synthesis of 1,4-hexadiene, and give, in suitable reaction conditions, good yields of 1,3-hexadiene, 3-methyl-1,4-pentadiene and 1,5-hexadiene.

Other olefins such as propylene and other dienes such as isoprene, pentadiene and dimethylbutadiene have been also studied.

1) *Catalytic systems*

Co-dimerizations are promoted by two component iron or cobalt catalysts of the Ziegler-Natta type, or by three component systems in which numerous electron-donor additives have been claimed. In certain cases, well defined complexes used with or without a Lewis acid have been described. Table 7 gives the main systems cited in the literature.

2) *Influence of the composition of catalytic systems on selectivity to form hexadiene*

Owing to the potential industrial uses of 1,4-hexadiene, the simplest hydrocarbon with two double bonds of unequal reactivity, much work has been done to improve the selectivity of ethylene-butadiene co-dimerization.

With conventional Ziegler type catalytic systems, such as iron (III)acetylacetonate and triethylaluminium [98], the product obtained at 30 °C from butadiene and ethylene is composed of. 1,4-hexadiene (41.3%), 1,3,5-hexatriene (0.8%) 2,4-hexadiene (2.3%), 1,3-hexadiene (traces), unreacted butadiene (10.5% and higher oligomers (45.1%).

The higher oligomers are chiefly dimers, trimers and high polymers of butadiene. The amount of 2,4-hexadiene increases in the later stages of the reaction.

Modifications of the nature of the iron compound, of the aluminium compound and use of solvents such as chlorobenzene do not increase the yield of 1,4-hexadiene above a value of around 60%.

The most important effect was observed with the use of chelating diphosphines, especially $Ph_2PCH_2CH_2PPh_2$. It is thought that diphosphines may fill the extra co-ordination vacancy of the metal, prevent the reaction between two moles of butadiene and thus suppress the formation of higher oligomers.

Table 7

ETHYLENE-BUTADIENE CO-DIMERIZATION – IRON AND COBALT CATALYSTS

metal complex	reducing agent or Lewis acid	additives	references
$Fe(acac)_3$	AlR_3		[76, 80, 86, 90]
$Fe(acac)_3$	$AlEt_2OEt$		[86]
$Fe(acac)_3$ or $FeCl_3$	BuLi		[88]
$Fe(acac)_3$ or $FeCl_3$	AlR_3	R_3N, C_6H_5CN, $CH_2=CHCN$	[89]
$FeCl_3$	$AlEt_3$	$P(C_6H_5)_3$	[77, 85]
$FeCl_3$	$AlEt_3$	diphosphines	[78, 106]
$FeCl_3$	$AlEt_3$	$P(OR)_3$	[77]
FeX_3	AlR_3	diarsines	[82]
$FeCl_3$	RMgX		[87]
$Fe(COT)_2$			[79]
$FeC_2H_4(DPE)_2$			[81, 91]
$FeC_2H_4(DPE)_2$	Lewis acids		[81, 91]
CoX_2	AlR_3		[99]
$Co(acac)_3$	AlR_3	diphosphines	[93]
$Co(acac)_3$	$AlEt_2Cl$ or $AlEtCl_2$		[98]
$Co(acac)_3$	AlR_2Cl	phosphines	[100]
$CoCl_2$	AlR_3	thiophosphines	[103]
CoX_2	AlR_3	diarsines	[102]
$CoCl_2$	$AlEt_3$	$(RO)_nPCl_{(3-n)}$	[97]
Co stearate	$AlEt_3$	$P(OR)_3$	[105]
$CoCl_2DPE$	$AlEt_3$		[92, 98, 99]
$Co(DPE)_2$	$Al(IsoBu)Cl_2$		[101]
$HCo(DPE)_2$	$AlEt_3$ or $AlEt_2Cl$ or $AlCl_3$		[96, 101]
$HCo(DPE)_2$	AlR_3	ROH, phenols	[104]

DPE means 1,2-bis(diphenilphosphino)ethane, COT means cyclooctatetraene.

In the case of cobalt, with the system $CoCl_2(Ph_2PCH_2CH_2PPh_2)$ + $AlEt_3$, Hata and co-workers [98] obtained at 70° a 96% selectivity of 1,4-hexadiene for an 80% conversion of butadiene.

The same catalyst with iron instead of cobalt gives the same selectivity but its activity is about ten times lower. For example, the rate of reaction for

FeCl$_3$, Ph$_2$PCH$_2$CH$_2$PPh$_2$, AlEt$_3$ is 20 kg of 1,4-hexadiene produced per 1 mole of transition metal in one hour. In the same conditions 1 mole of CoCl$_2$Ph$_2$PCH$_2$CH$_2$PPh$_2$ produces 200 kg of 1,4-hexadiene.

With the zerovalent iron complex Fe(Ph$_2$PCH$_2$CH$_2$PPh$_2$)$_2$C$_2$H$_4$, obtained by reducing iron (III) acetylacetonate with ethoxydiethylaluminium in the presence of 1,2-bis(diphenylphosphine)ethane the following hexadienes are obtained: 1,4-hexadiene, 1,3-hexadiene, 2,4-hexadiene and 1,5-hexadiene in addition to 2 : 1 adducts in large quantities [81]. The combination of the above iron complex with a Lewis acid (AlEt$_3$, AlEt$_2$Cl, AlEtCl$_2$ or BF$_3$ etherate) gives 95% 1,*cis* 4-hexadiene.

The function of the Lewis acid is not clearly understood. In fact, diethylaluminium chloride which is a stronger Lewis acid than triethylaluminium is more effective for the selective synthesis of 1,4-hexadiene. Ethylaluminium dichloride, the strongest Lewis acid, decomposes the catalyst during the reaction but it is more effective than AlEt$_2$Cl at a low molar ratio (2 : 1).

By the use of AlEt$_3$ as Lewis acid, and with a molar ratio AlEt$_3$/Fe-complex of 4, the content of 1,5-hexadiene may reach 31% of the hexadienes, but no catalytic system has been described with a better selectivity.

Until now, no catalytic system has allowed the exclusive production of 1,3-hexadiene. However, with conventional Ziegler type systems, such as FeCl$_3$, P(C$_6$H$_5$)$_3$, AlEt$_3$ [77] or Co(acac)$_2$, AlEt$_3$' [94] [94'], the content of 1,3-hexadiene in the C$_6$—dienes fraction may be as high as 45,8% (P/Fe = 1,5; Al/Fe = 1; t = 20°; reaction time = 5 hrs); but in that case a great quantity of butadiene dimers and polymers is obtained.

With the catalyst consisting of a cobaltous chloride-1,2-bis(diphenylphosphino) ethane complex and triethyl aluminium, when a vigorous reaction takes place at the initial stages of the reaction, higher oligomers are obtained as the main products, besides 1,4-hexadiene and its rearranged product 2,4-hexadiene. It has been shown [92] that between 80 and 100 °C the formation of 1,4-hexadiene is carried out selectively. But, at temperatures above 100° 1,4-hexadiene is isomerized to the conjugated diene to a considerable extent. The composition of the C$_8$-dienes depends on the reaction temperature. A comparative study of the dienes resulting from the reaction of ethylene with 1,4 or 2,4 hexadiene was carried out in order to elucidate the origin of the C$_8$-dienes [92].

3) *Addition of ethylene to other dienes*

The following 1,3 dienes have been particularly investigated:

isoprene [76, 80, 95, 97, 98, 106];

1,3-pentadiene [76, 80, 95, 97, 98];

2,3-dimethylbutadiene [80, 98];

2-methyl-1,3-pentadiene [80, 95, 98];

2-phenyl-1,3-butadiene [80].

and the effects of the diene structure in the iron (III) acetylacetonate-triethyla-
luminium catalyzed addition of ethylene have been studied. The reaction gives
excluvely *cis* isomers of 1-4 dienes. Methyl substitution control the orientation
of ethylene as reported in table 8 [98].

Table 8

1,3 diene	mode of addition %	total yield of (*) 1,4 diene %
butadiene	 $CH_2=CH-CH=CH_2$ ↑ 100	36
1,3-pentadiene	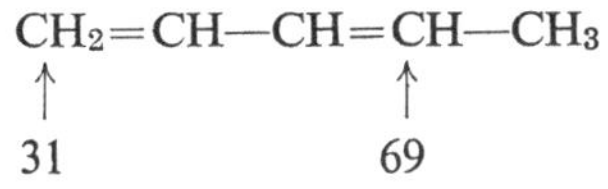 $CH_2=CH-CH=CH-CH_3$ ↑ ↑ 31 69	46
isoprene	CH_3 | $CH_2=C-CH=CH_2$ ↑ ↑ 57 43	70
2,3-dimethylbutadiene	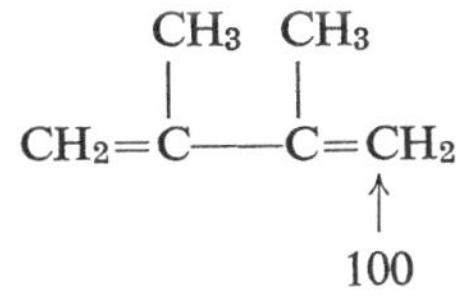 CH_3 CH_3 | | $CH_2=C----C=CH_2$ ↑ 100	78
2-methyl-1,3-pentadiene	CH_3 | $CH_2=C-CH=CH-CH_3$ ↑ ↑ 25 57	89

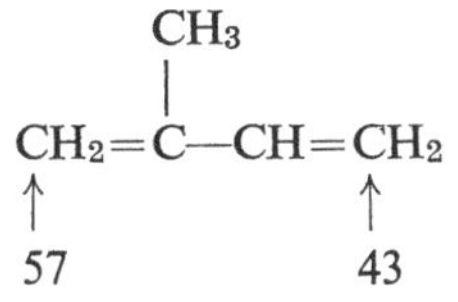

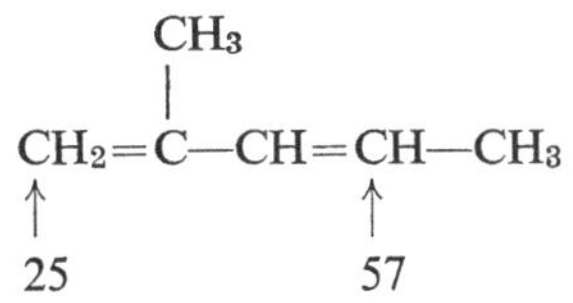

(*) Yield based on the amount of 1,3 diene consumed.

Once again, as in the case of butadiene, in the presence of a diphosphine
and using diethylaluminium chloride instead of triethylaluminium, the addition
of ethylene is more specific as shown in table 9

Table 9

1,3 diene	mode of addition %

isoprene

$$\overset{\displaystyle C}{\underset{\displaystyle \underset{75}{\uparrow}}{\overset{|}{C}}=\overset{|}{C}-C=\underset{\underset{25}{\uparrow.}}{C}}$$

1,3-pentadiene

$$\underset{\underset{15}{\uparrow}}{C}=C-C=\underset{\underset{85}{\uparrow}}{C}-C$$

2-methyl 1,3-pentadiene

$$\overset{\displaystyle C}{\underset{\underset{8}{\uparrow}}{\overset{|}{C}}}=\overset{|}{C}-C=\underset{\underset{92}{\uparrow}}{C}-C$$

1,3 dienes, in which *cisoid* conformations are sterically unfavoured, do not react with ethylene.

For example 1,*cis* 3-pentadiene and 4-methyl-1,3-pentadiene:

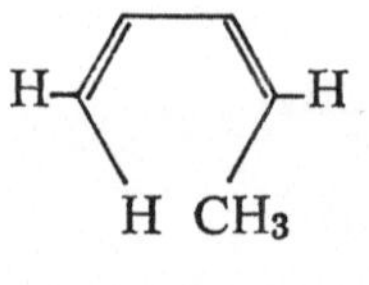

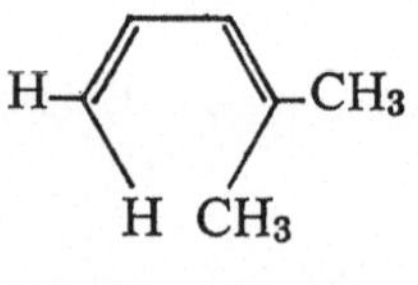

which do not react with ethylene, have steric hindrance in their *cisoid* conformations. However, 1,*trans* 3-pentadiene and 2-methyl 1,*trans* 3-pentadiene:

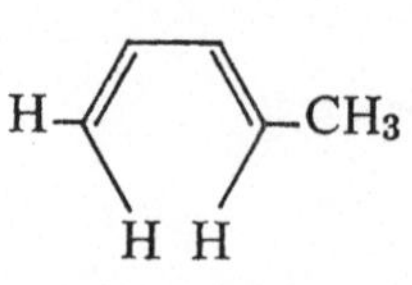

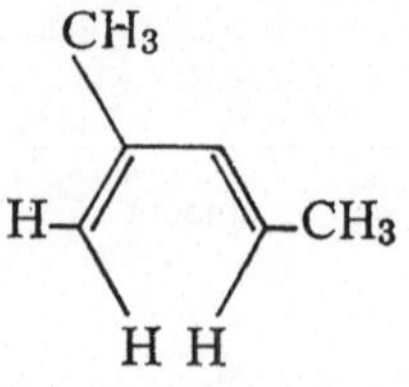

which easily react with ethylene have no such steric hindrance in their *cisoid* conformations.

4) *The mechanism of the 1,4 diene synthesis*

In the addition of ethylene to butadiene a hydrogen atom migrates from the ethylene to the butadiene. One of the possible mechanisms for this shift postulates a metal hydride as a reaction intermediate [78, 97, 98]. Butadiene is co-ordinated to this hydride in its *cisoid* configuration. Then, the adduct changes to a butenyl complex to which an ethylene molecule becomes co-ordinated. Interaction between the co-ordinated ethylene and the butenyl group leads to a 4-hexenyl residue bonded to the metal by a σ and a π bond. The hexenyl-metal intermediate finally leaves as 1,*cis* 4-hexadiene and the complex is free for further reaction.

Hata [98] depicted this mechanism in the case of a cobalt-diphosphine complex in the following way:

$$CH_2=CH-CH_2-CH=CH-CH_3$$

Recently, Dall'Asta and coll. [9] postulated a mechanism in which oligomerization occurs without a hydrogen or alkyl carrier from outside the complexed molecules that form the oligomer molecule.

To support their mechanism, they give evidence of the impossibility of the existence of an active π-crotyl iron bond as postulated by Iwamoto and Yuguchi [78].

For instance, starting from bis(cyclo-octatetraene) iron (0) and adding butadiene at —30° until the NMR spectrum indicated the displacement of all the cyclo-octatetraene by butadiene, they removed all the unreacted butadiene and treated the remaining complex with methanol. No butene was observed in addition to butadiene. Also, no hydrogen was observed after treatment with methanol or hydrogen chloride.

On the other hand, the zero valent complex of iron, $Fe(Ph_2PCH_2-CH_2PPh_2)_2 \cdot CH_2=CH_2$ [81], in the absence of an organometallic compound, gives, in the co-dimerization of butadiene and ethylene an appreciable amount of 1,5-hexadiene. This diene must be formed by a migration of hydrogen from ethylene to an internal butadiene carbon atom by a 1-2 type addition. The presence of σ-alkyl or π-crotyl iron is inconsistent with the formation of such a diene having any methyl end group.

Finally, by co-dimerizing 1,*trans* 3-pentadiene with perdeuterated ethylene in the presence of the tris(acetylacetonate)iron/AlEt$_3$ catalyst the autors obtained 3-methyl 1,*cis* 4-hexadiene containing exclusively the isomer (II) instead of an equimolecular mixture of isomers (I) and (II) as required by the symmetric structure of a π-allyl intermediate:

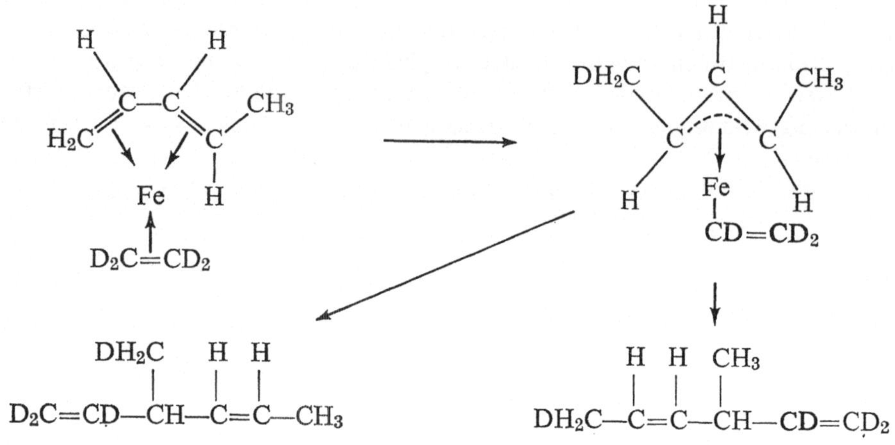

$$D_2C=CD-CH-C=C-CH_3$$

(I)

$$DH_2C-C=C-CH-CD=CD_2$$

(II)

The mechanism of the co-dimerization involves the co-ordination of one monomer; i. e. the butadiene, which may be mono or dico-ordinated and the oxidative addition of the other monomer to the metal by the splitting of a vinylic C—H bond:

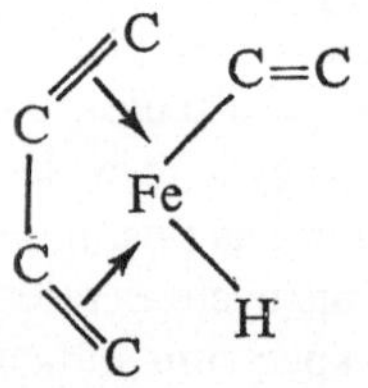

The complex, by rearrangement gives the 1,4-(or 1,5-)diene and leaves the metal free for further complex formation and addition of reactants.

The simultaneous formation of three different hexadienes (1,*cis* 4- hexadiene, 1,*trans* 3-hexadiene and 3-methyl-1,4-butadiene) may thus be explained by the different behaviour of the butadiene molecule towards the iron.

2.4.3. DIMERIZATION OF VINYL MONOMERS

Iron or cobalt complexes have been studied for the dimerization of such vinyl monomers as acrylonitrile but, until now, no catalytic dimerization has been described. Attempts to regenerate the active complex in a cyclic process have been, up to now, unsuccessful.

Most of the work achieved in this way uses iron carbonyl [107] [108] [109] [111] [112]. The carbonyl is treated with a mineral or organic base, with water, or water in the presence of hydrogen and caused to eliminate carbon monoxide. Polymerization inhibitors are often used in the "recipes".

For instance, 53 g acrylonitrile (1 mole), in the presence of 19.6 g $Fe(CO)_5$ (0.1 mole) and 0.3 g methylene blue under 50 atmospheres of hydrogen, gives, after 16 hours at 117 °C, 8.7 g of a fraction containing 85% adiponitrile.

Some of the iron carbonyl complexes presumably formed as intermediates in these reactions have been synthesized and subjected to reactions with acrylonitrile.

In the case of an alkali and $Fe(CO)_5$, a hydride is produced to which a mole of acrylonitrile is added. Two cyanoethyl complexes have been isolated [109]:

$$Fe(CO)_5 + MOH \rightarrow (CO)_4Fe\begin{smallmatrix} \diagup H \\ \diagdown M \end{smallmatrix} + CO_2$$

$$MHFe(CO)_4 + CH_2{=}CHCN \rightarrow (CO)_3Fe\begin{smallmatrix} \diagup CH_2{-}CH_2CN \\ \diagdown M \end{smallmatrix} + (CO)_3Fe\begin{smallmatrix} \overset{\displaystyle CH_3}{|} \\ \diagup CH{-}CN \\ \diagdown M \end{smallmatrix}$$

In the presence of a weak acid ($CO_2 + H_2O$), the hydride may be obtained:

$$(CO)_3FeM(ethylcyanide) + CO_2 + H_2O \rightarrow (CO)_3Fe\begin{smallmatrix} \diagup CH_2CH_2CN \\ \diagdown H \end{smallmatrix} + MHCO_3$$

This hydride adds a second mole of acrylonitrile:

$$(CO)_3Fe\begin{smallmatrix} \diagup CH_2{-}CHCN \\ \diagdown H \end{smallmatrix} + CH_2{=}CHCN \rightarrow (CO)_3Fe\begin{smallmatrix} \diagup CH_2{-}CHCN \\ \diagdown CH_2{-}CHCN \end{smallmatrix}$$

The same di-cyanoethyl iron tricarbonyl may be obtained in the following way:

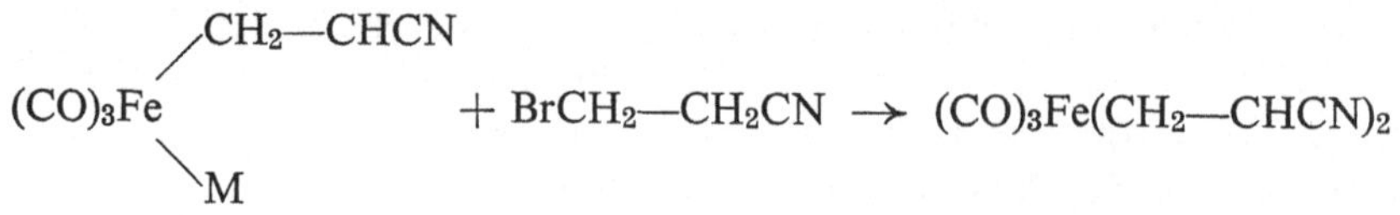

$$(CO)_3Fe\overset{\displaystyle CH_2-CHCN}{\underset{\displaystyle M}{\diagup}} \quad + BrCH_2-CH_2CN \rightarrow (CO)_3Fe(CH_2-CHCN)_2$$

With an organic base, the following reaction leads to the above complex:

$$[NEt_3H]\,[HFe_3(CO)_{11}] + CH_2 = CHCN \rightarrow (CO)_3Fe(CH_2-CHCN)_2$$

The decomposition of the different tricarbonyl iron cyanoethyl complexes has been studied by Thiers and co-workers [109] at 100 °C; $(CO)_3Fe(M)(CH_2-$ $-CHCN)$ gives a mixture of dimers and iron carbonyls; $(CO)_3FeH(CH_2-CHCN)$ gives mainly propionitrile; and $(CO)_3Fe(CH_2-CHCN)_2$ gives dimers. This last complex in the presence of CO or $CO + H_2O$ leads to dimers and $Fe(CO)_5$. In the presence of water alone, dimers are formed along with iron oxides and in the presence of hydrogen and acrylonitrile, polyacrylonitrile is formed.

Other patents claimed the use of irradiated $Fe(CO)_5$ in the presence of water. For instance [111], 96 g of acrylonitrile in the presence of water and $Fe(CO)_5$ (40 g) yield after 3 hrs of irradiation at 110 °C, 20 g of adiponitrile, 0.93 g of propionitrile and 0.37 g of methylene glutaronitrile.

A somewhat similar process has been claimed [110] to dimerize ethyl acrylate to diethyl adipate. Tetracarbonyl cobalt hydride $HCo(CO)_4$ in the presence of water and carbon monoxide has been patented for preparing 2-methylglutaroni-trile from acrylonitrile [112].

Another stoichiometric procedure for dimerizing acrylonitrile is based on the use of powdered manganese and cobalt chloride in dimethylformamide ac-cording to the following two-step reaction [113] [114]:

$$2CH_2=CH-CN + CoCl_2 + Mn \xrightarrow{\text{THF}} C_6H_6N_2Co(DMF)_n + MnCl_2$$

$$C_6H_6N_2Co(DMF)_n + H_2S \rightarrow nDMF + CoS + NC-(CH_2)_4-CN$$

The yield, based on converted acrylonitrile, exceeds 95%. Adiponitrile contains about 0,1% of 2-methylglutaronitrile. The following intermediate is probably formed in solution:

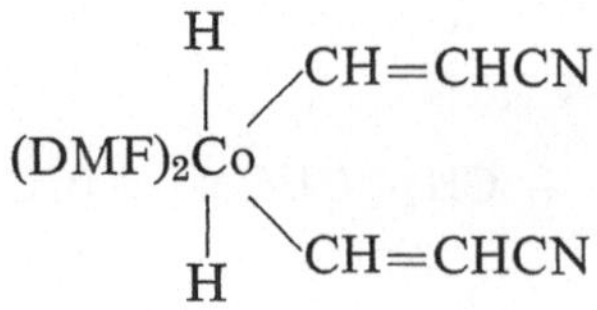

$$(DMF)_2\overset{\displaystyle \overset{H}{|}}{\underset{\displaystyle \overset{|}{H}}{Co}}\overset{\diagup CH=CHCN}{\diagdown CH=CHCN}$$

In addition to acrylonitrile, few acrylic monomers have been studied. However, acrylic and methacrylic acids, and methyl and ethyl acrylates have been reported to form dimers in the presence of pentacyanocobaltate (II) [115]. Co-dimers of butadiene and acrylates have been shortly reported by Wittenberg [4].

Acrylates may be co-dimerized with dienes (butadiene or isoprene) [116] [4] [159], by the use of a binary system consisting of Co(III) or Fe(III) acetylacetonate and triethylaluminium, in the following way:

$$CH_2=CR-CH=CH_2 + CH_2=CH-COOR' \rightarrow CH_3-CH=CR-CH_2-CH=$$

$$=CH-COOR' + CH_3-CR=CH-CH_2-CH=CH-COOR'$$

3-methylhepta-1,4,6-triene, which is the main product of the butadiene dimerization using cobalt catalysts, was not detected.

In both iron and cobalt catalytic systems, the activity of the catalyst was suppressed when it was prepared in the presence of an acrylate without butadiene.

2.5. Rhodium

It is well known that rhodium complexes catalyse homogeneously a great number of reactions of olefins, under mild conditions and with a high specific activity. They also catalyse the oligomerization of various olefins [117] although their activity is rather low in comparaison with other catalytic systems.

With the hydrated chloride ("RhCl₃ 3H₂O") Alderson [118] [119] has investigated the catalytic dimerization and co-dimerization of a wide variety of unsaturated compounds, such as ethylene, α-olefins, dienes, acrylates, styrene, halogenated olefins, etc. Among the most general characteristics of this reaction, the following may be mentioned,

a) Its relative insensitivity to solvents, although even small amounts of hydroxylic solvents such as alcohols promote its activity.

b) The wide temperature range in which the reaction can be carried out (from 20 °C to 200 °C). An increase of temperature accelerates the rate of isomerization more than the rate of dimerization.

c) Much greater activity for co-dimerization than for homodimerization reactions.

As already pointed out many olefins have been studied and they show a different selectivity.

Ethylene gives, with almost complete selectivity, a mixture of n-butenes, in which the content in 1-butene is higher at low temperature and for short reaction times (low conversion). Under certain experimental conditions 1-butene may then be the major product [120].

Propylene yields a mixture of n-hexenes and iso-hexenes. The co-dimerization of propylene with ethylene gives, with much greater activity, a mixture of C_4, C_5 and C_6 olefins, the C_5 fractions containing a nearly equal amount of n-pentenes and methylbutenes.

Methylacrylate yields mainly dimethyl α-dihydromuconate (dimethyl 2-hexenedioate) as follows

$$CH_2=CH—COOCH_3 \xrightarrow{\text{RhCl}_3} COOCH_3—CH=CH—CH_2—CH_2—COOCH_3$$

Likewise, acrylamide and its N-monosubstituted derivatives are transformed into the corresponding *trans*-α-dihydromuconamides (but this is not the case with N,N-disubstituted derivatives: cleavage of the amide group produces the corresponding secondary amines, the olefinic part being polymerized) [121]. On the other hand, acrylonitrile is not dimerized by rhodium, as observed with ruthenium.

Methyl acrylate-ethylene co-dimerization mainly produces methyl 3-pentenoate.

The addition of ethylene to butadiene occurs with great facility to the almost complete exclusion of the dimerization of either component [118] [119] [122] [125]:

$$CH_2=CH_2 + CH_2=CH—CH=CH_2 \rightarrow CH_2=CH—CH_2—CH=CH—CH_3$$

CH$_2$=CH—CH=CH—CH$_2$—CH$_3$ CH$_3$—CH=CH—CH=CH—CH$_3$

$$\downarrow +C_2H_4 \qquad\qquad\qquad\qquad \downarrow +C_2H_4$$

CH$_3$—CH=CH—CH—CH$_2$—CH$_3$ CH$_3$—CH$_2$—CH=CH—CH—CH$_3$

           | |

           CH   CH

           || ||

           CH$_2$ CH$_2$

The reaction occurs rapidly at 50 °C to give over 90% yield of a mixture of 1,4- and 2,4-hexadiene. Use of mild reaction conditions give almost exclusively the 1,4 isomer (the primary product) but at a somewhat higher temperature further reactions take place to give chiefly C_8 diolefins (3-ethyl-1,4-hexadiene and 3-methyl-1,4-heptadiene) formed by the addition of ethylene to 1,3-hexadiene and 2,4-hexadiene formed by isomerization. The isomerization appears to proceed to a much greater extent in the presence of ethylene than in the ethylene-free system.

The addition of ethylene to 1,3-pentadiene is extremely easy and results in the selective formation of 3-methyl-1,4-hexadiene.

The addition of propylene to conjugated dienes proceeds through the secondary carbon. Butadiene gives *trans*-2-methyl-1,4-hexadiene and isoprene 2,4-dimethyl-1,4-hexadiene.

The presence of chloride ion seems to be a requirement for the selective dimerization of dienes. A reasonable assumption is that co-ordinated chloride stops polymer chain growth by assisting hydrogen transfer [117].

Ethylene can also be added to styrene, 2-chloro-1,3-butadiene, and vinyl chloride, giving 2-phenyl-2-butene, chlorohexadienes and chlorobutenes respectively.

The addition of AlEtCl$_2$ to rhodium chloride increases the rate of reaction to a considerable extent, and it has been shown that the dimerization selectivity is improved by adding ligands such as triphenylbismuth [123]. In this way propylene, which had proved to be one of the less reactive olefins, is dimerized with more than 75% selectivity into a mixture containing 15% n-hexenes, the rest being mainly 4-methyl 1- and 2-pentene. The activation by a Lewis acid may probably be interpreted in a similar way as that observed with nickel.

A detailed mechanism, supported by spectral and kinetic data, has been suggested by Cramer, taking into account the dimerization of ethylene [124a], the co-dimerization of ethylene with butadiene [124b] and the isomerization of n-butenes [124c] by rhodium trichloride. These three reactions follow similar kinetic laws at limiting concentrations. So, the rate of dimerization of ethylene is described by the following equation:

$$\frac{d[C_4H_8]}{dt} = k[C_2H_4]\,[H^+]\,[Cl^-]\,[Rh]$$

and the formation of 1,4-hexadiene by:

$$\frac{d[C_6H_{10}]}{dt} = k[C_2H_4]\,[C_4H_6]^x\,[H^+]\,[Cl^-]\,[Rh]$$

with $x = 1$ at 50 °C and $x = 0.34$ at 30 °C.

Under otherwise identical conditions, the pseudo-first-order rate constant for ethylene absorption is three times higher in the presence of an excess of butadiene (1,4-hexadiene synthesis) than in its absence (butenes synthesis), the dimerization of butadiene being much slower.

The kinetics of the isomerization of butenes, for the first 80% of the reaction follows the equation:

$$-\frac{d[C_4H_8]}{dt} = k\,[C_4H_8]\,[H^+]\,[Cl^-]\,[Rh]$$

whilst the kinetics of the isomerization of 1,4-hexadiene are much more complicated.

Cramer has interpreted these data by assuming that the activation of rhodium trichloride consists of a reduction of rhodium(III) to rhodium(I), oxidation of rhodium(I) by protons to rhodium(III) hydride (hypothetical) and reaction of this hydride with a co-ordinated olefin to give a rhodium(III) alkyl, as follows

$$Rh_2Cl_2(C_2H_4)_4$$

$$\updownarrow \ +Cl^-$$

$$RhCl_3 \cdot 3H_2O + C_2H_4 \xrightarrow{\ EtOH\ } [RhCl_2(C_2H_4)_2]^- + CH_3CHO + 2H_2O + HCl$$

$$\updownarrow \ +HCl$$

$$[HRhCl_3(C_2H_4)_2]^- \rightleftarrows [C_2H_5RhCl_3(C_2H_4)]^-$$

(*hypothetical*)

1) *The dimerization of ethylene* is a "formal insertion" of a molecule of ethylene into the carbon-rhodium bond of the ionic complex (irreversible rate-determining chain-growth reaction), followed by the elimination of a hydrogen and reduction of rhodium(III) to give a 1-butene complex of monovalent rhodium(I). The co-ordinated butene is rapidly displaced by ethylene, reforming the initial complex:

$$[C_2H_5RhCl_3(C_2H_4)] \rightarrow [CH_3-CH_2-CH_2-CH_2-RhCl_3]^- \rightleftarrows$$

$$+ 2C_2H_4$$
$$[Cl_2Rh(C_4H_8)]^- + HCl \rightleftarrows [RhCl_2(C_2H_4)_2]^- + 1-C_4H_8$$

The 1-butene may be isomerized in a subsequent reaction.

The selectivity in butenes is attributed to the fact that all the equilibria are readily established and that the equilibrium constant highly favours the displacement of butene by ethylene [124d].

All the intermediate rhodium complexes are anions (in a cyclohexane water medium, rhodium is found only in the aqueous phase). It is also possible to dimerize ethylene by a non-ionic reaction by using $(\pi\text{-}C_5H_5)Rh(C_2H_4)_2$ as catalyst, but the reaction is not so fast.

2) In the *co-dimerization of butadiene with ethylene* the catalytic species are uncharged (rhodium is concentrated in the organic phase) and the reaction cycle does not invoke a rhodium hydride complex and reversible oxidation of rhodium(I) by HCl.

In the ethyl-rhodium complex, ethylene is displaced by butadiene. This is due to the greater stability of π-allyl complexes compared with σ-allyl ones:

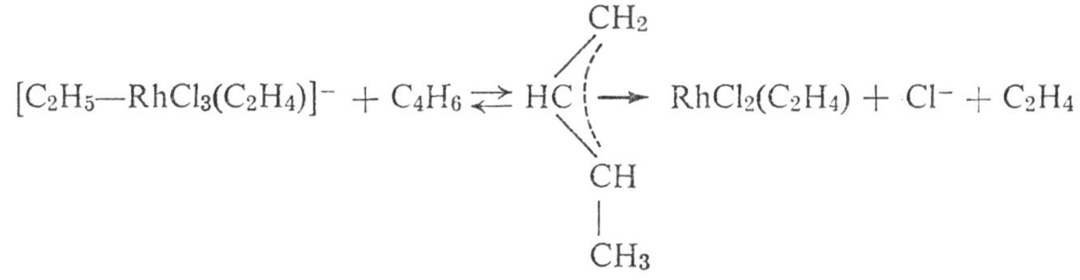

$$[C_2H_5\text{—}RhCl_3(C_2H_4)]^- + C_4H_6 \rightleftarrows \underset{\substack{\text{CH}_2 \\ \text{//} \\ \text{HC} \\ \backslash \\ \text{CH} \\ | \\ \text{CH}_3}}{} \longrightarrow RhCl_2(C_2H_4) + Cl^- + C_2H_4$$

and the preferential formation of 1,4-hexadiene is a result of the relatively high stability of the ethylene π-allyl rhodium complex compared to the butadiene π-allyl rhodium complex in which the equilibrium is displaced far to the left.

The reaction mechanism for the synthesis of 1,4-hexadiene is the following (non essential ligands are omitted):

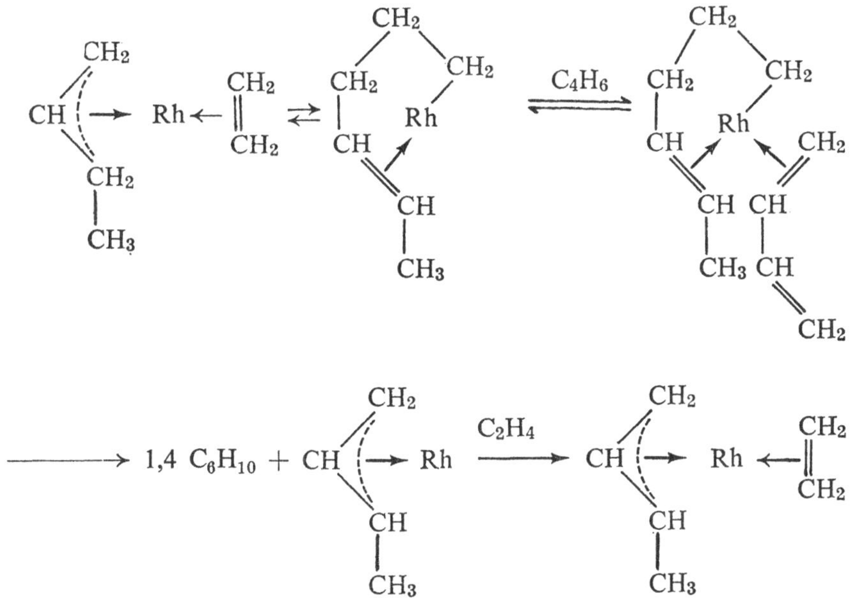

At 30 °C the rate determining step is the insertion of ethylene. But at 50 °C the rate determining step is the release of 1,4-hexadiene from the complex.

A similar hydride transfer inside the olefin complex has already been tentatively suggested by Bestian and co-workers in the case of the dimerization of olefins by titanium complexes (see titanium).

With respect to dimerization, isomerization is a fast reaction, but it is slowed down in the presence of more tightly co-ordinating olefins (e.g. ethylene or butadiene) than butene. This means that it is possible to increase the proportion of unisomerized products in a dimerization reaction by increasing the concentration of reagents and limiting the conversion.

2.6. Ruthenium

In the presence of various olefinic compounds, the catalytic behaviour of ruthenium is quite similar to that of rhodium, although under more extreme conditions ruthenium trichloride has lower activity and selectivity for dimerization [119].

In addition to the dimerizations and co-dimerizations already described in the case of rhodium, it has been reported that the co-trimerization of one allyl-acetate molecule with two butadiene molecules is catalyzed by a methanolic solution of ruthenium chloride [118].

However, the main difference between the two metals arises from their behaviour in the dimerization of acrylates and especially acrylonitrile. The dimerization of methylacrylate into dimethyl α-dihydromuconate by rhodium chloride is a sluggish reaction (34 mM of monomer per g $RhCl_3 3H_2O$, at 140 °C for 10 hours). Ruthenium chloride gives the same product but with a higher activity, especially in the presence of ethylene (which at the same time co-dimerizes with the methylacrylate forming methyl 3-pentenoate) (800 mM of monomer per g $RuCl_3$ at 150 °C for 16 hours) [119].

While rhodium does not dimerize acrylonitrile, ruthenium, provided that the reaction is performed under hydrogen pressure, dimerizes acrylonitrile into 1,4-dicyano-1-butene and adiponitrile with appreciable activity (800 mM of acrylonitrile per g $RuCl_3$ at 110 °C for 6.5 hours) [126a] [126b].

Under the same conditions it also catalyses the co-dimerization of acrylonitrile and methylacrylate [126c]. Besides the homodimerization products of each monomer, up to 50% of the head-to-head co-dimerization products are formed, i. e. methyl cyanovalerate $N \equiv CCH_2CH_2CH_2CH_2COOCH_3$ and methyl 5-cyano-4-pentenoate, $N \equiv CCH = CHCH_2CH_2COOCH_3$.

The bifunctional products obtained in this way are important industrial intermediates. The achievement of the inexpensive head-to-head coupling of readily available functional unsaturated compounds, has been sought after in many laboratories in recent years. The only one process now used on a semi-practical scale is the reductive duplication by cathode electrolysis or alkali metal, for which a mechanism has been postulated [127]. It consists of formation of the radical-anion of acrylonitrile, addition of an acrylonitrile molecule onto the radical-ion thus formed, capturing of an electron to form the di-anion and re-action of the di-anion with protons.

Besides *cis*- and *trans*-1,4-dicyanobutenes and adiponitrile which are the main products of the ruthenium catalyzed dimerization of acrylonitrile, varying amounts of propionitrile are formed (generally not less than half the amount of the dimers) and also a small amount of 2-methylene glutaronitrile, probably formed by a head-to-tail reaction of acrylonitrile under the influence of bases (phosphines, amines) eventually present in the medium.

Hydrogen seems to play an important role. The reaction does not proceed catalytically in its absence or when its pressure is below 1 or 2 atmospheres. Nonetheless, under an ethylene atmosphere a low yield of dimers is obtained [128a]. On the other hand, under a nitrogen atmosphere and in the presence of various organic compounds a minor quantity of 1,4-dicyano-1,3-butadiene, which is the products of an oxidizing coupling, is formed [129].

An increase in hydrogen pressure favours dimerization, particulary the formation of adiponitrile, and at the same time formation of propionitrile, but there is a pressure beyond which the dimerization selectivity decreases. The optimum appears to be between 5 and 20 atmospheres.

The most favourable temperature lies between 100 and 150 °C. Below 100 °C the reaction rate is low and above 180 °C the yield of dimers strongly decreases. This has been attributed to the thermal instability of the intermediates [128].

The reaction is hardly affected by the addition of various solvents like cyclohexane, benzene, tetrahydrofuran, ethylene glycol, propionitrile or dimethylformamide [126a] [126b] [130] which are either non-complexing or of too "hard" character to enter into competition with the acrylonitrile. So, dimerization is usually carried out in pure acrylonitrile or sometimes in the presence of alcohols which promote the reaction. Nevertheless, small amounts of certain "bases" favour or hinder the reaction, e.g. $SnCl_3^-$ anion together with N-methylmorpholine and "methylcellosolve" [131] or N-methylpyrrolidine [132] have a promoting effect and in the latter case this may arise from its ability to neutralize the hydrochloric acid formed by the reduction of ruthenium trichloride. Pyridine, carbon monoxide, phosphites, and triphenylphosphine, which may add to the metal atom, slow down the dimerization rate and favour the formation of propionitrile, (it is known that $RuCl_2(PPh_3)_3$ is an effective homogeneous catalyst for the hydrogenation of olefins) [133].

It follows that the form in which ruthenium is used seems to be of little importance provided that it does not contain too complexing ligands. In addition to the hydrated trichloride $RuCl_3 3H_2O$, the acetylacetonate $Ru(acac)_3$ [126a] [126b] [128b] [128c] [132] and the dichloro (dodeca-2,6,10-triene-1,2-diyl) ruthenium (IV) $RuCl_2(C_{12}H_{18})$ [126d] [128b] may also be used.

It is not yet clear what is the real catalytic complex, but several acrylonitrile ruthenium complexes provide greater activity and can be considered as the intermediates in the dimerization reaction. Some of these complexes were isolated, characterized and used as catalysts: $RuCl_2(acrylonitrile)_4$ obtained by refluxing mixture of acrylonitrile and $RuCl_3 3H_2O$ in 2-methoxyethanol [126e] [132], $RuCl_2(acrylonitrile)_3$ prepared in a similar way in methanol [128a] [128d] the i. r. spectrum of which reveals the co-ordination of acrylonitrile through the nitrogen lone pair; $RuCl_2(acrylonitrile)_3(H_2O)$ [132].

Various improvements for increasing the yield and the selectivity have been proposed, the most effective being the addition to the conventional system of oxides or hydroxides such as RuO_2, $Ru(OH)_3$ [126d]. Other patents claim the dimerization of acrylonitrile by various ruthenium catalysts [134] [135].

The hydrogen requirement of the dimerization suggests that ruthenium hydride complexes are involved as intermediates. In spite of the fact that phosphines decrease the yield and the selectivity of the reaction, ruthenium complexes with triphenylphosphine as ligand were taken as models for the catalytic species because complexes such as $RuClH(PPh_3)_3$, $[RuCl(PPh_3)_2(acrylonitrile)]_2$ and

$RuCl_2(PPh_3)_2(acrylonitrile)_2$ may be isolated and catalyse the dimerization [128] [132].

Two general mechanistic schemes have been formulated.

a) The first one [132] involves the three steps of a "degenerated" polymerization, i. e. "formal insertion" of an acrylonitrile molecule into a ruthenium-hydrogen bond, and then "formal insertion" of a second molecule into the ruthenium-carbon bond. Propionitrile and adiponitrile are formed by hydrogenolysis of the ruthenium-carbon bond by means of hydrogen or ruthenium hydride:

$$L_2RuCl_2(CH{=}CHCN) \xrightarrow{\text{H}_2} L_2Ru(CH_2{=}CHCN)HCl + HCl \xrightarrow{\text{C}_2\text{H}_3\text{CN}}$$

$$L_2Ru(CH_2{=}CHCN)HCl + CH_3CH_2CN$$

$$L_2ClRuCH_2CH_2CN \overset{\text{H}_2}{\nearrow}$$

$$\underset{CH_2{=}CHCN}{\uparrow}$$

$$\underset{\text{C}_2\text{H}_3\text{CN}}{\searrow}$$

$$\overset{CN}{\underset{|}{}} \\ L_2ClRuCHCH_2CH_2CH_2CN$$

$$\underset{CH_2{=}CHCN}{\uparrow}$$

$$\swarrow \qquad \searrow \text{H}_2$$

$$L_2ClRuH(CH_2{=}CHCN) \qquad L_2ClRuH(CH_2{=}CHCN)$$

$$+1,4\text{-dicyano-1-butene} \qquad + \text{adiponitrile}$$

A similar mechanism has been recently proposed, in which the manner of attachement of the acrylonitrile changes from the co-ordination through the double bond to that through the nitrogen lone-pair [128c].

b) The second scheme [131] involves the activation of a vinyl-hydrogen bond, followed by coupling. This mechanism is consistent with experimental findings obtained under a deuterium atmosphere.

$$Ru^{II}(CH_2{=}CHCN) \rightleftarrows HRu^{IV}CH{=}CHCN \xrightarrow{\text{C}_2\text{H}_3\text{CN}}$$

$$\xrightarrow{} NCCH_2CH_2Ru^{IV}CH{=}CHCN \xrightarrow{\text{C}_2\text{H}_3\text{CN}}$$

$$\xrightarrow{} NCCH{=}CHCH_2CH_2CN + Ru^{II}(CH_2{=}CHCN)$$

Propionitrile formation may be explained in the following way:

$$NCCH_2CH_2Ru^{IV}CH=CHCN + HRu^{IV}CH=CHCN \longrightarrow$$

$$\longrightarrow NCCH=CHRu—RuCH=CHCN + CH_3CH_2CN$$

The primary role of hydrogen is for the regeneration of the active catalyst:

$$NCCH=CHRu—RuCH=CHCN + H_2 \rightarrow 2HRu^{VI}CH=CHCN$$

2.7. Palladium

Palladium complexes have not yet reached the importance of those of such elements as nickel, cobalt or iron in the field of olefin dimerization and, unlike rhodium, they have not contributed extensively to the understanding of olefin dimerization mechanisms. Moreover, palladium is not, as is ruthenium, the metal of a specific dimerization (acrylonitrile). However, halides and complexes of palladium catalyze some reactions of olefins and conjugated dienes especially their dimerization and the migration of double bond. The understanding of the catalytic action of this element was helped by the isolation and characterization of the complexes which it forms with olefins.

2.7.1. OLEFIN DIMERIZATION

The dimerization of ethylene into n-butenes by means of tetrachlorobis (ethylene) dipalladium $(PdCl_2C_2H_4)_2$ in non hydroxylic media (benzene or dioxane) has been pointed out by Van Gemert and Wilkinson in 1964 [142]. However, Alderson [118] had already described the dimerization of ethylene into 2-butene by $PdCl_2$ in methanol and HCl (800 atm, 200 °C).

Since then, this investigation has been returned to and expanded, using palladium chloride which, in the presence of an olefin leads to the olefin complex [144] [145] [146] [147] [148]. Moreover, other palladium salts (fluoride, bromide, iodide, nitrate) were tested in the dimerization of olefins. With them, no formation of the complex $(C_2H_4)_2Pd_2X_4$ was observed. Palladium cyanide dimerizes ethylene twice as slowly as $PdCl_2$, probably on account of the deactivation of the catalyst by a polyethylene deposit formed at the same time as the dimer. The activating action of $AlEt_2Cl$ in the dimerization of propylene by $PdCl_2$ has also been described [123]. In this work, the authors have shown the influence of various ligands (triphenylphosphine, triphenylarsine and triphenylbismuth) on the structure of the hexenes formed.

The nature of the solvent of the reaction is of great importance. Besides benzene and dioxane, already mentioned, acetic acid [144] [145], halogenated hydrocarbons [143] [144] [147], nitro derivatives [143] [146] and sulfones [148] have been used. A comparative study of the dimerization of ethylene by $PdCl_2$ in various solvents [144] showed that acetic acid gave the best yield of butenes (about 1500 moles for mole of palladous chloride) whereas alcohol gave a yield of only 95 moles. The conditions under which ethylene is dimerized by $PdCl_2$ are variable, for instance: temperature (70 to 150 °C), ethylene pressure (50 to 100 atm), and time (10 to 24 hrs). A small percentage of high oligomers are formed. The butenes formed include *trans* 2-butene *cis* 2-butene and 1-butene with the proportions corresponding to the thermodynamic equilibrium.

An important contribution to the understanding of the mechanism of dimerization by palladium is due to Ketley and co-workers [143]. They demonstrated that the active solvents catalyze the formation and isomerization of the bridged complexes formed between the olefin and the palladium chloride. Therefore, the palladium chloride placed in chloroform suspension under 10 atm of ethylene at 50 °C is entirely transformed in 30 minutes into an insoluble yellow complex identified as di-μ-chloro dichlorobis (ethylene) dipalladium prepared according to the Kharasch method. The reaction forming the Kharasch complex is much slower and more incomplete in benzene, carbon tetrachloride, methylene chloride, nitrobenzene and nitroderivatives. On the contrary, in ethyl chloride and t-butyl chloride it takes place in less than one minute at ordinary temperature. By using chloroform containing traces of ethanol (0.03 M), the initial yellow complex goes into solution in about 24 hours. Under the conditions described above, the homogeneous red solution produces 0.1 mol/hr of butene.

The ethyl alcohol can be replaced by other alcohols or by water, and the concentration has no effect on the rate as long as it is between 0.02 and 0.1 M. For higher concentrations than 0.1 M, i. e. when $ROH/PdCl_2 > 1$ the catalyst is deactivated and metallic palladium is formed.

In methylene chloride, nitrobenzene and nitroethane, the transformation of the Kharasch complex into the soluble red catalyst takes place without any hydroxylic additive. On the other hand, in carbon tetrachloride and halogenated hydrocarbons, the yellow complex remains unchanged in the presence of ethylene for several days, even in the presence of alcohols or water.

According to Ketley [143], olefin dimerization may be illustrated as follows: in the first step, the solvent acting as weak ligands, or additives such as ethanol would cause the opening of the chlorine bridges of the Kharasch complex (I) and formation, under ethylene pressure, of complex (II).

(I)   (II)

Recently it was clearly demonstrated [141] the action of the co-catalyst as a base which splits the chlorine bridges.

Then, complex (II) would undergo a geometric isomerization bringing the two ethylene molecules into a position next to each other:

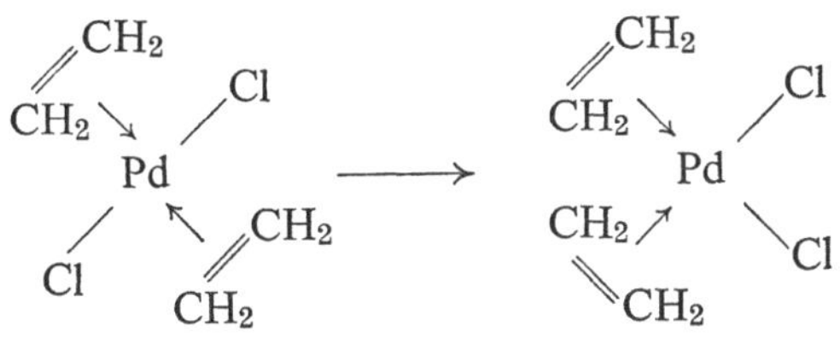

The detailed mechanism of the following dimerization step:

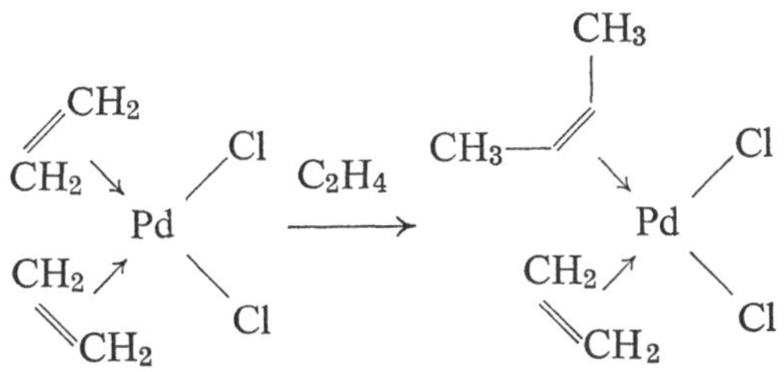

is far from obvious, no evidence having been obtained by the author of an intermediate metal hydride such as:

In fact, analyses of reaction solutions show no trace of vinyl chloride and the NMR spectra of the concentrated reaction solutions show no high-field signal characteristics of metal hydrides. Addition of sources of hydride (H$_2$, NaBH$_4$) inhibits the reaction. However, Ketley and co-workers postulated the transient formation of a Pd—H species. This hydride could arise from a vinylic hydrogen abstraction by the metal as recently postulated by Dall'Asta and others [9].

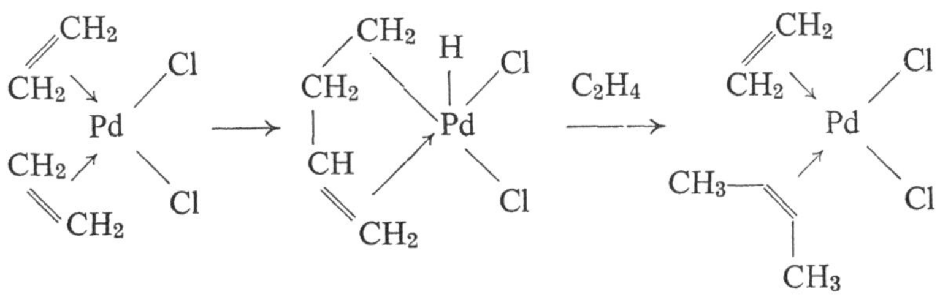

Ketley and coworkers also succeded in isolating and identifying the complex which is formed when the catalyst, after remaining at 40 °C for several days becomes inactive. The structure which appears to fit the data (NMR spectrum, and chemical analysis) most satisfactorily is that of a π-allyl compound.

Other authors [149], by reacting propylene with $PdCl_2$ in glacial acetic acid, have isolated in benzene a yellow crystalline complex with a composition $C_6H_{11}PdCl$ as a dimer.

In the case of propylene, using the conditions described above for ethylene, Ketley and co-workers succeeded in obtaining, in chloroform or methylene chloride, 65% of 2- and 3-hexenes and 35% of 2-methyl and 4-methylpentene. By replacing $PdCl_2$ with the propylene complex, up to 100% of normal hexenes can be obtained. The branched dimers are attributed to $PdCl_2$ acting as a weak Friedel-Crafts catalyst. The final composition of the olefin dimers depends on the isomerization of the primary products. In fact, isomerization of olefinic compounds by palladium salts and complexes is well known and has been extensively studied [136] [137] [138] [139] [140]. In a recent paper [160] some other examples of dimerization and codimerization of olefins, catalysed by $PdCl_2$, have been reported.

2.7.2. Butadiene-ethylene co-dimerization

The $PdCl_2$, $AlEt_2Cl$, $2PR_3$ system catalyses the butadiene-ethylene co-dimerization into pure *trans* 1,4-hexadiene [150].

However, the metallic palladium formed from the partial decomposition of the active complex also catalyses the isomerization of 1,4-hexadiene to 2,4-hexadiene.

Without thoroughly investigating the reaction mechanism of the *trans* 1,4-hexadiene synthesis, Schneider [150a] proposed the formation of a π-allyl complex of butadiene of the type synthesized by Slade and Jonassen [151]:

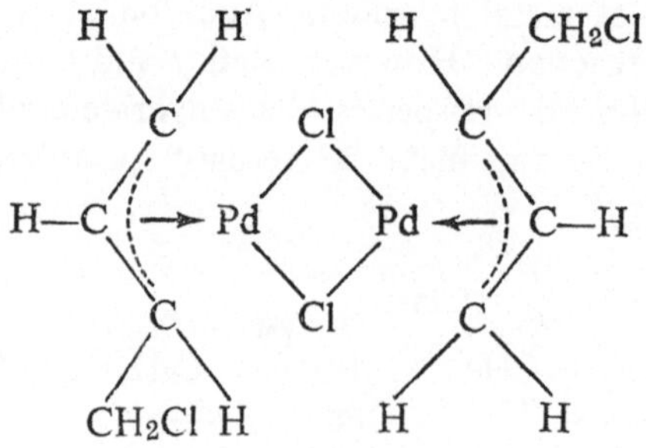

This compound has a π-methallyl structure and can occur in the *syn* and *anti* forms. The proposed structure of the intermediate state is:

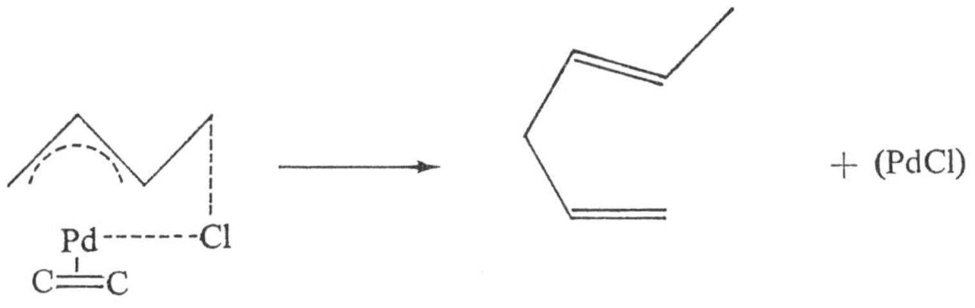

The unoccupied co-ordination sites of palladium could be occupied by phosphines and/or free ethylene or butadiene and the complex could be a dimer. Owing to steric hindrance, the butadiene may form, with the bulky palladium atom, a *syn* π-allyl complex rather than an *anti* π-allyl complex.

A further argument for steric hindrance being a major factor is obtained from the case of isoprene and ethylene. The preferential formation of *trans* 4-methyl-1,4-hexadiene implies the intervention of the *syn* form of the π-allyl complex. However, trace amount of 5-methyl-1,4-hexadiene may result from the unfavourable *anti* form:

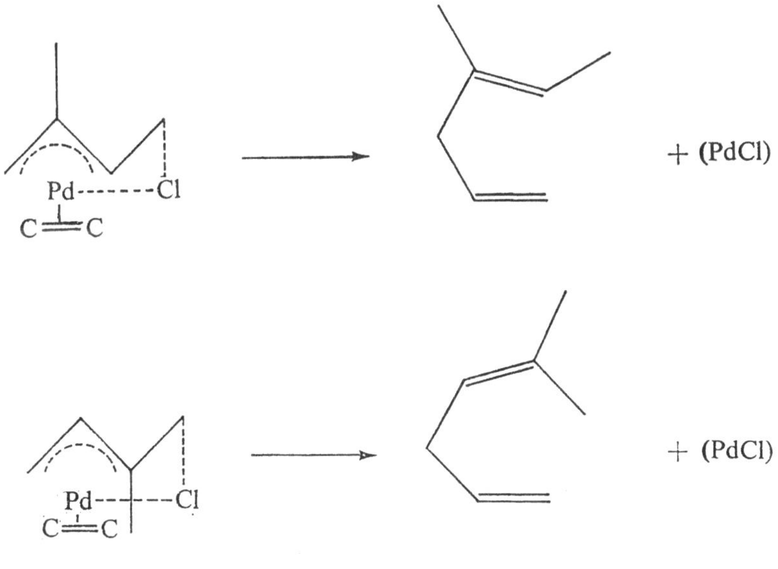

Phosphines have a marked effect on the selectivity of the reaction. The table 10 illustrates this influence in the case of triphenyl and tributyl phosphine. The reducing agent used was di-isobutyl aluminium chloride. No reaction between butadiene and ethylene occurs when a trialkyl aluminium is used. In the case of tributyl phosphine, the first experiment was run for 17 hours (a), and the second for 40 hours (b). The increase of 2,4-hexadiene content with time results from the decomposition of the palladium catalyst.

Table 10

| | PdCl$_2$ | PdCl$_2$(PPh$_3$)$_2$ | PdCl$_2$(PBu$_3$)$_2$ | |
			(a)	(b)
1,4-*trans* hexadiene	14	51	77	33
2,4-hexadiene	7	2	traces	55
High boiling fractions	79	47	23	13

2.8. Miscellaneous

Some other transition metals are active in dimerization reactions.

Thus, a binary system of cuprous oxide and isocyanide dimerizes methyl crotonate, [152] as follows:

$$2CH_3CH=CHCOOCH_3 \longrightarrow \begin{array}{c} CH_3CH=CCOOCH_3 \\ | \\ CH_3CHCH_2COOCH_3 \end{array}$$

Dimerization was also observed with crotonitrile and *trans* 3-pentene-2-one, but acrylates are not dimerized by this system. However, a patent [153] claims the dimerization of acrylonitrile in dicyanobutene by means of cuprous chloride.

In the same manner as rhodium trichloride, iridium trichloride dimerizes ethylene to a mixture of butenes, and propylene in a mixture of iso-hexenes [118]. The activity of iridium is greatly enhanced in the presence of AlEtCl$_2$ [123]. Likewise, a mixture of platinum chloride and organoaluminium compounds transforms propylene to dimers.

2.9. Heterogeneous dimerization catalysts related to co-ordination catalysis

It is now generally assumed that there is a close relationship between homogeneous and heterogeneous catalysed reactions, the support of the active site on a surface playing the part of the ligand on a co-ordinated metal atom in solution. The same intermediate species and similar mechanisms are invoked in both catalyses. Just as ligands can modify the electronegativity and the polarizability of a metal atom and hence its reactivity, so changes in the nature of the support drastically alter the behaviour of an active site.

As it has been pointed out above, mechanisms of co-ordination dimerizations and factors responsible for the dimer distribution are not fully understood.

Of course, features of heterogeneous dimerization are still more difficult to visualize. In this latter case, the selectivities depend on the nature and the acidity of the support, methods of pretreatment etc. The study is all the more complicated as, at high temperatures, the acidic part of the heterogeneous phase may be responsible by itself for the dimer formation. It is not the purpose of this review to discuss the heterogeneous dimerization catalysis. Nonetheless some typical examples must be given.

Special attention has been given to the cobalt oxide-on-carbon catalyst which is able to dimerize ethylene, propylene, butenes and hexenes [154]. For instance, propylene could be dimerized between 25 °C and 85 °C with more than 99% selectivity, into a mixture of about 50% n-hexenes and 50% 2-methylpentenes by means of a catalyst obtained by impregnating an ammoniated carbon with cobalt nitrate, then activating this precatalyst at 300 °C under a nitrogen flow. By modifying the process of catalyst activation, the isomer distribution may be slightly modified so that the major parts of the dimer fraction are 2-hexene and 4-methyl-2-pentene, which are the primary products of the reaction. A mechanism has been suggested which is of the "degenerated" polymerization type, involving a cobalt hydride intermediate. In contrast to co-ordination catalysis, nickel has been reported to be inactive or only slightly active in the same conditions. However, when a cobalt-on-carbon catalyst is activated by organoaluminium compounds [155] the isomer distribution is nearly the same as that of the dimers obtained by means of a nickel co-ordination catalyst.

Much work has also been devoted to catalysts consisting of nickel oxideon silica and alumina, or nickel oxide and molecular sieves, which dimerize olefins with high selectivity. Choosing suitable supports and special treatment, it is possible to obtain relatively high proportions of either end-to-end olefin dimers or head-to-end olefin dimers. In contrast to other catalysts, the nickel-oxide-silica-alumina may have a good activity for dimerizing higher olefins such as butenes, hexenes and so on. The primary products of propylene dimerization are 2-methyl pentenes, n-hexenes and the unexpected 3-methylpentenes [156]. To account for the formation of this unusual dimer a reaction scheme involving a "quasi cyclobutane" ring as intermediate has been tentatively suggested:

$$
\begin{array}{c}
\mathrm{C} \\
| \\
\mathrm{C{=}C{-}C{-}C{-}C}
\end{array}
\;+\;
\begin{array}{c}
\mathrm{C} \\
| \\
\mathrm{C{-}C{=}C{-}C{-}C}
\end{array}
\qquad
\begin{array}{c}
\mathrm{C} \\
| \\
\mathrm{C{-}C{-}C{=}C{-}C}
\end{array}
$$

$$
\begin{array}{c}
\mathrm{C}\quad\mathrm{C} \\
|\quad\; | \\
\mathrm{C{=}C{-}C{-}C}
\end{array}
$$

Such a "quasi cyclobutane" ring has also been invoked for other reactions, e. g. olefin metathesis.

In the two preceeding cases, thermal treatment was an essential requirement for catalytic activity. A more interesting approach may be achieved by the use of a catalyst composition produced by adding well defined organo-metallic complexes and an acidic support without any pretreatment prior to contact with the olefin. As has been seen earlier, Lewis acidity is an essential part of nickel based catalysts for attaining high activity and selectivity at moderated temperatures. This acidity may be brought about by insoluble inorganic oxides. For example, reacting π-allyl nickel chloride and silica-alumina in an hydrocarbon medium affords a catalyst which dimerizes propylene with poor activity. However, the catalytic activity may be enhanced by means of organoaluminium compounds [157]. In the same way treating protonated molecular sieves with $Ni(CO)_2(PR_3)_2$ in toluene produces a catalyst which dimerizes propylene at room temperature (64. c). None the less in this case, the "phosphine effect" discussed in the co-ordinated dimerization by nickel catalysts is not observed. Another patent describes the production of a catalyst for olefin dimerization [158] obtained by reacting nickelocene, silica-alumina and hydrogen in heptane.

2.10. Conclusion

As it has been achieved in the polymer field, co-ordination catalysis has opened a very attractive way to new products by dimerization. However many theoretical and practical important problems still remain to be solved, even if transition metals of the second series such as rhodium or palladium give well defined and effective complexes and if the mechanism of their catalytic action is now relatively clear.

However, in many cases, the catalytic complexes of the metal of the first series are not yet identified. Thus, need for a Lewis acid in the catalytic complex is not always clarified; moreover, the factors responsible for the mode of linking of the monomer remain to be explained.

Although many dimerization reactions between different olefinic compounds have been achieved, their industrial extension is still dependant of the improvement of their rate and selectivity. For instance, production of 4-methyl-1-pentene, r.-hexene 1,5-hexadiene and α,ω unsaturated dimers have not yet reached the state of industrial development.

In the case of highly specific dimerization and co-dimerization, the products obtained will undoubtedly be, in the near future, the new intermediates of industrial processes.

3. APPENDIX
3.1. Summary of main dimerized olefins

(when given, the percentage of main dimers is calculated on the basis of the dimer fraction, other products such as higher oligomers are not included).

Table 11

ETHYLENE

Main products	Catalytic systems			References
	Transition metal compounds	Co-catalysts	Additives or solvents	
Titanium				
1-butene (60 to 99%) the remainder being 2-butene, small amounts of n-hexenes	$Ti(OR)_4$	$AlEt_3$		[11, 12, 13, 14, 15]
prevailingly 1-butene	$Ti(OR)_4$	H_2AlNR_2		[17]
1-butene (> 99%) small amounts of n-hexenes	$(\pi\text{-}C_5Me_5)\,Ti(OR)_3$	$AlEt_3$		[16]
1-butene small amounts of 1-hexene	$(\pi\text{-}C_5H_5)\,TiCl_3$	Na/Hg or RMgX		[18, 19, 20]
prevailingly 1-butene	$Zr(OR)_4$	$AlEt_3$		[11, 12]
1-butene, 2-ethyl-1-butene and 2-ethyl-1-hexene in variable amounts according to the temperature	$TiCl_4$ MeTiCl_3	$Al_2Cl_3Me_3$ $AlCl_2Me$		[21.a]
Cobalt				
2-butene (95 to 99%) 1-butene (1 to 5%)	$H(N_2)\,Co(PPh_3)_3$ $Co(acac)_3$	$AlEt_3$		[75] [71]
Nickel				
mixed 1- and 2-butene in a variable ratio according to the conversion (prevailingly 2-butene)	$(\pi\text{-allyl Ni halide})_2$ $(\pi\text{-}C_5H_5)_2Ni$ $(\pi\text{-}C_5H_7)(\pi\text{-}C_5H_5)Ni$ $(\pi\text{-allyl Ni halide})_2$	$AlCl_3$, $AlBr_3$ or $AlRCl_2$	PR_3	[28.a] [62] [63] [25, 28], [55]
	$(\pi\text{-allyl Ni halide})_2$ $Ni(acac)_2$	$TiCl_4$ or $VOCl_3$ $Al_2R_{6-x}Cl_x$ $(x = 2,3$ or 4)	PPh_3	[54] [30, 33b, 33c, 44.g]
	$NiX_2\,(PR_3)_2$ NiX_2 hexamethyl-phosphoramide	» »		[46, 47] [50]
	Ni diisopropyl-salicylate	»		[41]

Table 11 (*continued*)

ETHYLENE

Main products	Catalytic systems			References
	Transition metal compounds	Co-catalysts	Additives or solvents	
mixed 1- and 2-butene and high amounts of n-hexene and 3-methylpentene (resulting from particular experimental conditions)	$Ni(CO)_x(PR_3)_{4-x}$ $NiCl_2$ $Ni(PR_3)_2(ethylene)$ $Ni(PCl_3)_4$ $Ni(fluoro-\beta-diketo$ enolate)	$Al_2R_{6-x}Cl_x$ $AlCl_3$ $AlCl_3$ $AlBr_3$ $AlEt_2(OEt)$	PR$_3$-amine BuLi	[44.*i*, 59] [36] [57] [58] [38]
	$NiCl_2$tetramethyl-cyclobutadiene $NiX_2(PR_3)_2$	$Al_2R_{6-x}Cl_x$ ($x = 2,3$ or 4) $Al_2R_{6-x}Cl_x$ ($x = 2,3$ or 4)	PR$_3$	[44.*a,c,e,f,h*]
	Ruthenium			
mixed 1- and 2-butene	$RuCl_3$		alcohol	[118, 119]
	Rhodium			
mixed 1- and 2-butene in a variable ratio	$RhCl_3$ 3 H_2O (and various rhodium complexes) $RhCl_3$ 3 H_2O		alcohol CHCl$_3$	[117, 118, 119 120, 124.*a*, 145] [143, 147]
	Palladium			
95 to 99 % 2-butene and 1 to 5 % 1-butene	$PdCl_2$ » » » $PdCl_2$(ethylene) $PdCl_2$(benzonitrile)$_2$		CHCl$_3$ or CH$_2$Cl$_2$ + EtOH acetic acid+HCl RNO$_2$ sulfones dibutylphtalate	[143, 144, 147] [144, 145] [143, 146, 147] [148] [142, 147] [160]
	Iridium			
95 to 99 % 2-butene and 1 to 5 % 1-butene	$IrCl_3$		alcohol	[118]

Table 12

PROPYLENE

Main products	Catalytic systems			References
	Transition metal compounds	Co-catalysts	Additives or solvents	
	Titanium			
unspecified dimers	Ti(OR)$_4$	AlEt$_3$		[14]
small amounts of 4-methyl 1- and 2-pentenes	Ti(OR)$_4$	AlEt$_3$		[15]
	Cobalt			
2-methyl 1-pentene	HCoN$_2$(PPh$_3$)$_3$			[75]
dimer distribution similar to that obtained with nickel	CoCl$_2$	Al$_2$Et$_{6-x}$Cl$_x$ (x = 2,3 or 4)		[35, 39.a,c,f, 44.h,i, 72, 74]
	Nickel			
n-hexene (20 to 30%) 2-methylpentene (68 to 78%) and 2,3-dimethylbutene (2 to 6%) (the "phosphine effect" is not discussed in these references)	(π-allyl Ni halide)$_2$ (π-allyl Ni halide)$_2$ (π-allyl Ni halide)$_2$	AlX$_3$ TiCl$_4$, MoCl$_5$ VOCl$_4$, WCl$_6$		[28] [39.a,g,e, 55] [39.a,g,e,f, 54]
	Ni(acac)$_2$	Al$_2$R$_{6-x}$Cl$_x$ (x = 2,3 or 4)	(PPh$_3$)	[30, 33a,b,d, 34, 35.b]
	Ni oleate Ni diisopropylsali- cylate or Ni alkyl- benzene sulfonate or Ni salt of mo- noalkyl ester of sulfuric acid	»		[39.a,b,c,d] [41, 43]
	Ni naphtenate	»	(PPh$_3$)	[49]
	NiCl$_2$(PPh$_3$)$_2$	»		[46]
	Ni(CO)$_2$(PPh$_3$)$_2$ or Ni(PPh$_3$)$_4$ or NiCl(PPh$_3$)$_3$	AlCl$_3$ or BF$_3$Et$_2$O		[56]
	Ni(PR$_3$)$_2$(ethylene)	AlCl$_3$		[57]
	NiBr(NO)(PPh$_3$)	AlEtCl$_2$		[53]
n-hexene (1 to 25%) 2-methylpentene (30 to 82%) and 2,3-dimethylbutene (4 to 68%) (the "phosphine effect" on the dimer distribution is exempli- fied in these references)	(π-allyl Ni halide)$_2$	Al$_2$R$_{6-x}$Cl$_x$ (x = 2, 3, 4, 6)	PR$_3$	[25, 28]
	π-allyl Ni halide (PR$_3$)	Al$_2$R$_{6-x}$Cl$_x$ (x = 2, 3, 4, 6)		[25, 28]
	Ni(acac)$_2$	Al$_2$R$_{6-x}$Cl$_x$ (x = 2,3 or 4)	PR$_3$	[33.c, 35.a]
	Ni oleate	»	PR$_3$	[39.e]
	NiCl$_2$tetramethyl- cyclobutadiene	»	PR$_3$	[44.a,b,e,f]
	NiX$_2$(PR$_3$)$_2$	»		[44.g, 45, 47, 48, 49]
	[PR$_4$][NiBr$_3$PR$_3$]	»		[51]

Table 12 (*continued*)

PROPYLENE

Main products	Catalytic systems			References
	Transition metal compounds	Co-catalysts	Additives or solvents	
»	Ni(acrylonitrile)$_2$	Al$_2$R$_{6-x}$Cl$_x$	PR$_3$	[61]
»	NiX$_2$	AlCl$_3$	PR$_3$-amine	[36]
»	NiX(PR$_3$)$_3$ or Ni(PR$_3$)$_4$	AlCl$_3$		[57]
»	Ni(CO)$_2$(PR$_3$)$_2$	AlCl$_3$, BF$_3$, GaBr$_3$, InBr$_3$		[60]
n-hexene (68%) 4-methylpentene (28%)	NiBr$_2$ (hexamethyl-phosphoramide)$_2$	Al$_2$Et$_3$Cl$_3$		[50]
n-hexene (75 to 80%) and 2-methylpentene (20 to 25%)	Ni(β-diketo enolate)	AlEt$_3$ or AlEt$_2$(OEt)		[37]
n-hexene (72 to 80%)	Ni(fluoro-β-diketo enolate)	AlEt$_2$(OEt)		[38]
n-hexene (40 to 51%) 2-methylpentene (45 to 57%) 2,3-dimethylbutene (2 to 4%)	Ni diisopropyl salicylate	AlEt$_2$F		[42]
	Ni oleate	AlEt$_2$F		[39.f]
	Ruthenium			
unspecified dimers	RuCl$_3$	alcohol		[119]
	Rhodium			
mixed dimers	RhCl$_3$ 3 H$_2$O		alcohol	[118, 119]
n-hexene (43%) 2-methylpentene (57%)	RhCl$_3$ 3 H$_2$O		CHCl$_3$, PhNO$_2$	[143, 147]
n-hexene (15%) 2-methylpentene (85%)	RhCl$_3$ 3 H$_2$O		BiPh$_3$ or SbPh$_3$	[123]
	Palladium			
n-hexene (65%) 2-methylpentene (35%)	PdCl$_2$		CHCl$_3$ or CH$_2$Cl$_2$	[143, 147]
n-hexene (90%)	PdCl$_2$		anisole	[143]
mixed hexenes	PdCl$_2$		acetic acid + HCl	[145]
n-hexene (100%)	PdCl$_2$(olefin)			[143]
n-hexene (25%) 2-methylpentene (75%)	PdCl$_2$	AlEtCl$_2$	PPh$_3$, AsPh$_3$ or SbPh$_3$	[123]
unspecified dimers	(π-allyl PdCl)$_2$	AlEtCl$_2$	P(OR)$_3$	[28.a]
	Iridium			
mixed dimers	IrCl$_3$		alcohol	[118]
	Platinum			
n-hexene (40%) 2-methylpentene (60%)	PtCl$_2$	AlEtCl$_2$		[123]

Table 12.1

PROPYLENE-ETHYLENE

Main products	Catalytic systems			References
	Transition metal compounds	Co-catalysts	Additives or solvents	
	Titanium			
2-methyl-1-butene (65 to 72%) and 3-methyl-1-butene (28 to 35%)	Ti(OR)$_4$	AlEt$_3$		[14, 15]
2-methyl-1-butene (95%)	MeTiCl$_3$	MeAlCl$_2$		[21a]
	Nickel			
n-pentene (19 to 63%) and methylbutene (37 to 81%)	NiCl$_2$tetramethyl-cyclobutadiene	Al$_2$R$_{6-x}$Cl$_x$ (x = 2,3 or 4)	PR$_3$	[44a,d,e,f]
unspecified co-dimers	Ni(acac)$_2$	»		[33c]
»	Ni diisopropyl salicylate	»		[41]
»	NiCl$_2$ (PR$_3$)$_2$	»		[47]
n-pentene (10%) and methylbutene (90%)	Ni acetate PR$_3$	»		[44j]
mixed oligomers	Ni(acac)$_2$	AlEt$_2$OEt		[37]
	Rhodium			
mixed co-dimers	RhCl$_3$ 3 H$_2$O		alcohol	[118, 119]
	Palladium			
n-pentene (97%) and methylbutene (3%)	PdCl$_2$(benzonitrile)$_2$		dibutyl phtalate	[160]

Table 13

n-BUTENES

Main products	Catalytic systems			References
	Transition metal compounds	Co-catalysts	Additives or solvents	
	Titanium			
2-ethyl-1-hexene (91%) and n-octenes (9%)	MeTiCl$_3$	MeAlCl$_2$		[21b]
	Nickel			
mixed dimers	Ni oleate	Al$_2$R$_{6-x}$X$_x$ (x = 2,3 or 4)		[39h]
»	Ni(acac)$_2$	»		[30]
»	NiCl$_2$ (PPh$_3$)$_2$	»		[47]
	Palladium			
octenes	PdCl$_2$(benzonitrile)$_2$		dibutyl phtalate	[160]

Table 13.1

n-BUTENE-ETHYLENE

Main products	Catalytic systems			References
	Transition metal compounds	Co-catalysts	Additives or solvents	
unspecified co-dimers	Ti(OR)$_4$	AlR$_3$		[14]
2-ethyl-1-butene (91%) and n-hexene (9%)	MeTiCl$_3$	MeAlCl$_2$		[21a]
3-methylpentene (86 to 90%) and n-hexene (10 to 14%)	NiX$_2$(PR$_3$)$_2$	Al$_2$R$_{6-x}$Cl$_x$		[44g,j]
mixed co-dimers	Ni(acac)$_2$	AlEt$_2$(OEt)		[37]

Table 13.2

n-BUTENE-PROPYLENE

Main products	Catalytic systems			References
	Transition metal compounds	Co-catalysts	Additives or solvents	
unspecified co-dimers	Ni naphtenate	Al$_2$R$_{6-x}$Cl$_x$ (x = 2,3 or 4)	PCl$_2$Ph	[40]
»	NiCl$_2$(PR$_3$)$_2$	Al$_2$R$_{6-x}$Cl$_x$ (x = 2,3 or 4)		[49a]
»	Ni(acac)$_2$	AlEt$_2$(OEt)		[37]
»	PdCl$_2$(benzonitrile)$_2$		dibutyl phtalate	[160]

Table 14

OLEFINS HIGHER THAN BUTENE

Starting olefins and products	Catalytic systems			References
	Transition metal compounds	Co-catalysts	Additives or solvents	
1-pentene, 1-hexene, 1-heptene and 1-octene				
products: branched dimers	MeTiCl₃	MeAlCl₂		[21b]
1-octene and 1-decene				
products: linear dimers	TiCl₄	Al₂Et₃Cl₃	propylene oxide	[22]
n-pentene				
products: unspecified dimers straight chain dimers	NiCl₂(PPh₃)₂ Ni(fluoro-β-diketo enolate)	AlEtCl₂ AlEt₂(OEt)		[47] [38]
unbranched (39 %) and monobranched (59 %) dimers	Ni diisopropyl salicylate	AlEt₂F		[42]

Table 14.1

OLEFINS HIGHER THAN BUTENE-ETHYLENE

Starting olefins and products	Catalytic systems			References
	Transition metal compounds	Co-catalysts	Additives or solvents	
1-pentene, 1-hexene, 1-heptene and 1-octene				
products: 2-ethylated olefins	MeTiCl₃	MeAlCl₂		[21a]
cyclohexene				
product: ethylcyclohexene	(π-allyl NiBr)₂	AlBr₃		[28a]
cyclopentene				
product: ethylcyclopentene	NiCl₂(PPh₃)₂ PdCl₂	AlEtCl₂		[47] [160]
4-vinylcyclohexene				
product: butenylcyclohexene	Ni(acrylonitrile)₂	Al₂R₆₋ₓClₓ		[61]

Table 15.1

BUTADIENE-ETHYLENE

Main products	Catalytic systems			References
	Transition metal compounds	Co-catalysts	Additives or solvents	
	Iron			
1,4-hexadiene (*cis*) > 50%	Fe(acac)₃	AlR₃		[76, 90, 98]
1,4-hexadiene 60%, higher oligom. 40%	Fe(acac)₂ 2 Pyridine	AlEt₃		[86]
1,4-hexadiene > 90%	FeCl₃ or Fe(acac)₃	AlEt₃	diphosphines	[76, 78, 98, 106]
»	FeCl₃ or Fe(acac)₃	AlR₃ or AlR₂Cl	diarsines	[83]
»	FeCl₃	AlEt₂(OEt)		[86]
»	FeCl₃	RMgX		[87]
»	Fe(acac)₃	n BuLi		[88]
»	Fe(acac)₃	Al(i Bu)₃	RNH₂, RCN CH₂=CHCN	[89]
1,5-hexadiene and **1,4**-hexadiene	Fe DPE C₂H₄	AlEt₂Cl and various Lewis acids		[81, 91]
»	Fe(cyclooctatriene)₂			[9, 79]
1,4-hexadiene + **1,3**-hexadiene + 3-methyl-1,4-pentadiene	FeCl₃	AlEt₃	PR₃, P(OR)₃, AsR₃, SbR₃	[77]
»	FeCl₃	AlEt₃	PPh₃	[85]
1,3-hexadiene and C₈ dienes	Fe(DPE)₂C₂H₄			[81, 91]
	Cobalt			
1,4-hexadiene (66%) **2,4**-hexadiene (17%)	CoCl₂	AlEt₃	Et₂P—PEt₂	[106]
1,4-hexadiene (main product)	Co(acac)₃	Al(i Bu)₂Cl	Ph₂PR	[100]
»	CoCl₂ or Co(acac)₃	AlR₂Cl	triphosphines	[99]
»	CoX₂	Al(i Bu)₂Cl, RMgX Al(iBu)₃	diarsines	[102]
1,4-hexadiene > 90%	CoCl₂[DPE]₂	AlEt₃		[92, 93, 98, 106]
»	HCo DPE	AlR₃, AlCl₃, phenols, AlR₃₋ₙClₙ		[96, 101, 104]
»	CoCl₂	AlEt₃	(RO)ₙPCl₃₋ₙ	[97]
»	CoCl₂	AlEt₃	thiophosphines	[103]
»	HCo DPE	LiAlH₄		[104]
»	CoCl₂	RₙAlX₃₋ₙ	P(OR)ₙX₃₋ₙ	[105]
»	CoCl₂ or Co(acac)₃	AlR₃, Al(i Bu)₂Cl	diarsine	[102]

Table 15.1 (*continued*)

BUTADIENE-ETHYLENE

Main products	Catalytic systems			References
	Transition metal compounds	Co-catalysts	Additives or solvents	
	Nickel			
1,4-hexadiene (*cis* + *trans*) (70 to 80%)	NiCl$_2$(PR$_3$)$_2$	AlR$_2$Cl		[29]
3-methyl-1,4-pentadiene (10 to 15%)	Ni(CO)$_2$(PR$_3$)$_2$	AlR$_2$Cl		[29]
1,4-hexadiene (*cis*)	NiX$_2$(hexamethyl phosphoramide)$_2$	Al$_2$R$_{6-x}$X$_x$ (x = 2,3 or 4)		[50]
1,5-cyclodecadiene (31 to 96%) and 1,4,9-decatriene (4 to 69%)	Ni(acac)$_2$ Ni(acac)$_2$ Ni(CO)$_2$L$_2$ Ni(CH$_2$=CH—CN)$_2$	AlR$_2$(OR) AlR$_3$	(PR)$_3$	[26, 67, 68] [65, 66a] [65] [66b]
	Rhodium			
1,4-hexadiene (main product) »	[RhCl(C$_2$H$_4$)$_2$]$_2$ RhCl$_3$ 3H$_2$O (and various Rh complexes)		alcohol + HCl	[125] [117, 122, 124b]
1,4-hexadiene *cis* (22%), *trans* (70 to 75%), isomerized hexadienes (5 to 8%) 3-methylpentadiene (0 to 2%)	RhCl$_3$ 3 H$_2$O		alcohol	[118, 119]
1,4-hexadiene	PdCl$_2$ or Pd(acac)$_2$	AlEt$_2$Cl	P(Ph)$_3$	[150a,b]

Table 15.2

BUTADIENE-PROPYLENE

Main products	Catalytic systems			References
	Transition metal compounds	Co-catalysts	Additives or solvents	
	Iron			
2-methyl-1,4-hexadiene (50%) 1,5-heptadiene (50%)	Fe(acac)$_3$	AlEt$_3$		[80, 90]
	Cobalt			
2-methyl-1,4-hexadiene	CoCl$_2$[DPE]$_2$	Al(C$_6$H$_5$)$_3$		[92, 106]
»	Co(acac)$_3$	AlEt$_3$	diphosphino-ethane	[106]
»	Co(acac)$_3$	Al(i Bu)$_2$Cl	Ph$_2$P-allyl	[100]
»	HCo(DPE)$_2$	AlEt$_2$Cl		[101]
2-methyl-1,4-hexadiene (60%) 2-methyl-1,3-hexadiene (40%)	CoCl$_2$	AlEt$_3$	PCl$_3$	[95]
	Nickel			
2-methyl-1,4-hexadiene	NiCl$_2$(PR$_3$)$_2$	AlR$_2$Cl		[29]
2-butyl-1,4-hexadiene (from 1-hexene)	NiCl$_2$(PR$_3$)$_2$	AlR$_2$Cl		[29]
	Rhodium			
2-methyl-1,4-hexadiene	RhCl$_3$ 3 H$_2$O		ethanol	[118, 119]

Table 16.1

ISOPRENE-ETHYLENE

Main products	Catalytic systems			References
	Transition metal compounds	Co-catalysts	Additives or solvents	
Iron				
4-methyl-1,4-hexadiene (76,5%)	FeCl$_3$	AlEt$_3$	diphosphino ethane	[106]
4-methyl-1,4-hexadiene (60 to 70%)	Fe(acac)$_3$	AlEt$_3$		[76, 80, 98]
4-methyl-1,4-hexadiene	Fe(acac)$_3$	AlEt$_2$Cl	diphosphino ethane	[81]
4-methyl-1,4-hexadiene (60%) 5-methyl-1,4-hexadiene (40%)	Fe(acac)$_3$	AlEt$_2$OEt		[86]
4-methyl-1,4-hexadiene (50%) 5-methyl-1,4-hexadiene (50%)	Fe(acac)$_3$	n BuLi		[88]
4-methyl-1,4-hexadiene (50%)	Fe(acac)$_3$	Al(i Bu)$_3$	CH$_2$=CH—CN	[89]
Cobalt				
4-methyl-1,4-hexadiene	CoCl$_2$	AlEt$_3$	diphosphino ethane	[106]
4-methyl-1,4-hexadiene (80%) »	CoCl$_2$DPE CoCl$_2$	AlEt$_3$ AlEt$_3$	PCl$_3$	[92] [95]
4-methyl-1,4-hexadiene (84%)	CoCl$_2$	AlEt$_3$	(RO)$_n$PCl$_{3-n}$	[97]
4-methyl-1,4-hexadiene (79%) 5-methyl-1,4-hexadiene (21%)	Co(acac)$_3$	Al(i Bu)$_2$Cl	Ph$_2$P-allyl	[100]
4-methyl-1,4-hexadiene (96%)	HCo(DPE)$_2$	AlEt$_2$Cl		[104]
Nickel				
4-methyl-1,4-hexadiene	NiX$_2$(PR$_3$)$_2$	AlR$_2$Cl		[29]
dimethyl-1,4,9-decatriene + dimethyl-1,5-cyclodecadiene	Ni(acac)$_2$ Ni(fumaronitrile)$_2$	AlR$_2$OR		[67] [66.b]
Rhodium				
4-methyl-1,4-hexadiene	RhCl$_3$3 H$_2$O		alcohol	[119]
Palladium				
4-methyl-1,4-hexadiene (55%)	PdCl$_2$ or Pd(acac)$_2$	AlEt$_2$Cl	P(Ph)$_3$	[150.b]

Table 16.2

ISOPRENE-PROPYLENE

Main products	Catalytic systems			References
	Transition metal compounds	Co-catalysts	Additives or solvents	
2,4-dimethyl-1,4-hexadiene	$NiX_2(PR_3)_2$	AlR_2Cl		[29.b]
2,4-dimethyl-1,4-hexadiene (?)	$RhCl_3\ 3\ H_2O$		alcohol	[119]

Table 17

PENTADIENE-ETHYLENE

Main products	Catalytic systems			References
	Transition metal compounds	Co-catalysts	Additives or solvents	
Iron				
3-methyl-1,4-hexadiene (70 to 80%) + 1,4-heptadiene (20 to 30%)	$Fe(acac)_3$	$AlEt_2OEt$		[86, 98]
3-methyl-1,4-hexadiene (70%) 1,4-heptadiene (30%)	$Fe(acac)_0$	$AlEt_0$		[76, 80]
3-methyl-1,4-hexadiene 1,4-heptadiene	$FeCl_3$	$Al(i\ Bu)_3$	Ph-CN	[89]
Cobalt				
3-methyl-1,4-hexadiene (90%) »	$CoCl_2$ $CoCl_2$	$AlEt_3$ $AlEt_3$	PCl_3 $P(OR)_nCl_{3-n}$	[95] [97]
Nickel				
3-methyl-1,4-hexadiene	$NiX_2(PR_3)_2$	AlR_2Cl		[29]
Rhodium				
3-methyl-1,4-hexadiene	$RhCl_3\ 3\ H_2O$		alcohol	[119]

Table 18

CONJUGATED DIENES HIGHER THAN C_5

Starting olefins and products	Catalytic systems			References
	Transition metal compounds	Co-catalysts	Additives or solvents	
Iron				
2,3-dimethylbutadiene + **ethylene** products: 4,5-dimethyl-1,4-hexadiene	Fe(acac)$_3$	AlEt$_2$OEt		[98]
2-methyl-1,3-pentadiene + **ethylene** products: 3,5-dimethyl-1,4-hexadiene 3,5-dimethyl-1,4-hexadiene (75 %) + 4-methyl-1,4-hepta-diene (25 %)	Fe(acac)$_3$	AlEt$_3$		[98]
2,3-dimethyl-1,3-butadiene + **ethylene** products: 4,5-dimethyl-1,4-hexadiene	Fe(acac)$_3$	AlEt$_3$		[76, 80]
methyl-1,3-pentadiene + **ethylene** products: 3,5-dimethyl-1,4-hexadiene 4-methyl-1,4-heptadiene	Fe(acac)$_3$	AlEt$_3$		[80]
2-phenyl-1,3-butadiene + **ethylene** products: 4-phenyl-1,4-hexadiene (96 %)	Fe(acac)$_3$	AlEt$_3$		[80]
Cobalt				
2-methyl-1,3-pentadiene + **ethylene** products: 3,5-dimethyl-1,4-hexadiene (79 %) 4-methyl-1,4-hepta-diene (21 %)	CoCl$_2$	AlEt$_3$	PCl$_3$	[95]
2,3-dimethyl-1,3-butadiene + **ethylene** products: 4,5-dimethyl-1,4-hexadiene	HCo[DPE]$_2$	Al(i Bu)Cl$_2$		[101]

Table 18

CONJUGATED DIENES HIGHER THAN C5

Starting olefins and products	Catalytic systems			References
	Transition metal compounds	Co-catalysts	Additives or solvents	
Nickel				
1,3-cyclohexadiene + ethylene products: 3-vinylcyclohexene	NiX₂(PR₃)₂	AlR₂Cl		[29]
1,3-cyclooctadiene + ethylene products: 3-vinylcyclooctene	NiX₂(PR₃)₂	AlR₂Cl		[29]
2-chlorobutadiene + ethylene products: 4-chloro-1,4-hexadiene	NiX₂(PR₃)₂	AlR₂Cl		[29]
2,4-hexadiene + ethylene products: 3-ethyl-1,4-hexadiene + 3-methyl-1,4-heptadiene	NiX₂(PR₃)₂	AlR₂Cl		[29]
Rhodium				
2-chlorobutadiene + ethylene products: chlorohexadiene	RhCl₃ 3 H₂O		ethanol	[118, 119]
2,4-hexadiene + ethylene products: 3-methyl-1,4-heptadiene	RhCl₃ 3 H₂O		alcohol	[119]

Table 19

STYRENE

Main products	Catalytic systems			Reference
	Transition metal compounds	Co-catalysts	Additives or solvents	
1,3-diphenyl-1-butene	(π-allyl Ni I)₂			[70]

Table 19.1

STYRENE-ETHYLENE

Main products	Catalytic systems			References
	Transition metal compounds	Co-catalysts	Additives or solvents	
3-phenyl-1-butene (32%) and 2-phenyl-2-butene (51%)	Ni(PCl₃)₄	AlBr₃	BuLi	[58]
2-phenyl-2-butene	RhCl₃ 3 H₂O		alcohol	[118, 119]
1-phenyl-1-butene and 2-phenyl-2-butene	PdCl₂			[160]

Table 19.2

STYRENE-BUTADIENE

Main products	Catalytic systems			References
	Transition metal compounds	Co-catalysts	Additives or solvents	
1-phenyl-1,4,8-decatriene	Fe(acac)₂	AlEt₃		[159]
»	Co(acac)₂	AlEt₃		[159]
»	Ni(acac)₂	AlEt₃		[159]

Table 20

ACRYLATES

Main products	Catalytic systems			References
	Transition metal compounds	Co-catalysts	Additives or solvents	
diethyl adipate (from ethyl acrylate)	Fe(CO)₅ or Fe(CO)₄Pyridine	(non catalytic reaction)		[100]
dimethyl α-dihydromuconate (from methyl acrylate)	RuCl₃ RhCl₃ 3 H₂O		alcohol alcohol	[118, 119] [119]

Table 20.1

ACRYLATE-ETHYLENE

Main products	Catalytic systems			References
	Transition metal compounds	Co-catalysts	Additives or solvents	
methyl 3-pentenoate	RuCl$_3$		alcohol	[118, 119]
»	RhCl$_3$ 3 H$_2$O		alcohol	[119]
methyl 3-pentenoate and methyl 2-pentenoate	PdCl$_2$(benzonitrile)$_2$			[160]

Table 20.2

ACRYLATE-BUTADIENE

Main products	Catalytic systems			References
	Transition metal compounds	Co-catalysts	Additives or solvents	
ethyl 4,6-heptadienoate (from ethyl acrylate)	Co$_2$(CO)$_8$			[159]
methyl 2,5,10-undecatrienoate (from methyl acrylate)	Ni(acac)$_2$	AlEt$_2$(OEt)	PPh$_3$	[69]
co-dimers from various dienes	Fe(acac)$_2$	AlEt$_3$		[116]
»	Co(acac)$_2$	AlEt$_3$		[159]

Table 20.3

ACRYLATE-STYRENE

Main products	Catalytic systems			Reference
	Transition metal compounds	Co-catalysts	Additives or solvents	
methyl 5-phenyl-4-pentenoate	PdCl$_2$(benzonitrile)$_2$			[160]

Table 21

ACRYLONITRILE

Main products	Catalytic systems			References
	Transition metal compounds	Co-catalysts	Additives or solvents	
Iron				
adiponitrile (non catalytic reaction)	Fe(CO)$_5$ and other carbonyl compounds		bases	[108, 109.*a,c*, 111]
»	[HFe$_3$(CO)$_{11}$] [NEt$_3$H]		methanol	[108]
Cobalt				
adiponitrile (non catalytic reaction)	CoCl$_2$	Mn	dimethyl formamide	[113, 114]
»	Co$_2$(CO)$_8$+ CH$_2$=CH—CN		H$_2$	[109.*b*]
2-methylglutaronitrile (non catalytic reaction)	HCo(CO)$_4$		H$_2$O + CO	[112]
Ruthenium				
1,4-dicyano-1-butene and adiponitrile (in variable proportions)	RuCl$_3$ 3 H$_2$O (and other ruthenium complexes)		H$_2$	[126, 128, 130, 131, 132, 134, 135.*a*]
	RuCl$_3$ (and other ruthenium complexes) on carbon		H$_2$	[135.*b*]
1,4-dicyano-1,3-butadiene	RuCl$_3$ (and other ruthenium complexes)		various organic bases	[129]

Table 21.1

ACRYLONITRILE-ACRYLATE

Main products	Catalytic systems			Reference
	Transition metal compounds	Co-catalysts	Additives or solvents	
methyl 5-cyano-4-pentenoate methyl *ω*-cyanovalerate	RhCl$_3$ 3 H$_2$O and various rhodium complexes		H$_2$	[126.*c*]

Table 22

MISCELLANEOUS

Starting olefins and products	Catalytic systems			References
	Transition metal compounds	Co-catalysts	Additives or solvents	
Isobutene				
products: 2,5-dimethyl-1- and 2-hexene	PdCl$_2$		PhNO$_2$	[147]
Acrylamides				
products: trans-α-dihydromuconamides	RhCl$_3$ 3 H$_2$O		alcohol	[121]
Allyl acetate-butadiene				
products: CH$_3$COOC$_{11}$H$_{17}$	RuCl$_3$		alcohol	[118]
Vinyl chloride-ethylene				
products: chlorobutenes	RhCl$_3$ 3 H$_2$O			[118]
α-methyl styrene-ethylene				
products: 3-phenyl-3-methyl-1-butene	Ni(PCl$_3$)$_4$	AlBr$_3$	BuLi	[58]
Various vinyl aromatics-butadiene	Ni(acac)$_2$	AlEt$_3$		[159]
Methyl crotonate				
products: dimethyl 3-pentene-1,3-dicarboxylate	Cu$_2$O		cyclohexyl-isocyanide	[152]
3-pentene-2-one				
products: 4-methyl-5-acetyl-5-hepten-2-one	Cu$_2$O		cyclohexyl-isocyanide	[152]

4. REFERENCES

[1] V. SH. FEL'DBLYUM and N. V. OBESHCHALOVA, "Russ. Chem. Reviews", *37*, 789, (1968).

[2] C. W. BIRD, *Transition metals intermediates in Organic Synthesis*, (1967), Logos Press, (London).

[3] G. WILKE and al.,
(*a*) "Angew. Chem. Int. Ed.", *2*, 105, (1963).
(*b*) "Angew. Chem. Int. Ed.", *5*, 151, (1966).

[4] H. MÜLLER, D. WITTENBERG, H. SEIBT and E. SCHARF, "Angew. Chem. Int. Ed.", *4*, 327, (1965).

[5] P. HEIMBACH, *Proc. I.U.P.A.C. Symposium Eastbourne*, (1968).

[6] D. J. CRAM, *Fundamentals of Carbanion Chemistry* (in "Organic Chemistry" Vol. 4), (1965), Academic Press, (New York).

[7] J. K. HAMBLING, "Chem. in Britain", *5*, 354, (1969).

[8] J. P. KENNEDY and R. M. THOMAS, "Makromol. Chem.", *64*, 1, (1963).

[9] A. CARBONARO, A. GRECO and G. DALL'ASTA, "J. Organometal. Chem.", *20*, 177, (1969).

[10] M. L. H. GREEN, *Organometallic Compounds*, Vol. 2, (1968), Methuen, (London).

[11] H. MARTIN, "Angew. Chem.", *68*, 306, (1956).

[12] Brit. Pat. to K. ZIEGLER, 787, 435, (1957).

[13] G. NATTA, "J. Polym. Sci.", *34*, 151, (1959).

[14] (*a*) Ital. Pat. to MONTECATINI, 586, 452, (1957).
(*b*) M. FARINA and M. RAGAZZINI, "Chimica e Industria" (Milan), *40*, 816, (1958).

[15] Ger. Offen. to TOYO SODA MANUFG. Co., 1, 803, 434, (1967).

[16] Belg. Pat. to CHEMISCHE WERKE HUELS, 634, 232, (1962), "Chem. Abst.", *61*, 4211.

[17] S. CESCA, W. MARCONI and M. L. SANTOSTASI, "J. Polym. Sci. B", *7*, 547, (1969).

[18] K. SHIKATA, Y. MIURA, S. NAKAO and K. AZUMA, "Kogyo Kagaku Zasshi", *68*, 2266, (1965), "Chem. Abst.", *64*, 15, 721.

[19] Jap. Pat. to TOKUYAMA SODA Co., 4, 961/69, (1965).

[20] Jap. Pat., to TOKUYAMA SODA Co., 4, 962/69, (1965).

[21] (*a*) H. BESTIAN, K. CLAUSS, H. JENSEN and E. PRINZ, "Angew. Chem. Int. Ed.", *2*, 32, (1963).
(*b*) H. BESTIAN and K. CLAUSS, "Angew. Chem. Int. Ed.", *2*, 704, (1963).

[22] (*a*) D. H. ANTONSEN, P. S. HOFFMAN and R. S. STEARNS, "Ind. Eng. Chem. Prod. Res. Develop.", *2*, 224, (1963).
(*b*) D. H. ANTONSEN, R. W. WARREN and R. H. JOHNSON, "Ind. Eng. Chem. Prod. Res. Develop.", *3*, 311, (1964).

[23] A. JOB and R. REICH, "Compt. Rend.", *177*, 330, (1924).

[24] K. ZIEGLER, H. G. GELLERT, E. HOLZKAMP, G. WILKE, E. W. DUCK and W. R. KROLL, "Annalen", *629*, 172, (1960).

[25] G. WILKE, "Proc. R. A. WELCH Foundation", *Conference on Chem. Res. IX. Organometal. Compounds*, (1966).

[26] P. HEIMBACH and G. WILKE, "Annalen", *727*, 183, (1969).

[27] L. PORRI, G. NATTA and M. C. GALLAZZI, "J. Polym. Sci. C", *16*, 2525, (1967).

[28] (a) Brit. Pat. to STUDIENGESELLSCHAFT KOHLE, 1,058,680, (1963).
(b) B. BOGDANOVIC and G. WILKE, *VIIth World Petroleum Congress*, (1967), Mexico.
(c) B. BOGDANOVIC and G. WILKE, "Brennstoff Chem.", *49*, 323, (1968).
(d) U. BIRKENSTOCK, H. BONNEMANN, B. BOGDANOVIC, D. WALTER and G. WILKE, "Adv. Chem. Ser.", *70*, 250, (1968).

[29] (a) R. G. MILLER, T. J. KEALY and A. L. BARNEY, "J. Amer. Chem. Soc.", *89*, 3756, (1967).
(b) French Pat. to E. I. DU PONT DE NEMOURS and Co., 1,388,305, (1963).

[30] Y. CHAUVIN, N. H. PHUNG, N. GUICHARD and G. LEFEBVRE, "Bull. Soc. Chim. France", 3223, (1966).

[31] L. FARADY, L. BENCZE and L. MARKO, "J. Organometal. Chem.", *17*, 107, (1969).

[32] (a) R. G. MILLER, "J. Amer. Chem. Soc.", *89*, 2785, (1967).
(b) R. G. MILLER and P. A. PINKE, "J. Amer. Chem. Soc.", *90*, 4500, (1968).

[33] (a) J. EWERS, "Angew. Chem. Int. Ed.", *5*, 584, (1966).
(b) French Pat. to SHOLVEN-CHEMIE AKTIENGESELLSCHAFT, 1,497,673, (1965).
(c) J. EWERS, "Erdòl Kohle Erdgas. Petrochemie", 763, (1968).
(d) Neth. Appl. to SHOLVEN-CHEMIE AKTIENGESELLSCHAFT, 68,12618, (1967).

[34] Brit. Pat. to THE BRITISH PETROLEUM Co., 1,106,734, (1966).

[35] (a) French Pat. to ESSO RESEARCH and ENGINEERING Co., 1,533,588, (1966).
(b) French Pat. to ESSO RESEARCH and ENGINEERING Co., 1,560,865, (1967).

[36] Neth. Appl. to FARBWERKE HOECHST, 68,09282, (1967).

[37] Brit. Pat. to THE BRITISH PETROLEUM Co., 1,101,498, (1965).
Brit. Pat. to THE BRITISH PETROLEUM Co., 1,123,474, (1966).
Neth. Appl. to THE BRITISH PETROLEUM Co., 68,00291, (1967).
Neth. Appl. to THE BRITISH PETROLEUM Co., 68,00288, (1967).
Neth. Appl. to THE BRITISH PETROLEUM Co., 68,05854, (1967).
Neth. Appl. to THE BRITISH PETROLEUM Co., 68,18508, (1967).

[38] U.S. Pat. to SHELL OIL Co., 3,424,815, (1967).

[39] (a) V. SH. FEL'DBLYUM and N. V. OBESHCHALOVA, "Dokl. Akad. Nauk SSSR", *172*, 368, (1967).
(b) V. SH. FEL'DBLYUM, N. V. OBESHCHALOVA and A. I. LESHCHEVA, "Dokl. Akad. Nauk SSSR", *172*, 111, (1967).
(c) French Pat. to NAUCHNO-ISSLEDOVATELSKY INST. MONOMEROV, 1,420,952, (1965).
(d) V. SH. FEL'DBLYUM, N. V. OBESHCHALOVA, A. I. LESHCHEVA and T. I. BARANOVA, "Neftekhimiya", *7*, 379, (1967).
(e) V. SH. FEL'DBLYUM, *IVth Internat. Congress on Catalysis Moscow*, (1968).
(f) Neth. Appl. to NAUCHNO-ISSLEDOVATELSKY INST. MONOMEROV, 67,05681, (1967).
(g) N. V. OBESHCHALOVA, V. SH. FEL'DBLYUM and N. M. PASHCHENKO, "Zh. Org. Khim.", *4*, 1014, (1968).

(*h*) V. SH. FEL'DBLYUM, A. I. LESHCHEVA, N. V. OBESHCHALOVA, P. P. YABLONSKII and N. M. PASHCHENKO, "Neftekhimiya", *8*, 533, (1968).

[40] Neth. Appl. to IMPERIAL CHEMICAL INDUSTRIES, 68,12220, (1967).

[41] French Pat. to SHELL INTERNATIONAL RESEARCH MAATSCHAPPIJ, 1,385,503, (1962).

[42] U.S. Pat. to SHELL OIL Co., 3,355,510, (1965).

[43] U.S. Pat. to SHELL OIL Co., 3,327,015, (1963).

[44] O. T. ONSAGER, H. WANG and U. BLINDHEIM
  (*a*) "Helv. Chim. Acta", *52*, 187, (1969).
  (*b*) "Helv. Chim. Acta", *52*, 196, (1969).
  (*c*) "Helv. Chim. Acta", *52*, 215, (1969).
  (*d*) "Helv. Chim. Acta", *52*, 224, (1969).
  (*e*) "Helv. Chim. Acta", *52*, 230, (1969).
  (*f*) Brit. Pat. to SENTRALINSTITUTT FOR INDUSTRIELL FORSKNING, 1,124,123, (1965).
  (*g*) French Pat. to SENTRALINSTITUTT FOR INDUSTRIELL FORSKNING, 1,519,181, (1966).
  (*h*) French Pat. to SENTRALINSTITUTT FOR INDUSTRIELL FORSKNING, 1,535,201, (1966).
  (*i*) French Pat. to SENTRALINSTITUTT FOR INDUSTRIELL FORSKNING, 1,537,550, (1966).
  (*j*) Neth. Appl. to SENTRALINSTITUTT FOR INDUSTRIELL FORSKNING, 68,10033, (1967).

[45] (*a*) M. UCHINO, *Thesis*, (1968), Paris.
  (*b*) M. UCHINO, Y. CHAUVIN and G. LEFEBVRE, "Compt. Rend.", *265*, 103, (1967).

[46] French Pat. to GELSENBERG BENZIN AKTIENGESELLSCHAFT, 1,547,921, (1966).

[47] Belgian Pat. to PHILLIPS PETROLEUM Co., 707,477, (1966).

[48] French Pat. to SUN OIL Co., 1,528,160, (1966).

[49] (*a*) Brit. Pat. to IMPERIAL CHEMICAL INDUSTRIES, 1,131,146, (1966).
  (*b*) Brit. Pat. to IMPERIAL CHEMICAL INDUSTRIES, 1,140,821, (1967).

[50] Neth. Appl. to FARBWERKE HOECHST, 67,06533, (1966).

[51] (*a*) U.S. Pat. to SUN OIL Co., 3,459,825, (1967).
  (*b*) U.S. Pat. to SUN OIL Co., 3,472,911, (1968).

[52] G. DESGRANDCHAMPS, H. HEMMER and M. HAURIE, Private communication.

[53] U.S. Pat. to PHILLIPS PETROLEUM Co., 3,427,365, (1966).

[54] Brit. Pat. to FARBWERKE HOECHST, 1,146,190, (1965).

[55] T. ARAKAWA and K. SAEKI, "Kogyo Kagaku Zasshi", 71, 1028, (1968,) "Chem. Abst.", *69*, 110, 199.

[56] G. HATA and A. MIYAKE, "Chem. and Ind." (London), 921, (1967).

[57] Neth. Appl. to FARBWERKE HOECHST, 68,10848, (1967).

[58] Neth. Appl. to the INTERNATIONAL SYNTHETIC RUBBER Co., 68,13476, (1967).

[59] U.S. Pat. to PHILLIPS PETROLEUM Co., 2,969,408, (1955).

[60] M. BORN, Y. CHAUVIN, G. LEFEBVRE and N. H. PHUNG, "Compt. Rend.", *268*, 1600, (1969).

198

[61] Brit. Pat. to THE B. F. GOODRICH Co., 1,138,575, (1966).

[62] M. TSUTSUI and T. KOYANO, "J. Polym. Sci. A 1", 5, 681, (1967).

[63] U.S. Pat. to SHELL OIL Co., 3,424,816, (1967).

[64] (a) M. BORN, Y. CHAUVIN, J. GAILLARD and G. LEFEBVRE, Unpublished results.
(b) G. LEFEBVRE and Y. CHAUVIN, VIIth World Petroleum Congress, (1967), Mexico.
(c) M. BORN, Y. CHAUVIN, N. H. PHUNG and G. LEFEBVRE, Unpublished results.

[65] U.S. Pat. to COLUMBIAN CARBON Co., 3,420,904, (1966).

[66] (a) French Pat. to THE B. F. GOODRICH Co., 1,494,631, (1965).
(b) French Pat. to THE B. F. GOODRICH Co., 1,494,632, (1965).

[67] French Pat. to STUDIENGESELLSCHAFT KOHLE, 1,351,938, (1962).

[68] French Pat. to CHEMISCHE WERKE HUELS, 1,424,314, (1964).

[69] French Pat. to SOCIÉTÉ DES USINE CHIMIQUES RHÔNE-POULENC, 1,433,409, (1966).

[70] L. I. REDKINA, K. L. MARCOVECKII, E. I. TINYAKOVA and B. A. DOLGOPLOSK "Dokl. Akad. Nauk. SSSR", 186, 397, (1969).

[71] G. HATA, "Chem. and Ind." (London), 223, (1965).

[72] Brit. Pat. to THE BRITISH PETROLEUM, 1,101,657, (1965).

[73] U.S. Pat. to PHILLIPS PETROLEUM Co., 1,129,463.

[74] Neth. Appl. to THE BRITISH PETROLEUM Co., 66,18467, (1966).

[75] LYONG SUN PU, A. YAMAMOTO, S. IKEDA, "J. Amer. Chem. Soc.", 90, 7170, (1968).

[76] G. HATA, "J. Amer. Chem. Soc.", 86, 3903, (1964).

[77] M. IWAMOTO and S. YUGUCHI, "Bull. Soc. Chem. Japan", 39, 2001, (1966).

[78] M. IWAMOTO and. S. YUGUCHI, "J. Org. Chem.", 31, 4290, (1966).

[79] A. CARBONARO, A. GRECO and G. DALL'ASTA, "Tetrahedron Lett.", 22, 2037, (1967).

[80] G. HATA and DAN AOKI, "J. Org. Chem.", 32, 3754, (1967).

[81] G. HATA and A. MIYAKE, "Bull. Soc. Chem. Japan", 41, 2762, (1968).

[82] G. HATA, H. KONDO and A. MIYAKE, "J. Amer. Chem. Soc.", 90, 2278, (1968).

[83] U.S. Pat. to E. I. DU PONT DE NEMOURS AND Co., 3,407,245, (1967).

[84] U.S. Pat. to TOYO RAYON, 3,408,418, (1965).

[85] French Pat. to TOYO RAYON, 1,446,790, (1964).

[86] French Pat. to TOYO RAYON, 1,417,455, (1963).

[87] French Pat. to MONTECATINI-EDISON, 1,544,365, (1966).

[88] French Pat. to MONTECATINI-EDISON, 1,561,984, (1966).

[89] U.S. Pat. to THE B. F. GOODRICH Co., 3,441,627, (1965).

[90] Ital. Pat. to MONTECATINI-EDISON, 724,306, (1964).

[91] French Pat. to TOYO RAYON, 1,548,453, (1966).

[92] M. IWAMOTO, K. TANI, H. IGADI and S. YUGUCHI, "J. Org. Chem.", 32, 4148, (1967).

[93] M. IWAMOTO and S. YUGUCHI, "Kogyo Kagaku Zasshi", 70, 1505-8, (1967).

[94] T. SAITO, Y. UCHIDA and A. MISONO, "Bull. Chem. Soc. Japan", 38, 1397, (1965).

[94'] D. WITTENBERG, "Angew. Chem. Int. Ed.", *3*, 153, (1964).

[95] Y. TAJIMA and E. KUNIOKA, "Chem. Comm.", 603, (1968).

[96] M. IWAMOTO and S. YUGUCHI, "Chem. Comm.", 28, (1968).

[97] G. HATA and A. MIYAKE, "Bull. Chem. Soc. Japan", *41*, 2443, (1968).

[98] A. MIYAKE, G. HATA, M. IWAMOTO and S. YUGUCHI, *VIIth World Petroleum Congress*, (1967), Mexico.

[99] U.S. Pat. to E. I. DU PONT DE NEMOURS AND CO., 3,445,540, (1968).

[100] French Pat. to E. I. DU PONT DE NEMOURS AND CO., 1,561,485, (1967).

[101] French Pat. to E. I. DU PONT DE NEMOURS AND CO., 1,536,670, (1966).

[102] U.S. Pat. to E. I. DU PONT DE NEMOURS AND CO., 3,407,244, (1967).

[103] French Pat. to TOYO RAYON, 1,538,950, (1966).

[104] French Pat. to TOYO RAYON, 1,487,354, (1965).

[105] French Pat. to TOYO RAYON, 1,481,339, (1965).

[106] French Pat. to TOYO RAYON, 1,462,308, (1965).

[107] A. MISONO, Y. UCHIDA, T. SAITO and K. UCHIDA, "Bull. Chem. Soc. Japan", *40*, 1889, (1967).

[108] A. MISONO, Y. UCHIDA, K. TAMAI and M. HIDAI, "Bull. Chem. Soc. Japan", *40*, 931, (1967).

[109] (*a*) French Pat. to SOCIÉTÉ DES USINES CHIMIQUES RHÔNE-POULENC, 1,377,425, (1963),
French addition 85,320, (1964).
French addition 85,717, (1964).
French addition 87,045, (1964).
French addition 86,341, (1964).
(*b*) French Pat. to SOCIÉTÉ DES USINES CHIMIQUES RHÔNE-POUELNC, 1,381,511, (1963).
(*c*) French Pat. to SOCIÉTÉ DES USINES CHIMIQUES RHÔNE-POULENC, 1,453,988, (1965).

[110] French Pat. to TOYO RAYON, 1,542,199, (1966).

[111] French Pat. to TOYO RAYON, 1,524,921, (1967).

[112] U.S. Pat. to E. I. DU PONT DE NEMOURS AND CO., 3,206,498, (1963).

[113] G. AGNES, G. P. CHIUSOLI and G. COMETTI, "Chem. Comm.", 1515, (1968).

[114] G. AGNES, G. P. CHIUSOLI and G. COMETTI, *IVth Int. Conference on Organometallic Chemistry, Bristol*, (1969).

[115] J. KWIATEK, I. L. MADOR and J. K. SEYLER, "Advan. Chem. Ser.", *37*, 201, (1963).

[116] A. MISONO, Y. UCHIDA, T. SAITO and K. UCHIDA, "Bull. Chem. Soc. Japan", *40*, 1889, (1967).

[117] R. CRAMER, "Accounts Chem. Res.", *1*, 186, (1968).

[118] U.S. Pat. to E. I. DU PONT DE NEMOURS AND CO., 3,013,066, (1961).

[119] T. ALDERSON, E. L. JENNER and R. V. LINDSEY, "J. Amer. Chem. Soc.", *87*, 5638, (1965).

[120] French Pat. to IMPERIAL CHEMICAL INDUSTRIES Ltd., 1,521,991, (1966).

[121] Y. Kobayashi and S. Taira, "Tetrahedron", *24*, 5763, (1968).

[122] French Pat. to E. I. Du Pont de Nemours and Co. 1,319,578, (1961),

[123] N. H. Phung and G. Lefebvre, "Compt. Rend.", *265*, 519, (1967).

[124] (a) R. Cramer, "J. Amer. Chem. Soc.", *87*, 4717, (1965).
(b) R. Cramer, "J. Amer. Chem. Soc.", *89*, 1633, (1967).
(c) R. Cramer, "J. Amer. Chem. Soc.", *88*, 2272, (1966).
(d) R. Cramer, "J. Amer. Chem. Soc.", *89*, 4621, (1967).

[125] Jap. Pat. to Toyo rayon, "Chem. Abst.", *70*, 77283, 68,28447.

[126] (a) French Pat. to Société des Usines Chimiques Rhône-Poulenc, 1,472,033, (1965).
(b) French Pat. to Société des Usines Chimiques Rhône-Poulenc, 1,541,443, (1965).
(c) French Pat. to Société des Usines Chimiques Rhône-Poulenc, 1,493,068, (1966).
(d) French Pat. to Société des Usines Chimiques Rhône-Poulenc, 1,546,530, (1967).
(e) French Pat. to Société des Usines Chimiques Rhône-Poulenc, 1,505,334, (1965).

[127] M. Figeys and H. P. Figeys, "Tetrahedron", *24*, 1097, (1968).

[128] (a) A. Misono, Y. Uchida, M. Hidai, H. Shinohara and Y. Watanabe, "Bull. Chem. Soc. Japan", *41*, 396, (1968).
(b) A. Misono, Y. Uchida, M. Hidai and Y. Watanabe, "Chem. Comm.", 704, (1968).
(c) A. Misono, Y. Uchida, M. Hidai, I. Inomata, Y. Watanabe and M. Takeda, "Kogyo Kagaku Zasshi", *72*, 1801, (1969).
(d) A.Misono, Y.Uchida, M. Hidai and H. Kanai, "Chem.Comm.", 354, (1967).

[129] Neth. Appl. to Halcon International, 68,03864, (1967).

[130] French Pat. to Mitsubishi Petrochemical Co., 1,560,994, (1967).

[131] E. Billig, C. B. Strow and R. L. Pruett, "Chem. Comm.", 1307, (1968).

[132] J. D. McClure, R. Owyang and L. H. Slaugh, "J. Organometal. Chem.", *12*, p 8, (1968).

[133] P. S. Hallman, B. R. McGarvey and G. Wilkinson, "J. Chem. Soc. A", 3143, (1968).

[134] French Pat. to Imperial Chemical Industries Ltd., 1,519,113, (1966).

[135] (a) French Pat. to E. I. Du Pont de Nemours and Co., 1,520,883, (1966).
(b) French Pat. to E. I. Du Pont de Nemours and Co., 1,572,892, (1967).

[136] (a) N. R. Davies, "Nature", *201*, 490, (1964).
(b) N. R. Davies, "Australian J. Chem.", *17*, 212, (1964).

[137] (a) J. F. Harrod and A. J. Chalk, "J. Amer. Chem. Soc.", *86*, 1776, (1964).
(b) J. F. Harrod and A. J. Chalk, "J. Amer. Chem. Soc.", *88*, 3491, (1966).

[138] G. C. Bond and M. Hellier, "J. Catalysis", *4*, 1, (1965).

[139] M. B. Sparke, L. Turner and A. J. M. Wenham, "J. Catalysis", *4*, 332, (1965).

[140] (a) I. I. Moiseev, "Bull. Acad. Sc. U.S.S.R. Div. Chem. Sci.", n° *9*, 1690, (1965).
(b) I. I. Moiseev, "Bull. Acad. Sc. U.S.S.R. Div. Chem. Sci.", n° *10*, 1609, (1966).
(c) I. I. Moiseev, "Zh. Org. Khim.", *4*, 354, (1968).

[141] F. CONTI, G. PREGAGLIA and R. UGO, *Am. Chem. Soc. Minneapolis Meeting, Petr. Div. Abstr.*, Paper B 39, (1969).

[142] J. T. VAN GEMERT and P. R. WILKINSON, "J. Phys. Chem.", *68*, 645, (1964).

[143] A. D. KETLEY, L. P. FISHER, A. J. BERLIN, C. R. MORGAN, E. H. GORMAN and T. R. STEADMAN, "Inorg. Chem.", *6*, 657, (1967).

[144] Y. KUSUNOKI, R. KATSUNO, N. HASEGAWA, S. KUREMATSU, Y. NAGAO, K. ISHII and J. TSUTSUMI, "Bull. Chem. Soc. Japan", *39*, 2021, (1966).

[145] Ger. Pat. to CONSORTIUM FÜR ELEKTROCHEMISCHE INDUSTRIE, 1,193,934, (1957).

[146] U.S. Pat. to SHELL OIL Co., 3,361,840, (1963).

[147] French Pat. to W. R. GRACE AND Co., 1,499,833, (1965).

[148] U.S. Pat. to SHELL OIL Co., 3,354,236, (1967).

[149] I. I. MOISEEV, A. P. BELOV and G. YU. PEK, "Russ. J. Inorg. Chem.", *10*, 180, 1965.

[150] (a) W. SCHNEIDER, *Am. Chem. Soc. Minneapolis Meeting*, *14*, n° 2 B, 89, (1969).
(b) Brit. Pat. to THE B. F. GOODRICH Co., 1,153,336, (1965).

[151] P. E. SLADE and H. B. JONASSEN, "J. Amer. Chem. Soc.", *79*, 1277, (1957).

[152] T. SAEGUSA, Y. ITO, S. KOBAYASHI and S. TOMITA, "Chem. Comm.", 273, (1968).

[153] French Pat. to IMPERIAL CHEMICAL INDUSTRIES, 1,519,114, (1966).

[154] (a) R. G. SCHULTZ, J. M. SCHUCK and B. S. WILDI, "J. Catalysis", *6*, 385, (1966).
(b) R. G. SCHULTZ, R. M. ENGELBRECHT, R. N. MORRE and L. T. WOLFORD, "J. Catalysis", *6*, 419, (1966).
(c) R. G. SCHULTZ, "J. Catalysis", *7*, 286, (1967).

[155] Neth. Appl. to THE BRITISH PETROLEUM Co., 66,09513, (1965).

[156] H. IMAI, T. HASEGAWA and H. UCHIDA, "Bull. Chem. Soc. Japan", *41*, 45, (1968).

[157] Neth. Appl. to SHELL INTERNATIONAL RESEARCH MAATSCHAPPIJ, 68,13667, (1967).

[158] U.S. Pat. to SHELL OIL Co., 3,459,826, (1968).

[159] French Pat. to BADISCHE ANILIN UND SODA-FABRIK, 1,337,558.

[160] M. G. BARLOW, M. J. BRYANT, R. N. HASZELDINE and A. G. MACKIE, "J. Organometal. Chem.", *21*, 215, (1970).

[161] B. BOGDANOVIC, personal communication.

Chapter 4

Selective Homogeneous Hydrogenation of Dienes and Polyenes to Monoenes

A. ANDREETTA, F. CONTI and G. F. FERRARI

Centro Ricerche Montecatini-Edison, Bollate (Milan) Italy

1. INTRODUCTION

In the last ten years there has been a rapid increase in the number of papers concerning the homogeneous catalytic hydrogenation of polyenes to monoenes; this increase has been parallel to the extensive development of transition metal complexes as catalysts for a large number of homogeneous reactions. Studies in this field have originally been devoted to the production of edible fatty esters (containing one double bond in *cis*-configuration) from the corresponding natural polyenes.

Besides the great efforts in this direction, much work of theoretical interest has been done which has resulted in substantial progress in the knowledge of reaction mechanisms. In our opinion some of these studies will be of great importance in the near future for a satisfactory correlation between homogeneous and heterogeneous hydrogenation of polyenes to monoenes.

Iron, cobalt and nickel carbonyls were the first complexes used in the selective homogeneous hydrogenation, the substrates being polyunsaturated natural products such as soybean and linseed oils; subsequently, great progress has been made in this field, with the discovery of the pentacyanocobaltate catalytic system and with studies of its reaction mechanism.

Since 1964 a great number of papers has appeared, thus emphasizing the ever growing interest in the field. Many reviews concerning the activation of hydrogen and the catalytic homogeneous hydrogenation of alkenes have been published; one of these is also given in this series [1]. Moreover some reviews including also the homogeneous hydrogenation of polyenes have appeared [2-6].

The aim of this review is to describe the existing catalytic systems and to emphasize their outstanding properties; in this context some space has been given to the description of the available mechanism of the reactions. Moreover, tables summarising the most significant results have been included to point out the best performances of every catalytic system. It is hoped that people concerned with organic syntheses will take advantage of the contents of this review whenever the appropriate catalyst is to be chosen; for this reason a general description of factors governing the selectivity in the hydrogenation of polyenes to monoenes has been included.

2. GENERAL CONSIDERATIONS

2.1. Selectivity

In some favourable cases selective hydrogenation of diolefins gives monoenes in yields as high as 100%; for example, this is the case when π-Arene-Cr(CO)$_3$, Co(CN)$_5$H^{3-}, Cp$_2$VCl$_2$—LiC$_4$H$_9$ or LiAlH$_4$ are employed as catalysts. In these instances the exceptionally good results are ascribed to the complete inertness of the catalytic systems with respect to the hydrogenation of the resulting monoenes.

However, in general the hydrogenation of dienes to monoenes takes place with simultaneous formation of the corresponding saturated compounds.

Although the direct path from dienes to saturated products has been described as a side reaction [5], as a rule the reduction proceeds stepwise according to the following scheme:

$$\text{diene } (d) \xrightarrow{\ k'\ } \text{monoene } (m) \xrightarrow{\ k''\ } \text{saturated } (s) \qquad (1)$$

Generally speaking the "selectivity" in the case of these consecutive reactions indicates the extent to which intermediate monoenes, as opposed to saturated compounds (*), are formed:

$$S_m = \frac{\text{monoenes} \times 100}{\text{monoenes} + \text{saturated}}$$

_____

(*) The formation of by-products different from saturated compounds (such as polymers etc.) has been neglected.

In the hypothetical case of a non-catalyzed sequence of type (1), with $k' \simeq k''$ (case a of fig. 1), selectivity depends only upon the concentrations of diene and monoene in the course of the reaction and decreases with conversion. In this case a "true selective reaction" does not take place; conversely this event

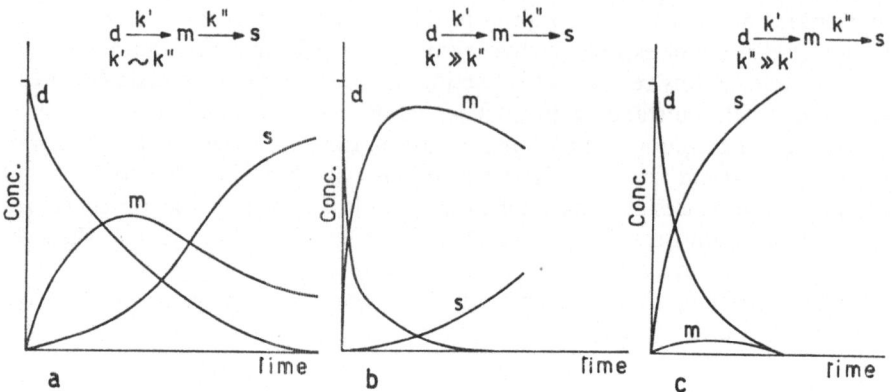

Figure 1. Typical concentration curves for the consecutive reactions type (1).

occurs whenever other factors (e.g. kinetic factors as in case b of fig. 1) contribute to keep the monoene concentration at a high level in a large range of conversions.

In order to determine whether a reaction is selective or not, let us suppose that an experiment is carried out, whereby a diene and an intermediate monoene, initially in equal concentrations, are reduced; selective hydrogenation takes place whenever a ratio higher than unity is obtained between the initial rates of reduction of the diene and monoene.

While in the hypothetical case of a non-catalyzed sequence of reactions only kinetic factors must be considered (case b of fig. 1), in the presence of the catalyst, contribution of both thermodynamic and kinetic factors to the selectivity must be taken into account.

This effect is derived from the existence of complex intermediates of different stabilities, containing diene or monoene moieties.

Many studies have been made to identify the intermediate species and to clarify the mechanisms of the reactions. Significant results have been obtained, concerning the nature of the reaction intermediates: in fact hydride-, σ-alkyl-, σ-and π-alkenyl-, π-mono-and di-olefin-complexes have been isolated or identified in the reaction mixtures.

The reaction paths, proposed on the basis of the available evidence and especially on the characterization of the intermediates [1], are exemplified in scheme 1 where M is the metal atom and L_x or L_y is an indefinite number of ligands L.

In this scheme L_yMH and L_xM indicate active species, which are possibly different from the starting complexes, where the number of ligands is concerned; in fact, an activation process is generally necessary whereby some ligands must be displaced. Although these preliminary reactions are often very fast, in some

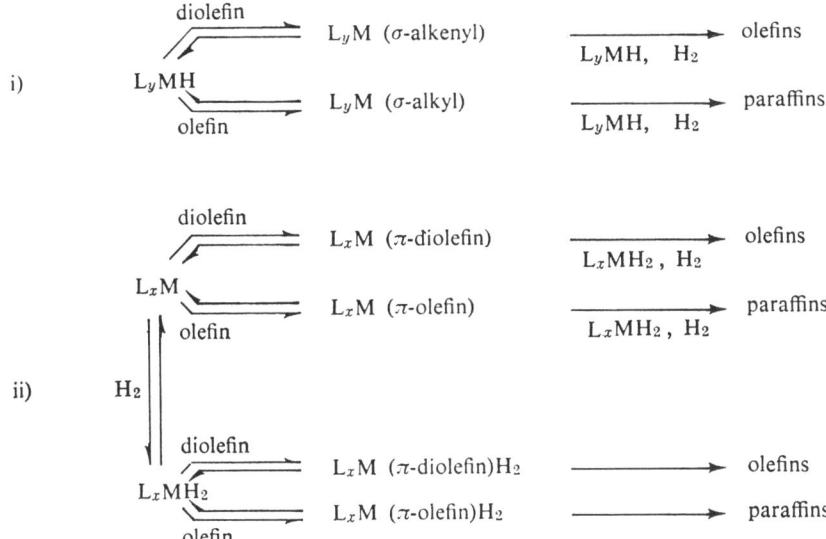

Scheme 1

cases slow displacement results in the observation of an induction period.

According to the mechanism i) of scheme 1, hydrogenation takes place through insertion of the substrate (diene or monoene) into a MH bond, an alkenyl or alkyl intermediate respectively being formed.

Conversely, uncertainty sometimes exists about the actual path (hydridic or olefinic) whenever the hydrogenation takes place according to mechanism ii).

There is enough preparative evidence to indicate that equilibria involving diolefins are very fast and largely shifted to the π-or σ-complexes; stable complexes such as π-diene-Fe(CO)$_3$, $(\sigma\text{-}C_4H_7)Co(CN)_5{}^{3-}$ and $(\pi\text{-}C_4H_7)Co(CO)_2P(n\text{-}C_4H_9)_3$ have been isolated in the course of hydrogenation. Conversely π-mono-olefin complexes of moderate stability are known only for a small number of metals [7]; moreover it is important to remember that Chatt et al. [8] have recently shown that the equilibrium (2) lies completely to the left.

$$HPtCl(PPh_3)_2 + RCH{=}CH_2 \rightleftharpoons RCH_2CH_2PtCl(PPh_3)_2 \qquad (2)$$

The complete set of reactions, which take place during the hydrogenation of dienes catalyzed by transition metal complexes (see scheme 1), can be represented in a general way as follows:

where Cat denotes the free catalyst which is related to the total amount by the following equation:

$$[Cat_{tot}] = [Cat-d] + [Cat-m] + [Cat] \qquad (5)$$

According to the observations set forth above, the case in which the equilibria concerning dienes and monoenes are slow with respect to their subsequent reactions, can be considered uncommon or unlikely; if it occurs, kinetic factors would be related to selectivity as illustrated in fig. 1. Conversely, emphasis must be given to the general path involving fast pre-equilibria followed by slow reactions. In this case, assuming a constant concentration of the hydrogenating species (H_2 or L_yMH or L_xMH_2) during the reaction, the rate of formation of monoenes and saturated compounds are as follows:

$$r_m = k_2 [Cat-d] - k_4 [Cat-m] \qquad (6)$$

$$r_s = k_4 [Cat-m] \qquad (7)$$

With reference to equations (3) and (4) when $k_2 \ll k_{-1}$ and $k_4 \ll k_{-3}$ the concentrations of intermediates Cat-d and Cat-m attain the equilibrium value independently of further reactions.

When this happens, let us consider the two cases in which [Cat-d] and [Cat-m] are or not negligible as compared to [d] and [m] respectively (case 1 and case 2).

Case 1. Assuming $[d] \gg [Cat-d]$ and $[m] \gg [Cat-m]$, Eqs (8) and (9) can be derived from the equilibria (3) and (4):

$$[Cat-d] = K_1 [Cat] [d] \qquad (8)$$

$$[Cat-m] = K_3 [Cat] [m] \qquad (9)$$

wherein $K_1 = \dfrac{k_1}{k_{-1}}$ and $K_3 = \dfrac{k_3}{k_{-3}}$.

By introducing (8) and (9) in (6) and (7), the following ratio can be obtained:

$$\frac{r_m}{r_s} = \frac{k_2}{k_4} \frac{K_1}{K_3} \frac{[d]}{[m]} - 1 \qquad (10)$$

The above equation shows the dependence of selectivity on both kinetic and thermodynamic factors; as a rule, the latter are probably of overwhelming importance. It is thus evident that when thermodynamic factors are operative, good selectivity can be obtained up to high conversions.

The concentration of Cat-d and Cat-m can also be related to the total concentration of the catalyst; in fact the following equations can be obtained from (5), (8) and (9):

$$[\text{Cat-}d] = \frac{K_1[d][\text{Cat}_{tot}]}{1 + K_1[d] + K_3[m]} \tag{11}$$

$$[\text{Cat-}m] = \frac{K_3[m][\text{Cat}_{tot}]}{1 + K_1[d] + K_3[m]} \tag{12}$$

It is worth noting that Eqs. (11) and (12), which are similar to those of some enzymic systems, correspond to the Eqs. (13) and (14) obtained for the chemisorption on heterogeneous catalysts in the gaseous phase [9] [10]:

$$\theta_d = \frac{b_d P_d}{1 + b_d P_d + b_m P_m} \tag{13}$$

$$\theta_m = \frac{b_m P_m}{1 + b_d P_d + b_m P_m} \tag{14}$$

wherein surface coverages θ_d and θ_m are related to the pressures P_d and P_m of diene and monoene respectively, and b_d and b_m are the equilibrium constants for adsorption. The similarity between Eqs. (11) (12) and (13) (14) is related to the fact that the active centers of heterogeneous catalysts are always present in negligible amounts, with respect to the substrate.

Case 2. When the catalyst is used in large amounts because of its low activity, which is a frequent case with homogeneous catalysts, the concentration of the intermediates Cat-d and Cat-m cannot be neglected. Under these conditions, the following equations must be written with reference to the equilibria which appear in (3) and (4):

$$K_1 = \frac{[\text{Cat-}d]}{\{[d] - [\text{Cat-}d]\}[\text{Cat}]} \tag{15}$$

$$K_3 = \frac{[\text{Cat-}m]}{\{[m] - [\text{Cat-}m]\}[\text{Cat}]} \tag{16}$$

These equations together with Eq. (5) can afford additional and useful considerations concerning the effect of thermodynamic factors. The ratio [Cat-d]/{[Cat-d] + [Cat-m]} (see the following figures) represents the pure contribution to the selectivity of the competition between the substrates on the catalyst.

By solving the system of Eqs. (5), (15), and (16) using some sets of values for K_1, K_3, $[d]$, $[m]$, and $[\text{Cat}_{tot}]$, chosen on the basis of qualitative experimental evidences, it appears that selectivity is strictly related to the ratio K_1/K_3 (fig. 2). When $K_1/K_3 = 1$ (which is rather improbable, dienes being co-ordinated more than monoenes) a non-selective reaction takes place; it is evident that the higher the above ratio, the better is the selectivity. Thus for example, when $K_1/K_3 = 100$ (fig. 2), the catalyst being almost completely tied up by the diene, negligible hydrogenation of monoenes takes place up to high conversions.

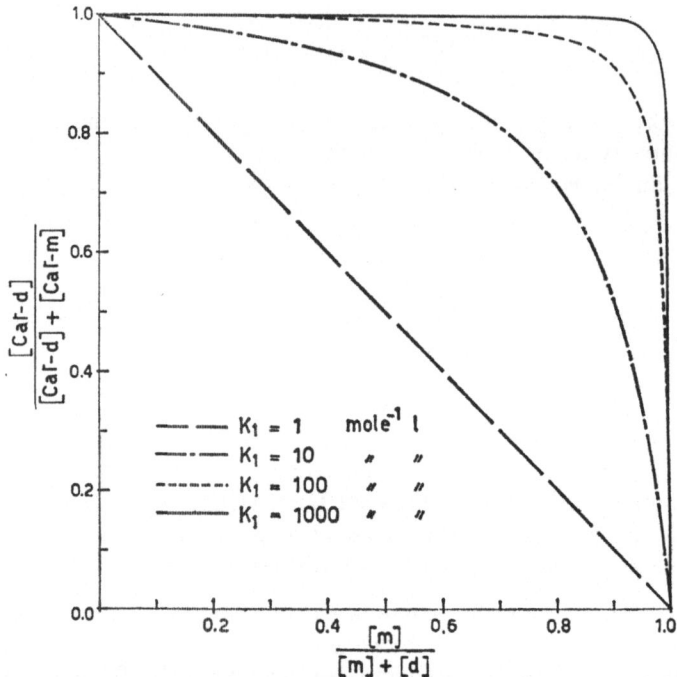

Figure 2. Factor of competition against composition of the unsaturates for different ratios K_1/K_3; $K_3 = 1$ mole^{-1} l, $[Cat_{tot}] = 10^{-2}$ mole^{-1} l, $[m] + [d] = 1$ mole^{-1} l.

Moreover the resolution of the system of Eqs. (5), (15), (16) shows (fig. 3) that lower selectivities are obtained at higher concentrations of the catalyst (and consequently of the species Cat-d and Cat-m), all the other parameters being constant. In order to explain this interesting feature, let us consider the following expression, obtained from Eqs. (15) and (16):

$$\frac{[Cat\text{-}d]}{[Cat\text{-}m]} = \frac{K_1\{[d] - [Cat\text{-}d]\}}{K_3\{[m] - [Cat\text{-}m]\}} \tag{17}$$

It appears that, an increase of the catalyst concentration does not affect $\{[m] - [Cat\text{-}m]\}$ appreciably (K_3 being generally low), but it affords a marked decrease of $\{[d] - [Cat\text{-}d]\}$ (K_1 being much larger) with the consequence that $[Cat\text{-}d]/[Cat\text{-}m]$ is lowered.

Therefore, where simple thermodynamic factors are concerned, it can be said that high selectivity is favoured by both large ratios of K_1/K_3 and by low

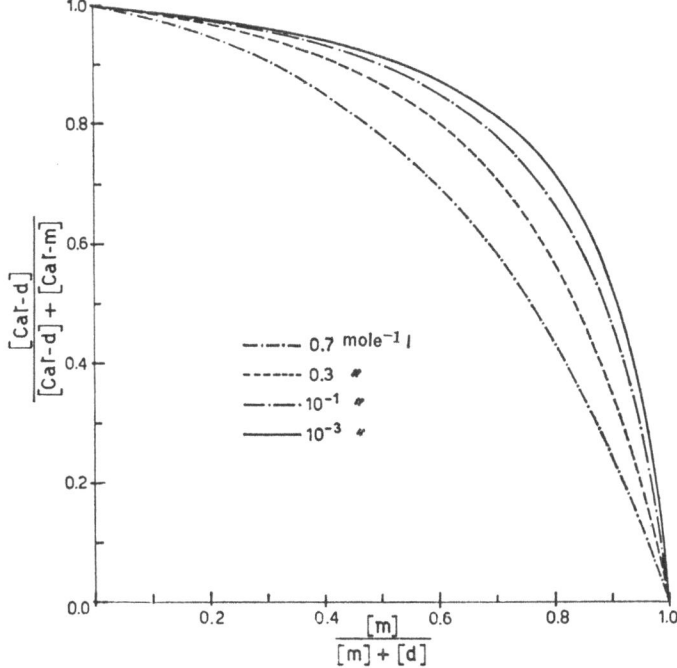

Figure 3. Factor of competition against composition of the unsaturates for different catalyst concentrations; $K_3 = 1$ mole^{-1} 1, $K_1 = 10$ mole^{-1} 1, $[m] + [d] = 1$ mole^{-1} 1.

concentrations of active intermediates. The latter condition (which is fulfilled whenever the catalysts are used in low concentration) is also operative when they have low equilibrium constants. Accordingly, higher selectivity corresponds to lower absolute values of K_1 and K_3, their ratio being constant (fig. 4).

However, even though it appears convenient to keep [Cat-d] and [Cat-m] at low levels, in practice the need of acceptable reaction rates for (6) and (7) can be a limiting factor. While the catalysts that give reasonable rates of hydrogenation can be profitably used at low concentrations, in many other cases this is not possible because of the low catalytic activity and consequently the selectivity is lowered.

Clearly the observed selectivity of the reaction is also comprehensive of the kinetic parameters (Eqs. (3) and (4) which can add positively or negatively to the contribution of thermodynamic factors.

212

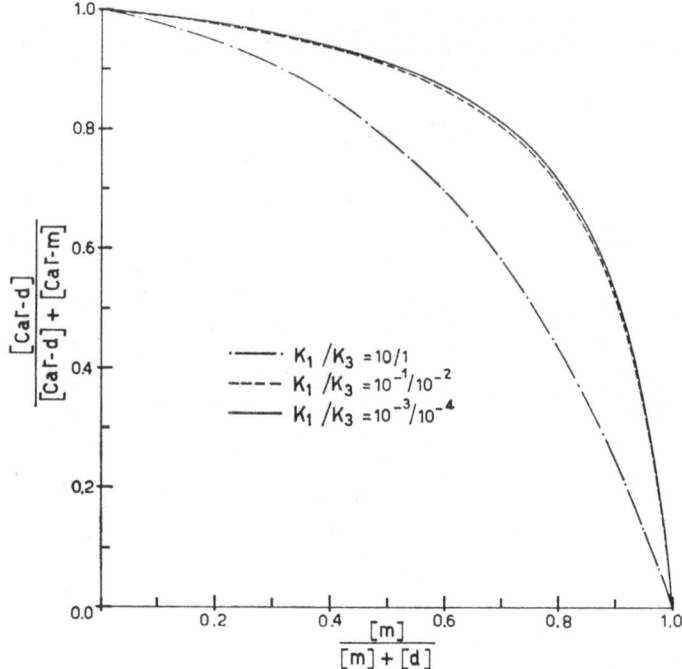

Figure 4. Factor of competition against composition of the unsaturates for different equilibrium constants; $[Cat_{tot}] = 0.7$ mole^{-1} 1, $[m] + [d] = 1$ mole^{-1} 1.

2.2. Stereoselectivity

In the homogeneous hydrogenation of dienes, the formation of only one olefinic isomer is seldom observed; as a rule a mixture of monoenes is obtained, whose composition does not correspond to the thermodynamic equilibrium. The prevalent formation of a specific isomer is very interesting from both a theoretical and practical point of view.

Whenever theoretical considerations on stereoselectivity are to be made, the important question that arises is whether the hydrogenation products can be considered as the initial ones. In fact, side reactions, such as isomerization or preferential subsequent hydrogenation of some isomers, can modify the initial distribution of monoenes.

In this context it can be said that while isomerization is always detrimental to the stereoselectivity, preferential hydrogenation of some of the resulting monoenes may modify the composition in both directions.

Generally speaking the composition of the initial products is strictly related to both the conformation of the hydrocarbon moiety in the intermediate catalytic species and to the type of attack of the reducing species. Usually, for many catalytic systems, the assumption has been made that σ-alkenyl and/or π-allyl species are involved as intermediates [1].

Evidence exists concerning the formation of these complexes by diene insertion into the metal-hydrogen bonds [11] [12]. σ-Alkenyl and π-allyl structures have been detected in the case of the pentacyanocobaltate catalyzed hydrogenation of dienes (see 3.1) and the existence of a σ-π equilibrium has been suggested to explain the distribution of the hydrogenation products; similar intermediates have been postulated for heterogeneous selective catalysts [10].

However, little is definitely known about the intimate mechanism of many selective hydrogenations; further studies will be necessary to shed light on this aspect of the problem.

σ-Alkenyl and π-allyl structures are probably not involved in other catalytic systems; the results obtained in π-Arene-Cr(CO)$_3$ catalyzed hydrogenation are better explained in terms of a 1,4 concerted *cis*-addition of a dihydrido species to the conjugated dienes (see 3.2.3.1.).

When σ-alkenyl and π-allyl species are involved, the stereoselectivity of the reaction can be controlled by the factors which affect the σ-π conversion, such as concentration and nature of the ligand, temperature, solvent etc. which can alter the conformation of the π-allyl intermediate (*syn* or *anti* for instance) and, as a consequence, the geometry of the final products.

It is well known that, in the heterogeneous hydrogenation of dienes on metal surfaces, different distributions of monoenes can be obtained by changing some parameters such as carrier, temperature and solvent, but the effect on the products composition is never so marked as for the homogeneous systems [2].

Although some interesting results have been obtained concerning the control of stereoselectivity, many aspects of the problem are still to be clarified and it is hoped that, when these difficulties are overcome, a better knowledge of all the factors governing stereoselectivity will help in obtaining the required product composition.

2.3. Catalytic systems and their classification

Among the catalysts used in the homogeneous selective hydrogenation of dienes, complexes of groups VI and VIII of transition metals are the most frequent, as shown in table 1, in which typical catalysts are listed. Several of these complexes have also been used in the catalytic hydrogenation of alkenes [1].

Lithium aluminium hydride is the only reported example of a non-transition metal hydride which homogeneously catalyzes selective hydrogenation of dienes.

Although many criteria can be adopted in classifying the available material, we have divided the various catalytic systems on the basis of their chemical properties. Moreover, while to LiAlH$_4$ has been given a separate section, it seemed convenient to collect transition metal complexes according to their ability to activate molecular hydrogen.

Table 1

TYPICAL TRANSITION METAL CATALYSTS FOR SELECTIVE HYDROGENATION

Group	Metal	Catalysts
VI b	Cr, Mo, W	$CpMH(CO)_3$, π-arene-$M(CO)_3$
VIII	Fe	$Fe(CO)_5$, π-diene-$Fe(CO)_3$, $[CpFe(CO)_2]_2$, $Fe(SaProR)$
	Ru	$RuHCl(PPh_3)_3$, $RuCl_2(PPh_3)_{3-4}$
	Co	$Co(CN)_5^{3-}$ $[Co(CO)_4]_2$, $[Co(CO)_3PR_3]_2$, $[Co(CO)_2PBu_3]_3$, $Co(SaProR)$
	Rh	$RhCl(PPh_3)_3$, $RhH(DMG)_2PPh_3$
	Ni	$[Ni_2(CN)_6]^{4-}$, $Ni I_2(PPh_3)_2$
	Pd	$Pd(CN)_2(PPh_3)_2$, $PdCl_2(PPh_3)_2 + SnCl_2$
	Pt	$PtCl_2(AsPh_3)_2$, $PtCl_2(PPh_3)_2 + SnCl_2$, $H_2PtCl_6 + SnCl_2$
I b	Cu	$Cu(SaProR)$

2.4. List of abbreviations

In this review the following abbreviations have been used:

Ph	for	$-C_6H_5$
Et	»	$-C_2H_5$
Pr	»	n-C_3H_7
Bu	»	n-C_4H_9
Py	»	pyridine
acac	»	acetylacetonate
Cp	»	cyclopentadienyl
S	»	sorbate
dipy	»	2,2'-dipyridyl
DMG	»	dimethylglyoximate
DPE	»	1,1'-bis(diphenylphosphino)ethane
DPP	»	1,1'-bis(diphenylphosphino)propane

M (SaProR)   for

3. COMPLEXES ACTIVATING MOLECULAR HYDROGEN

3.1. Cyano complexes

Among the great number of transition metal cyano complexes only cobalt and nickel derivates have been reported as being able to catalyze the selective hydrogenation of dienes. Although the two catalytic systems can be considered similar in some aspects, such as chemical structure, catalytic activity in mild conditions and high selectivity, they are different as far as hydrogen activation is concerned.

While cobalt-cyano complexes have been extensively studied for both practical interest and theoretical aspects the nickel system has been only partially investigated in the last few years.

3.1.1. COBALT-CYANO COMPLEXES

Excellent reviews on this catalytic system are available [2] [3]. This section will focus only on the catalytic hydrogenation of dienes.

The pentacyanocobaltate(II) ion reacts in aqueous solution reversibly with molecular hydrogen forming hydrido compounds [13-15]. The reaction takes place at room temperature and at hydrogen pressure of less than 1 atmosphere according to the following equilibrium:

$$2 \, Co(CN)_5^{3-} + H_2 \rightleftharpoons 2 \, Co(CN)_5H^{3-} \tag{18}$$

The hydrogen can be transferred to suitable unsaturated compounds containing conjugated double bonds, such as butadiene, cinnamic acid, sorbic acid, etc. [16-23].

The hydrogenation of organic substrates has been generally carried out in aqueous media at room temperature and atmospheric pressure but examples of catalytic hydrogenation under more extreme conditions have been reported [16].

In some cases the solvent (generally water) is the source of hydrogen *via* the formation of a hydrido-complex as shown in reaction (19):

$$2 \ Co(CN)_5^{3-} + H_2O \rightleftharpoons Co(CN)_5H^{3-} + Co(CN)_5OH^{3-} \tag{19}$$

In every case, the selectivity of the hydrogenation of dienes to monoenes is very high (table 2), and sometimes, very high stereoselectivity can be obtained, depending on the reaction conditions.

The inability of pentacyanocobaltate complexes to hydrogenate mono-olefins generally permits the quantitative reduction of several dienes to monoenes.

The conjugation is not the only condition for the reduction of dienes to monoenes but steric factors are also of great importance. In effect, *cisoid* conformation seems to favour the reduction of dienes: when this conformation is sterically hindered, as in the case of 2,5-dimethyl-2,4-hexadiene, the reduction does not occur [18]. Incidentally it must be pointed out that the same behaviour has been reported in the case of arene-chromium carbonyl catalyzed hydrogenation [2] [5] (see 3.2.3.1.). However, other factors are probably involved since for example 3-methylene-cyclohexene, in which a fixed s-*trans* conformation is undoubtedly present, can be reduced [2].

Many studies have been reported in the last few years aimed at elucidating the mechanism of hydrogenation with special reference to the selectivity and stereoselectivity of the catalytic system. The first mechanism was proposed by Kwiatek [17] [18], using butadiene as a model. It involves three steps: (20) reversible addition of the hydrido complex to butadiene, (21) cleavage of the resulting butenyl complex by the hydrido complex, and (22) hydrogen absorption by pentacyanocobaltate(II):

$$Co(CN)_5H^{3-} + C_4H_6 \rightleftharpoons Co(CN)_5(C_4H_7)^{3-} \tag{20}$$

$$Co(CN)_5H^{3-} + Co(CN)_5(C_4H_7)^{3-} \rightleftharpoons 2 \ Co(CN)_5^{3-} + C_4H_8 \tag{21}$$

$$2 \ Co(CN)_5^{3-} + H_2 \rightleftharpoons 2 \ Co(CN)_5H^{3-} \tag{22}$$

The reversibility of the first step has been demonstrated by the formation of mono-, di- and tri-deuterated 1-butene in the reduction of butadiene with D_2 using D_2O as a solvent [18]. The reversibility of equation (20) has been further confirmed by the observation that the butenyl cobalt complex disproportionates, in the absence of hydrogen, giving equal amounts of butenes and butadiene. This result can be explained only by taking into account the formation of $Co(CN)_5H^{3-}$ through the reverse of equation (20) followed by equation (21).

The relatively slow rates of the reverse reaction (20) and the forward reaction (21) have allowed a butenylcyanocobaltate complex to be isolated and characterized by n.m.r. [25]. A freshly prepared deuterium oxide solution of $Co(CN)_5(C_3H_5)^{3-}$ shows the signals of a σ-structure which gradually becomes rear-

Table 2
SELECTIVE HYDROGENATIONS CATALYZED BY CYANO COMPLEXES

Catalyst	Substrate (% Convn.)	Product(s) (% Compn.)		Conditions	Ref.
$K_3Co(CN)_5$	1,3-Butadiene (100)	1-Butene trans-2-Butene cis-2-Butene	(89.5) (8.4) (2.1)	$H_2 = 1$ atm, 20 °C Water, CN/Co = 10	[16]
$K_3Co(CN)_5$	1,3-Butadiene (100)	1-Butene trans-2-Butene cis-2-Butene	(12.5) (85.1) (2.4)	$H_2 = 1$ atm, 20 °C Water, $Co^{2+} = 0.08M$ CN/Co = 5	[16]
$K_3Co(CN)_5$	1,3-Butadiene (80)	1-Butene trans-2-Butene cis-2-Butene	(47.9) (30.3) (21.8)	$H_2 = 100$ atm, 100°C Water, CN/Co = 10	[16]
$K_3Co(CN)_5$	1,3-Butadiene	1-Butene trans-2-Butene cis-2-Butene	(84.0) (14.0) (2.0)	$H_2 = 1$ atm, Water, $Co^{2+} = 0.2M$ CN/Co = 5	[22]
$K_2Co(CN)_4(dipy)$	1,3-Butadiene	1-Butene trans-2-Butene	(86.0) (14.0)	$H_2 = 1$ atm, Water	[22]
$CoCl_2 + LiCN(1:4)$	Cyclopentadiene (100)	Cyclopentene Dicyclopentadiene	(95.0) (5.0)	$H_2 = 60$ atm, 15 °C Ethanol CN/Co = 4	[33]
$K_3Co(CN)_5$	Sodium Sorbate (100)	2-Hexenoate 3-Hexenoate 4-Hexenoate	(82.0) (17.0) (1.0)	$H_2 = 1$ atm, 25 °C Water	[34]
$K_3Co(CN)_5$	Sorbic Acid (100)	2-Hexenoic Acid 3-Hexenoic Acid	(96.2) (3.8)	$H_2 = 1$ atm, 25 °C Methanol	[34]
$K_3Co(CN)_5$	1,3-Butadiene	1-Butene trans-2-Butene cis-2-Butene	(10.0) (84.0) (6.0)	$N_2 = 1$ atm, 20 °C Glycerin-Methanol CN/Co = 5	[35]
$K_3Co(CN)_5$	1,3-Butadiene	1-Butene trans-2-Butene cis-2-Butene	(38.0) (7.0) (55.0)	$N_2 = 1$ atm, 20 °C Glycerin-Methanol CN/Co = 6	[35]
$K_4Ni_2(CN)_6 + NaBH_4$	1,3-Butadiene (100)	1-Butene trans-2-Butene cis-2-Butene	(14.0) (65.0) (21.0)	$N_2 = 1$ atm, 0 °C Water	[42]
$K_4Ni_2(CN)_6 + NaBH_4$	Isoprene (100)	2-Methyl-1-Butene 2-Methyl-2-Butene 3-Methyl-1-Butene	(21.0) (79.0) (traces)	$N_2 = 1$ atm, 0 °C Water	[42]

ranged to a π-structure, presumably by loss of a cyanide ligand [25]. The σ-structure can be immediately obtained by addition of cyanide ion, indicating a fast equilibrium (23).

Since the [CN]/[Co] ratio (where [ ] denotes a molar concentration) strongly affects the distribution of hydrogenation products of butadiene, the stereoselectivity has been correlated to the above σ-π equilibrium [18].

$$\begin{array}{c} \text{(structure: butenyl–Co(CN)}_4\text{ σ-complex)} \end{array} \underset{+CN^-}{\overset{-CN^-}{\rightleftarrows}} \begin{array}{c} \text{(structure: butenyl–Co(CN)}_3\text{ π-complex)} \end{array} \qquad (23)$$

The effect of the ratio [CN]/[Co] on the stereoselectivity has also been observed with other substrates, such as isoprene and 1,3-pentadiene (table 3), but not in the hydrogenation of homoconjugated norbornardiene [3].

A complete kinetic study of the hydrogenation of butadiene in aqueous solution has been recently reported [26]. The reversible reaction of butadiene

Table 3

DEPENDENCE OF PRODUCTS DISTRIBUTION ON THE CN/Co RATIO IN THE HYDROGENATION OF BUTADIENE, ISOPRENE AND 1,3-PENTADIENE

CN/Co	%	%	%
4.5	86	1	13
5.5	70	1	29
6.0	12	3	85
8.5	19	1	80
5.0	21	1	78
7.0	91	6	3
5.0	2	0	98
7.0	21	12	67

with the hydrido complex (scheme 2), which is the fast step of the overall reaction, is in accordance with the following rate law:

$$-d\,[C_4H_6]/dt = k_1\,[C_4H_6]\,[Co(CN)_5H^{3-}] \qquad (24)$$

Since the observed rate constant is independent of the cyanide ion concentration, the formation of a σ-complex as the initial intermediate has been suggested; it is probably the branched isomer σ_1 (scheme 2).

The composition of the resulting butenes and the kinetic of the overall reaction have been explained on the basis of equilibria between σ and π species at low [CN]/[Co] ratios (e.g. 5), and between the two species (σ_1, σ_2), at high [CN]/[Co] ratios (scheme 2):

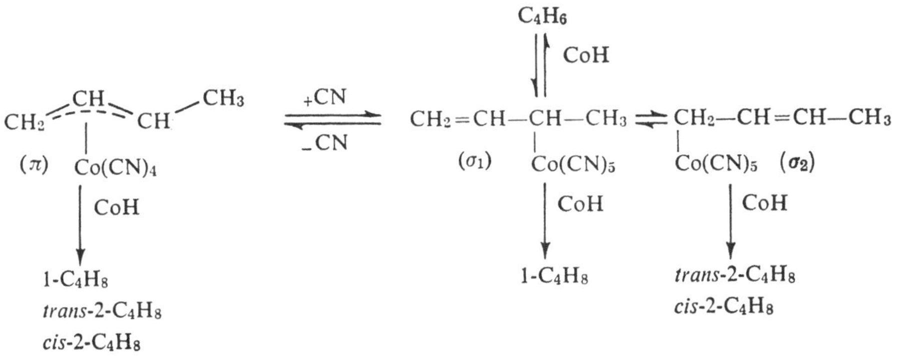

Scheme 2

The stereoselectivity of the reaction has been explained by considering the effects of the concentrations of both the cyanide ion and the hydrido complex [26]. It has been found that in the hydrogenation of butadiene the ratio between 1-and 2-butene increases with increasing $Co(CN)_5H^{3-}$ concentration. Since the exclusive formation of 1-butene is due to σ_1, the increase of $Co(CN)_5H^{3-}$ concentration conceivably enhances the amount of σ_1, relative to the π species at low [CN]/[Co] ratios (e.g. 5.2), and to the σ_2 species in the presence of excess cyanide.

While the method by which the hydrido complex changes the balance of this $\sigma_1 \rightleftharpoons \sigma_2$ equilibrium is not clear spectrophotometric evidence has been reported [27] according to which species lower in cyanide ligands are present in dilute solutions at [CN]/[Co] = 5. The effect of the hydrido complex concentration on product distribution has also been reported by Farcas et al [22] and Kwiatek and Seyler [3] (table 2).

Some recent studies have been concerned with the mechanism of hydrogen addition; according to Kwiatek and Seyler [3] the distribution of butenes obtained from the hydrogenation of butadiene can be explained by assuming a γ attack on the allylic system by the hydrido complex. At [CN]/[Co] > 5.5

the formation of 1-butene is prevalent (reaction (25)), while at $[CN]/[Co] < 5.5$ *trans*-2-butene is obtained preferentially (reaction (26)). According to the same authors the more stable *syn* form of the π-allyl complex can easily explain the prevalent formation of *trans*-2-butene:

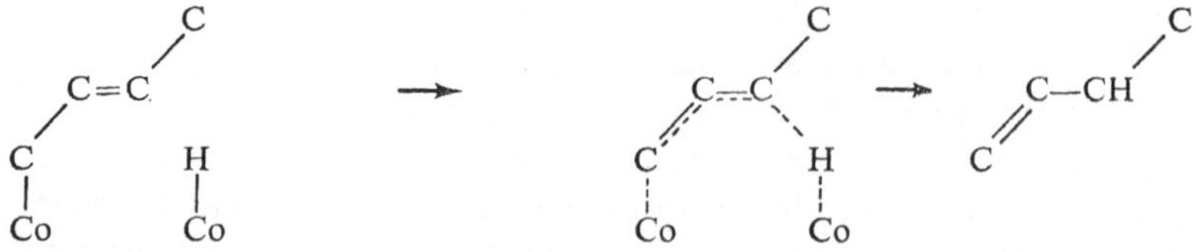

(25)

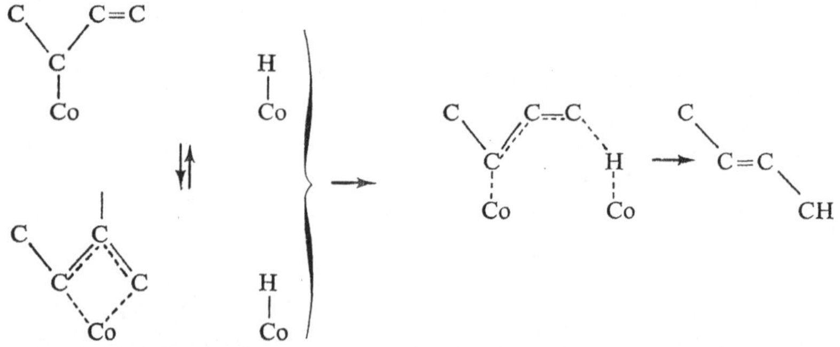

(26)

The validity of this general mechanism has been confirmed by other authors [26] but a different transition state has been proposed. It has been assumed that the reaction of the hydrido complex with the butenyl complex proceeds through a transition state in which the C—H bond, which is about to be formed, acts as a bridge between the two cobalt atoms [26].

The structural features of the substrates are important in determining the course of the cyanocobaltate catalyzed reactions. In fact the hydrogenation of sorbic acid to 2-hexenoic acid has been found to occur through a completely different mechanism. According to Simandi et al [28] the hydrogenation occurs via a two step hydrogen-atom transfer from $Co(CN)_5H^{3-}$ to the 4-5 double bond of sorbic acid with intermediate formation of a free radical (reactions (27), (28), (29) and (30) where S is the sorbate ion):

$$2\ Co(CN)_5^{3-} + H_2 \rightleftharpoons 2\ Co(CN)_5H^{3-} \tag{27}$$

$$Co(CN)_5H^{3-} + S \rightarrow Co(CN)_5^{3-} + \dot{S}H \tag{28}$$

$$Co(CN)_5H^{3-} + \dot{S}H \rightarrow Co(CN)_5^{3-} + SH_2 \tag{29}$$

$$Co(CN)_5^{3-} + \dot{S}H \rightleftharpoons Co(CN)_5SH^{3-} \tag{30}$$

The organometallic species $Co(CN)_5SH^{3-}$ has been shown to be present in significant amounts, depending on the reaction conditions; however, its role as an intermediate of the hydrogenation has been ruled out. The same mechanism has been proposed for the hydrogenation of cinnamic acid [29].

The reactions of hydridopentacyanocobaltate with the anions of α, β unsaturated acids have been recently studied by n.m.r. spectroscopy [30]. The results further support the mechanism previously proposed by Simandi and Nagy.

The effect of the pH on the distribution of hydrogenation products has also been considered [3] [19] [20] [31]. Addition of NaOH, $Ca(OH)_2$ or NH_3 to the cyanocobaltate(II) solution favours the formation of l-butene at low $[CN]/[Co]$ ratios [19]. This effect has only been observed when the alkali compounds have been added in the presence of hydrogen before the introduction of potassium cyanide [19] [20]. The reason for this effect is not completely understood: probably hydroxyl ions suppress the cyanide hydrolysis, so that the formation of π-allyl species is lowered.

The kinetics of hydrogen absorption have been reported for several cyclopolyolefins and the effect of their structure on the rate has been studied [32]. The rate decreases in the order: 1,3-cyclohexadiene > cyclopentadiene > cycloheptatriene > 1,3-cycloheptadiene > 1,3,5-cyclooctatriene > cyclooctatetraene. The reason for this effect has been tentatively attributed to the hybridization of the two external carbon atoms of the conjugated systems.

The catalytic hydrogenation of dienes and unsaturated systems has been generally carried out in aqueous solutions due to the very low solubility of the pentacyano complexes in organic solvents. However, two main difficulties arise when using aqueous solutions: the solubility of many organic substrates is very low and, on the other hand, low hydride concentrations are obtained due to the aging reaction [13] [23]. According to Kwiatek [2] this reaction is reversible: consequently the hydrido complex may be restored autocatalytically in the presence of molecular hydrogen. However, very slow inverse reaction is to be expected since this equilibrium, if it exists, lies well to the right and a low forward reaction has been observed [30] [33].

Some examples of aqueous-alcoholic or alcoholic solutions have been reported. Surprisingly the hydrogenation of sorbic acid [34] gives 2-hexenoic acid with higher selectivity in methanol solution (96.2%) than in water (81.7%). Hydrogenation of butadiene in a glicerine-methanol solution has been reported [35] to proceed in a nitrogen atmosphere, the solvent being the hydrogen source. At $[CN]/[Co] > 5.2$ the amount of resulting butenes decreases in the order cis-2-butene > 1-butene > trans-2-butene. The low 1-butene content can be tentatively explained by considering the very low concentration of the hydrido complex in accordance with the interpretations of many authors [3] [22] [26] [27]; in fact 1-butene is again formed as a major product in a hydrogen atmosphere.

The catalytic hydrogenation of cyclopentadiene to cyclopentene by an ethanolic solution of $Li_3Co(CN)_5$ has been recently reported [36]. No effect of the solvent on the selectivity has been found. The results of kinetic experiments seem to agree with the mechanism suggested by Kwiatek et al. [3].

A different catalytic system is obtained when an ethanolic solution of LiCN and anhydrous $CoCl_2$ (4 : 1 ratio) are mixed in the presence of cyclopentadiene [36]. The hydrogenation takes place at much higher rate (c.a. ten fold) than in the case of ethanolic solutions of $Li_3Co(CN)_5$ being used. This 4 CN : 1 Co catalytic system is active in the hydrogenation of butadiene provided that the substrate concentration is lower than that of the cobalt catalyst.

3.1.2. SUBSTITUTED COBALT-CYANO COMPLEXES

Many efforts have been made to improve the catalytic activity and to change the stereoselective course of the hydrogenation by replacing part of the cyanide ligands with other ligands [22] [37-39]. A number of amine complexes has been studied and their catalytic activity investigated, principally by Piringer and Farcas [40], and by Wymore [38]. The substitution of a cyanide ligand with groups which are weaker π-acids destabilizes the cobalt-hydrogen bond, and improves the catalytic activity [38] Ligands, such as 2,2'-dipyridyl, ethylendiamine, o-phenanthroline and, in general chelating amines, have been employed in the hydrogenation of butadiene and other unsaturated systems.

According to Wymore, $Co(dipy)(CN)_3^-$ is the most active catalyst in an ethanol-water medium. Unfortunately, kinetic data on the hydrogenation of dienes are still lacking. The catalytic activity of $Co(dipy)(CN)_2$, in the selective hydrogenation of 1,3-cyclohexadiene, sorbic acid and α,β unsaturated acids has also been recently reported [39]. The initial rate of hydrogenation of 1,3-cyclo-hexadiene is approximately three times the observed rate of hydrogenation catalyzed by $Co(CN)_5^{3-}$. The initially very rapid hydrogenation of sorbic acid appears to be "poisoned" after a few minutes. The decrease in the catalytic activity has been ascribed to the formation of Co(III) through oxidation by the substrate [39].

3.1.3. NICKEL-CYANIDE SYSTEMS

Aqueous solutions of potassium hexacyanodinickelate(I) $K_4Ni_2(CN)_6$, synthesized by reduction of $K_2Ni(CN)_4$ with sodium amalgam or sodium bo-rohydride, cause selective hydrogenation of diolefins. The reaction takes place at 0-25 °C, in an inert atmosphere and is catalytic in the presence of sodium bo-rohydride [41-43]. The distribution of hydrogenation products is very similar to that observed for the pentacyanocobaltate(II) catalyzed reaction (table 2). The selectivity is very high although a small amount of saturated and isomerized products has been found at long reaction times [42].

In the presence of CN^- pure $Ni_2(CN)_6^{4-}$ appears to cause hydrogenation of crotonic acid in an aqueous solution in an inert atmosphere [43]; therefore, the aqueous solvent must be the hydrogen source. This conclusion is further confirmed by the fact that the reduction of maleic acid in deuterium oxide gives 2,3-dideutero-succinic acid [43]. The catalytic system can be represented by (31)

$$Ni_2(CN)_6^{4-} + \text{substrate} \xrightarrow[2\ CN^-]{H_2O} 2\ Ni(CN)_4^{2-} + \text{products} \qquad (31)$$

$$\xrightarrow[-2CN^-]{+BH_4^-}$$

Some observations and experimental results which help in the elucidation of the reaction mechanism have been reported. $K_4Ni_2(CN)_6$ reacts with butadiene and other unsaturated substrates in aqueous solution to form relatively stable complexes [41]; the red colour of the solution of pure $K_4Ni_2(CN)_6$ ($\lambda = 285$ and $475\ m\mu$) turns yellow ($\lambda = 315\ m\mu$) after the addition of butadiene [42]. Cleavage of the intermediate complex to give monoenes takes place at a lower rate than absorption of butadiene [28].

Kinetic studies have shown that the rate of hydrogenation of crotonic acid increases with decreasing the pH and increasing the cyanide concentration [43] The formation of a paramagnetic Ni(I) complex, followed by disproportionation of the intermediate adduct to give Ni(II) and Ni(I) complexes, has been suggested [43] and butyrric acid results from a double protonation as indicated in the following equations:

$$Ni_2(CN)_6^{4-} \xrightarrow{CN^-} 2\ Ni(CN)_4^{3-} \xrightarrow{L} [Ni(I)(CN)_3]_2L^{4-} + 2\ CN^- \qquad (32)$$

$$[Ni(I)(CN)_3]_2L^{4-} \rightleftharpoons Ni(II)(CN)_3^- + Ni(I)(CN)_3L:^{3-} \qquad (33$$

$$Ni(I)(CN)_3L:^{3-} \xrightarrow{H^+} Ni(I)(CN)_3LH^{2-} \rightleftharpoons Ni(II)(CN)_3^- + :LH^{1-} \qquad (34)$$

$$:LH^{1-} \xrightarrow{H^+} LH_2 \qquad (35)$$

Since it is well known that structural features of the substrate are very important in the mechanism of catalytic hydrogenation by cyano complexes, the same mechanism cannot be reasonably invoked in the case of nickel cyanide catalyzed hydrogenation of dienes. A π-allyl intermediate such as:

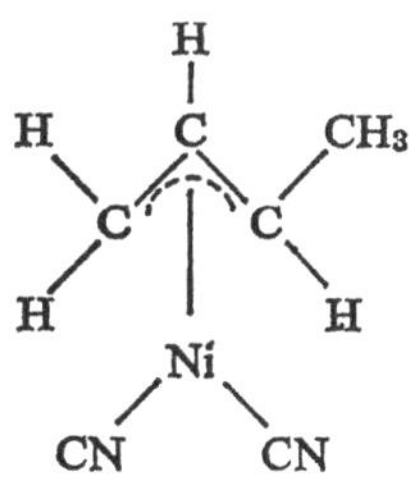

has been suggested without any experimental evidence [42]. Further studies will be necessary to elucidate some features of this interesting catalytic system.

3.2. Transition metal-carbonyls

Extensive use of transition-metal-carbonyls in the selective hydrogenation of polyenes to monoenes is quite recent and papers dealing with this topic have become more and more frequent, especially during the last three years. It is not surprising that the first studies have been concerned with iron, cobalt and nickel carbonyls [44], many laboratories having been acquainted with these catalysts for several years, due to their widespread use in the carbonylation reactions of unsaturated compounds. Subsequently, substituted metal carbonyls have been employed in which such ligands as tertiary phosphines, arenes, conjugated dienes and cyclopentadienyl are present (table 1).

Consequently better results have been reported for both selectivity and stereoselectivity of the reactions considered.

3.2.1. CARBONYL DERIVATIVES OF IRON

Iron containing catalysts have been widely used [5] [45] for the selective hydrogenation of polyunsaturated fatty acid esters, in particular simple $Fe(CO)_5$ which was the first catalyst to be used as a result of both its availability and stability at high temperatures even in the absence of carbon monoxide. As a rule all the complexes of this class exhibit low catalytic activity, so that generally they have been used in high concentrations.

In an attempt to discover more active catalysts than $Fe(CO)_5$, it has been found that other systems such as π-diene-iron(0)carbonyls and cyclopentadienyl-iron(I)carbonyls can be profitably used.

3.2.1.1. *Iron(0)carbonyls and π-diene-iron(0)carbonyls*

Detailed studies have been carried out on the hydrogenation of several fatty acid substrates (c.a. 1 M), such as methyl-linoleate (*cis*-9-, *cis*-12-octadecadienoate, [5] [45] methyl-linoleate (*cis*-9-, *cis*-12-, *cis*-15-octadecatrienoate) [5] [46] and soybean oil methyl ester [44], using simple iron pentacarbonyl as catalyst (substrate/catalyst molar ratio from 1 to 10). The reactions were carried out in cyclohexane solution at 150-180 °C and at 27 atmospheres of hydrogen pressure (table 4); in every case long reaction times (3-4 hrs.) were necessary in order to insure acceptable conversions. Due to the isomerizing properties of the catalytic system, a mixture of several isomers was always obtained (table 4). Experimental data clearly demonstrate [5] that higher selectivity is obtained in the presence of smaller amounts of catalyst; this effect has been explained on the basis of a competition between the diene and monoene on the catalyst (see fig. 3).

During the hydrogenation of methyl linoleate catalyzed by $Fe(CO)_5$, diene complexes were isolated and characterized as mixtures of isomeric, conjugated methyl octadecadienoate-iron tricarbonyl [5] [45] [47] [48]:

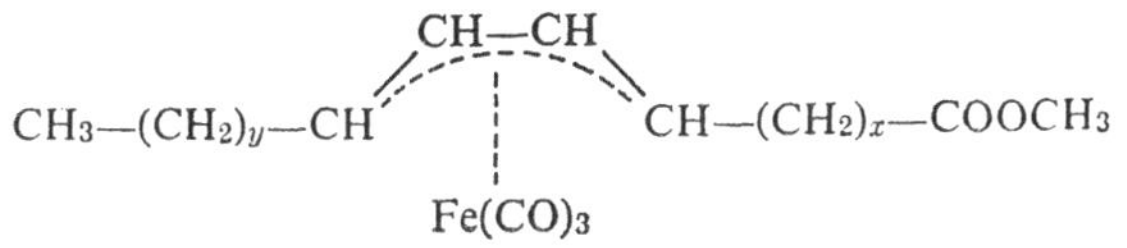

$$CH_3-(CH_2)_y-CH \underset{\underset{Fe(CO)_3}{\diagdown \quad \diagup}}{\overset{\overset{CH=CH}{\diagup \quad \diagdown}}{}} CH-(CH_2)_x-COOCH_3$$

in which $x,y = 7,5$; $8,4$; $6,6$; $9,3$; $5,7$; $10,2$; $3,9$; $4,8$.

Similarly triene complexes, consisting of mixtures of isomers containing a conjugated π-diene-Fe(CO)$_3$ unit and one free olefinic bond, have been isolated during the homogeneous hydrogenation of cis-9-, cis-12-, cis-15-octadecatri-enoate [5] [46] [48].

The concentration of the π-diene-Fe(CO)$_3$ species has been followed during the course of the hydrogenation [5]: when starting from Fe(CO)$_5$, the concentration of diene-Fe(CO)$_3$ reaches a maximum in the early stages of the reaction and then slowly decreases parallel to the decrease of diene [5]. On the other hand, whenever preformed diene-Fe(CO)$_3$ is used, its concentration is initially almost constant and then decreases with a similar trend as above.

The above results suggest that the conjugated diene-Fe(CO)$_3$ complexes are possible intermediates in the hydrogenation; since the same intermediate diene-Fe(CO)$_3$ has been isolated when starting from both unconjugated and conjugated dienes, preliminary isomerization seems to be a necessary step of the reaction.

It is significant that the use of preformed diene-Fe(CO)$_3$ species brings about higher reaction rates than Fe(CO)$_5$ [45]. Frankel et al [49] have pointed out that the catalytic activity in the hydrogenation of methyl sorbate is affected by substitution on the diene moieties.

Kinetic investigations and studies on ^{14}C labelled fatty esters [5] have been carried out in order to elucidate the mechanism of this catalytic hydrogenation.

During the reduction of a mixture of conjugated methyl trans, trans-octa-decadienoate and of labelled methyl linoleate, the same rate of reaction has been observed for the two substrates [5], suggesting that the isomerization of linoleate to a conjugated diene structure is not the rate determining step of the hydrogenation [5].

Ligand exchange (36) has been reported to be an important reaction taking place during the Fe(CO)$_5$ catalyzed hydrogenation [5] [50]

$$LFe(CO)_3 + S \rightleftharpoons SFe(CO)_3 + L \tag{36}$$

where L and S are conjugated dienes. In the case of methyl sorbate-Fe(CO)$_3$ $(2.2 \cdot 10^{-2}M)$ and diphenylbutadiene $(4.4 \cdot 10^{-2}M)$ the above reaction has been found to be first order in the metal complex and first order in the substrate [50]. Whenever the free diene is unconjugated (e. g. methyl linoleate), ligand exchange does not take place [5]; this is the reason why the isomerization of unconjugated to free conjugated dienes does not occur with $Fe(CO)_5$ to a great extent [48].

Kinetic data of methyl linoleate hydrogenation have been found, by means of an analog computer, to be in accordance with the following scheme [5]:

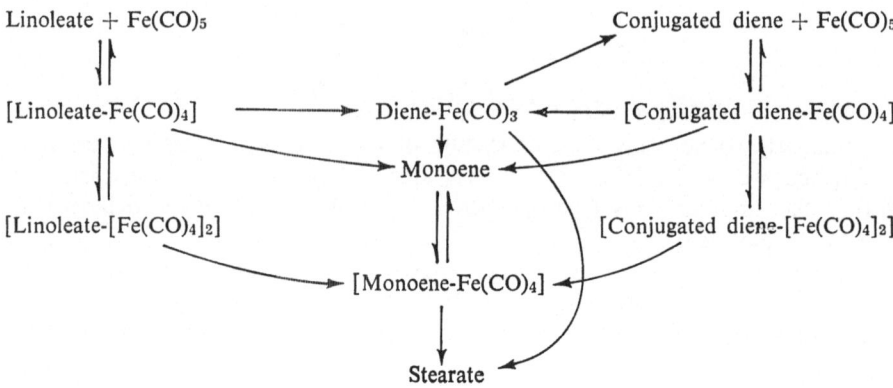

Scheme 3

Besides species of the type π-diene-$Fe(CO)_4$, complexes in which one double bond is involved in the co-ordination appear as reaction intermediates (scheme 3). Although these species have not been isolated during the hydrogenation, their role as intermediates has been postulated on the basis of the existence of similar compounds with butadiene [51], maleic anhydride [45] and dimethyl fumarate [45].

Kinetic evidence shows that the formation of monoene through an intermediate π-diene-$Fe(CO)_3$ is the most important path [5]; interestingly the direct formation of stearate from the above intermediate is significant [5]. This result explains the formation of saturated product at the initial stage of the reaction [5].

Unfortunately despite the isolation of intermediate species such as diene-$Fe(CO)_3$ and the kinetic approach, the intimate mechanism of the reaction is not completely understood. Some doubt exists about the mode of evolvement of intermediate iron complexes and the nature of the hydrogenating species, such as H_2 or $H_2Fe(CO)_3$.

3.2.1.2. *Cyclopentadienyl-iron*(I)*carbonyls*

Hydrogenation of cyclopentadiene catalyzed by cyclopentadienyl-iron(I) carbonyl has been reported by Sternberg et al. [52], who observed the formation of some cyclopentene and cyclopentane in the synthesis of $[CpFe(CO)_2]_2$ from cyclopentadiene and $Fe(CO)_5$.

Selective hydrogenation of 1,3-butadiene (1.7 M) to a mixture of butenes has been carried out using $[CpFe(CO)_2]_2$ (6.8 10^{-2} M) as catalyst [53]; a butene content as high as 94% with respect to the butene-butane mixture has been found (table 4). Possibly the reaction involves the formation of the hydridic species CpFeH(CO)$_2$, which has been invoked also in the hydrogenation of cyclopentadiene with iron carbonyls [52] and in the recently reported hydroformylation of propylene with [CpFe(CO)$_2$]$_2$ [54].

3.2.2. CARBONYL DERIVATIVES OF COBALT

Carbonyl complexes of cobalt have been extensively investigated as hydrogenation catalysts; as a rule, these complexes show higher activity than iron-carbonyls, so that lower temperatures have been used. When simple cobalt-carbonyls are employed, partial decomposition of the catalyst can take place; also the presence of carbon monoxide generally brings about the formation of some carbonylation products [55]. The advantage of employing phosphine substituted cobalt-carbonyls is due to their enhanced thermal stability; however, in this instance the reduction does not stop at the monoene stage as is the case when simple cobalt carbonyls are used.

3.2.2.1. *Cobalt*(0)*carbonyl*

Simple [Co(CO)$_4$]$_2$ (2 · 10^{-2} M) has been used in the selective hydrogenation of polyunsatured fatty acid esters (0.7-1.0 M) to the corresponding monoenes in cyclohexane solution at 75 °C and 204 atmospheres of hydrogen pressure [55] (table 4).

The reduction has been reported as finishing at the mono-unsaturated esters stage [56] [57]; extensive geometric and positional isomerization of double bonds takes place during the reaction, due to the well known isomerizing properties of HCo(CO)$_4$.

A complicating factor is introduced by the thermal instability of the catalyst, which brings about some decomposition when operating in the absence of carbon monoxide. However, the catalytic hydrogenation is still considered homogeneous, since any heterogeneous catalysis from finely divided cobalt would be poisoned by carbon monoxide evolved from the decomposition [56].

Reduction of polyenes to monoenes using cobalt carbonyl formed "in situ" from bivalent cobalt salts has also been reported [58].

Interestingly, Co$_4$(CO)$_{12}$ (2.6 · 10^{-3} M) hydrogenates 1,3-butadiene (2.0 M) at 80 °C and 80 atmospheres of hydrogen pressure without appreciable formation of butane [59].

3.2.2.2. *Substituted cobalt*(0)*carbonyls*

The use of substituted cobalt carbonyls in the homogeneous hydrogenation of linear and cyclic polyenes has been recently reported [60] [61] (table 4).

Table 4

SELECTIVE HYDROGENATION CATALYZED BY TRANSITION METAL CARBONYLS

Catalyst	Substrate (% Convn.)	Product(s) (% Compn.)	Conditions	Ref.
Fe(CO)$_5$	Methyl Linoleate (88.0)	Octadecanoate (15.0) Octadecenoates (74.7) Conj. Octadecadienoates (10.3)	H$_2$ = 27 atm, 180 °C Cyclohexane	[45]
DiphenylbutadieneFe(CO)$_3$	Methyl Sorbate (91.2)	Hexanoate (8.2) 2-Hexenoate (55.9) 3-Hexenoate (26.0) 4-Hexenoate (9.9)	H$_2$ = 13.5 atm, 165 °C Benzene	[50]
[CpFe(CO)$_2$]$_2$	1,3-Butadiene (77.0)	Butane (5.2) 1-Butene (27.6) cis-2-Butene (29.2) trans-2-Butene (22.6) Higher hydrocarbons (15.4)	H$_2$ = 150 atm, 150 °C Heptane	[53]
[Co(CO)$_4$]$_2$	Methyl Linoleate (40.0)	Octadecenoates (87.5) Conj. Octadecadienoates (12.5)	H$_2$ = 204 atm, 75 °C Cyclohexane	[56]
Co$_4$(CO)$_{12}$	1,3-Butadiene (85.3)	1-Butene (35.0) cis-2-Butene (39.3) trans-2-Butene (25.7)	H$_2$ = 80 atm, 84 °C Heptane	[59]
[Co(CO)$_3$P(n-C$_4$H$_9$)$_3$]$_2$	1,5,9-Cyclododecatriene (100)	Cyclododecane (2.7) Cyclododecene (97.3)	H$_2$ = 25 atm, 140 °C Benzene	[60]
[Co(CO)$_3$P(C$_6$H$_{11}$)$_3$]$_2$	1,5,9-Cyclododecatriene (100)	Cyclododecane (1.2) Cyclododecene (98.8)	H$_2$ = 25 atm, 110 °C Benzene	[60]
[Co(CO)$_3$P(n-C$_4$H$_9$)$_3$]$_2$	1,5-Cyclooctadiene (100)	Cyclooctane (13.0) Cyclooctene (87.0)	H$_2$ = 25 atm, 145 °C Benzene	[60]

Catalyst	Substrate	(%)	Products	Conditions	Ref.
[Co(CO)₃P(n-C₄H₉)₃]₂	1,5-Hexadiene	(100)	Hexane (15.6) Unspecified Hexene (84.4)	H₂ = 25 atm, 145 °C Benzene	[60]
[Co(CO)₃PPh₃]₂	1,4-Cyclohexadiene	(70.1)	Cyclohexane (4.8) Cyclohexene (92.2) 1,3-Cyclohexadiene (3.0)	H₂ = 30, 155-170 °C Benzene	[60]
[Co(CO)₃PPh₃]₂	1,4-Cyclohexadiene	(100)	Cyclohexane (2.5) Cyclohexene (97.5) Oxo products (not considered)	H₂ = 30 atm, CO = 5 atm, 170-175 °C, benzene	[60]
[Co(CO)₃PPh₃]₂	1,5,9-Cyclododecatriene		Cyclododecene (18.6) Cyclododecadiene (30.0) Unspecified Cyclododecatriene (51.4)	H₂ = 25 atm, 155 °C Benzene	[60]
[Co(CO)₂P(n-C₄H₉)]₃	1,3-Butadiene	(83.0)	Butane (14.8) 1-Butene (60.3) cis-2-Butene (15.7) trans-2-Butene (9.2)	H₂ = 30 atm, 66 °C Heptane	[61]
Cr(CO)₆	Methyl Sorbate	(100)	2-Hexenoate (3.2) 3-Hexenoate (96.8)	H₂ = 48 atm, 165 °C Acetone	[65]
CycloheptatrieneCr(CO)₃	Methyl cis-9,trans-11-Octadecadienoate (95.0)		Octadecenoate(s)	H₂ = 30 atm, 125 °C Cyclohexane	[24]
CycloheptatrieneCr(CO)₃	Methyl Sorbate	(100)	Hexanoate (1.0) 2-Hexenoate (1.0) 3-Hexenoate (98.0)	H₂ = 30 atm, 120 °C Cyclohexane	[66]
ChlorobenzeneCr(CO)₃	Methyl Sorbate	(100)	Hexanoate (4.3) 2-Hexenoate (0.1) 3-Hexenoate (95.6)	H₂ = 48 atm, 150 °C Cyclohexane	[67]

Table 4

SELECTIVE HYDROGENATION CATALYZED BY TRANSITION METAL CARBONYLS

Catalyst	Substrate (% Convn.)		Product(s) (% Compn.)		Conditions	Ref.
Methyl BenzoateCr(CO)₃	Methyl Sorbate	(100)	Hexanoate 2-Hexenoate 3-Hexenoate	(1.1) (0.1) (98.8)	H₂ = 48 atm, 150 °C Cyclohexane	[67]
Methyl BenzoateCr(CO)₃	Methyl linoleate	(94.8)	9-Octadecenoate 10-Octadecenoate 11-Octadecenoate 12-Octadecenoate Other Octadecenoates	(24.3) (24.1) (22.5) (22.3) (6.8)	H₂ = 30 atm, 175 °C Cyclohexane	[24]
Methyl BenzoateCr(CO)₃	Methyl trans-9,trans-11,trans-13-Octadecatrienoate	(100)	10-Octadecenoate 12-Octadecenoate 9,12-Octadecadienoate 10,13-Octadecadienoate Conj. Octadecadienoate Octadecatrienoate	(18.5) (71.0) (7.5) (3.0)	Not reported	[67]
Methyl BenzoateCr(CO)₃	1,3-Hexadiene	(100)	1-Hexene cis-2-Hexene trans-2-Hexene cis-,trans-3-Hexene	(6.0) (66.0) (10.0) (18.0)	H₂ = 30 atm, 160 °C Pentane	[24]
Methyl BenzoateCr(CO)₃	2,4-Hexadiene	(100)	2-Hexene 3-Hexene	(10.0) (90.0)	H₂ = 30 atm, 160 °C Pentane	[24]
Methyl BenzoateCr(CO)₃	(structure)	(5.0)	(structure)		H₂ = 30 atm, 160 °C Pentane	[24]

Catalyst	Substrate		Products		Conditions	Ref.
Methyl BenzoateCr(CO)₃	1,3-Cyclohexadiene	(100)	Cyclohexane Cyclohexene	(2.0) (98.0)	H₂ = 30 atm, 160 °C Hexane	[24]
Methyl BenzoateCr(CO)₃	1,4-Cyclohexadiene	(93.0)	Cyclohexene		H₂ = 30 atm, 160 °C Hexane	[24]
Methyl BenzoateCr(CO)₃	1,5-Cyclooctadiene	(100)	Cyclooctene 1,3-Cyclooctadiene 1,4 Cyclooctadiene	(70.0) (28.0) (2.0)	H₂ = 30 atm, 170 °C Hexane	[24]
BenzeneCr(CO)₃	Methyl Sorbate	(99.6)	Hexanoate 2-Hexenoate 3-Hexenoate	(2.1) (3.6) (94.3)	H₂ = 48 atm, 165 °C Cyclohexane	[67]
BenzeneCr(CO)₃	Methyl cis-9, trans-11-Octadecadie-noate	(100)	9-Octadecenoate 10-Octadecenoate Other octadecenoates	(42.0) (41.0) (17.0)	H₂ = 30-48 atm, 175 °C Cyclohexane	[24, 67]
BenzeneCr(CO)₃	Methyl trans-9, trans-11-Octadecadie-noate	(100)	10-Octadecenoate Other Octadecenoates	(92.0) (8.0)	H₂ = 30 atm, 175 °C Cyclohexene	[24]
PhenanthreneCr(CO)₃	Methyl Sorbate	(100)	2-Hexenoate 3-Hexenoate	(97.4) (2.6)	H₂ = 40 atm, 150 °C Cyclohexane	[50]
Stilbene[Cr(CO)₃]₂	Methyl Sorbate	(100)	2-Hexenoate 3-Hexenoate	(1.7) (98.3)	H₂ = 4.1 atm, 100 °C Acetone	[65]
1,4-Diphenylbutadiene-[Cr(CO)₃]₂	Methyl Sorbate	(98.4)	2-Hexenoate 3-Hexenoate	(1.8) (98.2)	H₂ = 4.1 atm, 100 °C Acetone	[65]
	Methyl Sorbate	(100)	3-Hexenoate Other Products	(99.0) (1.0)	H₂ = 4.8 atm, 115 °C Cyclohexane	[67]

Table 4

SELECTIVE HYDROGENATION CATALYZED BY TRANSITION METAL CARBONYLS

Catalyst	Substrate (% Convn.)	Product(s) (% Compn.)	Conditions	Ref.
CycloheptatrieneMo(CO)$_3$	Methyl Sorbate (100)	2-Hexenoate (10.0) 3-Hexenoate (90.0)	H$_2$ = 30 atm, 100 °C Cyclohexane	[66]
MesityleneMo(CO)$_3$	Methyl Sorbate (100)	Hexanoate (1.0) 2-Hexenoate (11.0) 3-Hexenoate (86.0) 4-Hexenoate (2.0)	H$_2$ = 30 atm, 100 °C Cyclohexane	[66]
MesityleneMo(CO)$_3$	Methyl Sorbate (100)	Hexanoate (9.0) 2-Hexenoate (8.0) 3-Hexenoate (28.0) 4-Hexenoate (55.0)	H$_2$ = 30 atm, 150 °C Cyclohexane	[66]
MesityleneW(CO)$_3$	Methyl Sorbate (77.0)	Hexanoate (15.6) 2-Hexenoate (55.8) 3-Hexenoate (16.9) 4-Hexenoate (11.7)	H$_2$ = 30 atm, 165 °C Cyclohexane	[66]
[CpCr(CO)$_3$]$_2$	Isoprene (ca. 100)	2-Methyl-2-Butene (95.0) 2-Methyl-1-Butene (3.0) 3-Methyl-1-Butene (2.0)	H$_2$ = 90 atm, 70 °C Benzene	[71]
[CpCr(CO)$_3$]$_2$	1,3-Pentadiene	trans-2-Pentene (88.0) cis-2-Pentene (12.0)	H$_2$ = 90 atm, 70 °C Benzene	[71]
[CpCr(CO)$_3$]$_2$	4-Methyl-1,3-Pentadiene	2-Methyl-2-Pentene (78.0) 4-Methyl-2-Pentene (22.0)	H$_2$ = 90 atm, 70 °C Benzene	[71]

Catalyst	Substrate		Products		Conditions	Ref.
[CpCr(CO)$_3$]$_2$	1,3-Cyclohexadiene	(91.0)	Cyclohexene	(100)	H$_2$ = 90 atm, 70 °C Benzene	[71]
[CpCr(CO)$_3$]$_2$	1,3,5-Cyclooctatriene		1,5-Cyclooctadiene 1,4-Cyclooctadiene	(74.0) (26.0)	H$_2$ = 90 atm, 70 °C Benzene	[71]
CpMo(CO)$_3$H	Isoprene		2-Methyl-2-Butene 2-Methyl-1-Butene 3-Methyl-1-Butene	(90.1) (6.6) (3.3)	N$_2$ = 1 atm, 59 °C Heptane	[72]
CpMo(CO)$_3$H	2,4-Hexadiene	(50.0)	trans-2, trans-3-Hexene cis-2,cis-3-Hexene	(79.0) (21.0)	N$_2$ = 1 atm, 59 °C Heptane	[72]
CpMo(CO)$_3$H	4-Methyl-1,3-Pentadiene	(50.0)	2-Methyl-2-Pentene 4-Methyl-2-Pentene	(76.0) (24.0)	N$_2$ = 1 atm, Heptane $\{$ t$^{1}/_2$ = 10 hrs at 20 °C $\{$ t$^{1}/_2$ = 0.2 hrs at 59 °C	[72]
CpMo(CO)$_3$H	2,4,6-Octatriene	(50.0)	Probably: 3,5-Octadiene 2,4-Octadiene	 (83.0) (17.0)	N$_2$ = 1 atm, 20 °C Heptane 0.1 hrs	[72]
CpMo(CO)$_3$H	1,3,5-Hexatriene	(50.0)	2,4-Hexadiene: trans, trans 2,4-Hexadiene trans, cis 2,4-Hexadiene cis, cis 2,4-Hexadiene	 (66.0) (31.0) (3.0)	N$_2$ = 1 atm, 20 °C Heptane 0.03 hrs	[72]
CpW(CO)$_3$H	1,3,5-Hexatriene	(50.0)	2,4-Hexadiene: trans, trans 2,4-Hexadiene trans, cis 2,4-Hexadiene cis, cis 2,4-Hexadiene	 (66.0) (30.0) (4.0)	N$_2$ = 1 atm, 20 °C Heptane 9 hrs	[72]
CpW(CO)$_3$H	4-Methyl-1,3-Pentadiene	(50.0)	2-Methyl-2-Pentene 4-Methyl-2-Pentene	(72.0) (28.0)	N$_2$ = 1 atm, 60 °C Heptane 70 hrs	[72]

Complexes of the type $[Co(CO)_3PR_3]_2$ and $[Co(CO)_3(PR_3)_2][Co(CO)_4]$ $(1.4 \cdot 10^{-2}$ M) (R = alkyl, cycloalkyl, aryl) selectively reduce polyenes (1.1-2.6 M) to monoenes in benzene solution at 110-155 °C and 25 atmospheres of hydrogen pressure [60] [62].

The catalytic activity depends strongly on the nature of the tertiary phosphine being in the order $P(C_6H_{11})_3 > P(n\text{-}C_4H_9)_3 > P(C_6H_5)_3$ and decreases parallel to the decrease of the σ-donor properties [60] [62]. It is interesting to note that the use of excess phosphine (n-butyl) results in a decrease in selectivity; this effect has been explained by the formation of a different catalytic species, i.e. $CoH(CO)_2[P(n\text{-}C_4H_9)_3]_2$ [60] [63].

The dependence of the selectivity on the structure of the olefin is not great [60] (1,5,9-cyclododecatriene > 1,5-cyclooctadiene > 1,5-hexadiene). The reason for the best selectivity being achieved with 1,5,9-cyclododecatriene is not yet clear, but it might be due to some steric factors: the alkyl complex produced from cyclododecene and the metal hydride may not be sufficiently stable to be attacked by hydrogen, and may prefer to eliminate the monoene again [60].

The rate of hydrogenation of olefins depends strongly on the presence of other substrates; e.g. in the $[Co(CO)_3Pbu_3]_2$ catalyzed hydrogenation of 1,5,9-cyclododecatriene, the yield of cyclododecene is sharply decreased (from 97.3 to 2.4%) in the presence of approximately equimolar amounts of 1,5-hexadiene under otherwise identical conditions. This result is tentatively explained by assuming that hydrogenation takes place on conjugated systems (*via* allyl intermediates) whose formation is more difficult in a C_{12}-ring than in the linear hexadiene.

A cluster compound of possible formula $[Co(CO)_2PBu_3]_3$ [61] $(1.9 \cdot 10^{-2}$ M) has been used in the selective hydrogenation of 1,3-butadiene (1.0 M) mainly to 1-butene in heptane solution (table 4). Mild conditions (66 °C and 15 atm.) are required for this reaction which shows a typical autocatalytic trend. The π-allyl complex (π-C_4H_7)Co-$(CO)_2PBu_3$, has been isolated from the reaction mixture, and the same catalytic activity has been obtained using this π-allyl complex as catalyst.

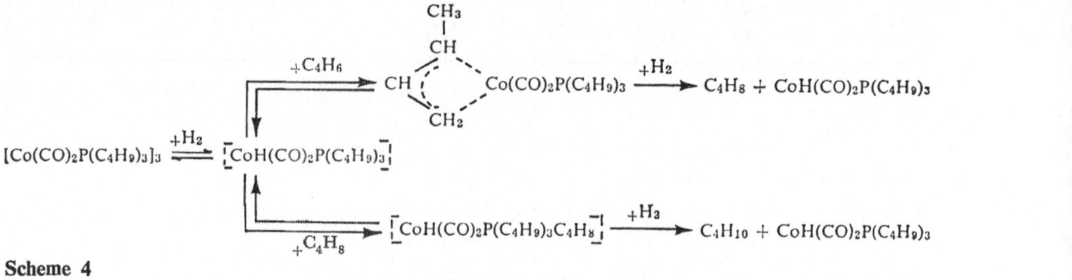

Scheme 4

High selectivity to butenes (c.a. 90%) has been obtained up to a conversion of 70% [61]; since 1,3-butadiene hydrogenates more slowly than butenes, a strong competition of the diene with respect to the monoene on the catalyst (scheme 4) obviously determines the favourable selectivity trend. Probably the actual catalyst of the reaction is a co-ordinatively unsaturated hydride, derived from the cluster compound [64], whose existence has been recently postulated [61].

3.2.3. CARBONYL DERIVATIVES OF CHROMIUM, MOLYBDENUM AND TUNGSTEN

Substituted carbonyl complexes of Cr, Mo and W have been used in the last three years as catalysts for the selective hydrogenation of several poly-unsaturated substrates.

Arene and cyclopentadienyl carbonyls of the above metals have also been studied as catalysts. Interestingly, simple carbonyls of this group have been investigated very little; only a short report has recently appeared [65] concerning the catalytic activity of $Cr(CO)_6$ in the hydrogenation of methyl sorbate to methyl 3-hexenoate (table 4).

A great deal of work in this area has been devoted to the investigation of the catalytic properties of π-arene-$Cr(CO)_3$; this is understandable in view of the outstanding stereoselective aspects of these catalytic systems.

3.2.3.1. π-arene-metal carbonyls

i) *General considerations.* Arene-metal tricarbonyl complexes (10^{-2} M) of formula π-arene-$M(CO)_3$ (M=Cr, Mo, W) are excellent catalysts in the selective homogeneous hydrogenation of 1,3- and 1,4-dienes ($2 \cdot 10^{-1}$ M) to monoenes [24] [50] [66-69] at 100-180 °C and 30-50 atmospheres of hydrogen pressure (table 4). The order of catalytic activity of several complexes of Cr, Mo, and W has been related to the thermal stability of the complexes during hydrogenation [66]. The activity of mesitylene-$M(CO)_3$ decreases in the order Mo < W < Cr, and selectivity in the order Cr < Mo < W [66]. While the catalytic activity is modified by the presence of substituents on the arene group, in the order [66]:

$$Cl > COOCH_3 > H > CH_3 > (CH_3)_3 > (CH_3)_6$$

any change of the arene moiety does not appear to greatly effect the selectivity of the system [66] [67].

It appears that electron-withdrawing substituents increase the catalytic activity, while the opposite is true for electron-repelling ones. It is interesting to note that cycloheptatriene-$Cr(CO)_3$ shows higher activity than benzene-$Cr(CO)_3$ [67].

Interestingly, the presence of two $Cr(CO)_3$ moieties per molecule of complex (e.g. α, β, γ) greatly improves the efficiency of the catalyst [65] [67]; thus for instance almost quantitative hydrogenation of methyl sorbate ($2 \cdot 10^{-1}$ M) to methyl-3-hexenoate has been obtained in 30 minutes by using complex γ ($2.7 \cdot 10^{-3}$ M) as the catalyst in hexane solution at 115 °C and 5 atmospheres of hydrogen pressure [67] (table 4).

The activity of arene-metal carbonyls has been reported to increase sharply parallel to the increase of the polarity of the reaction media [65] [67].

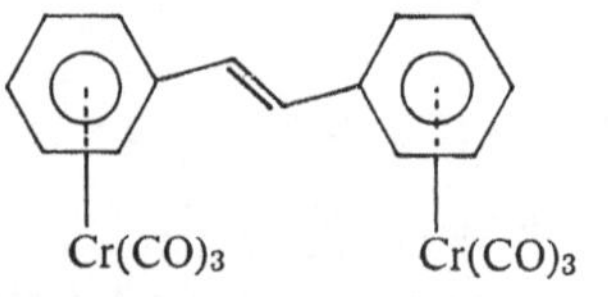

(α)

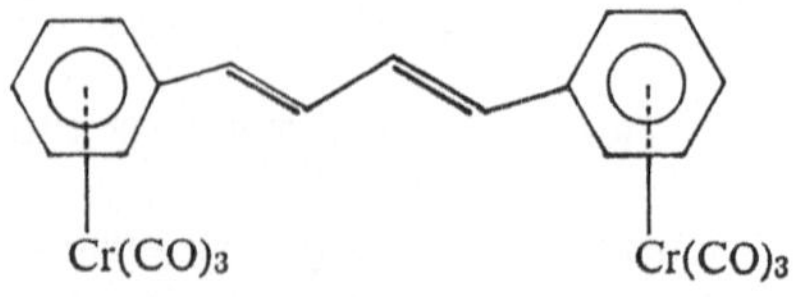

(β)

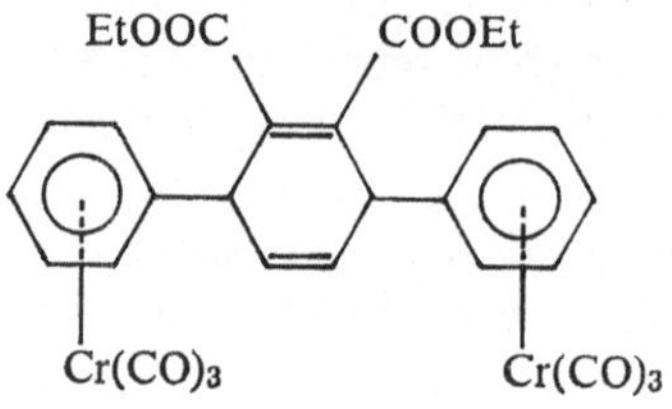

(γ)

An important feature of these catalytic systems is their ability to give mainly monoenes; the type of isomer which prevails depends strongly on the metal and on the reaction conditions (table 4); thus, for example, in the hydrogenation of methyl-sorbate, mainly methyl 4-hexenoate is obtained using molybdenum complexes, while methyl 2-hexenoate is the principal product in the presence of tungsten derivatives [66]. On the other hand, the reduction of the same substrate in the presence of chromium complexes gives almost exclusive formation of methyl 3-hexenoate [66].

Moreover, chromium catalyzed reactions are highly stereoselective, in that monoenes with high *cis* content (85 to 94%) can be obtained [24] [66-69] by hydrogenation of several dienes, the catalyst having negligible isomerising properties. This result is very important since other selective homogeneous or heterogeneous catalysts [66] generally give mainly *trans* monoenes.

For the above reasons great attention has been paid especially to the arene-chromium carbonyl system.

ii) *Mechanism of hydrogenation of conjugated dienes catalyzed by chromium complexes.* π-arene-chromium carbonyls show no activity in the hydrogenation of either internal or terminal mono-olefins [24] [69]; consequently, almost exclusive formation of monoenes is obtained over a wide range of reaction conditions.

These catalytic systems reduce dienes to monoenes by 1,4-hydrogen addition.

This mechanism has been clearly demonstrated by deuterium tracer studies in the case of arene-chromium carbonyl catalyzed hydrogenation of methyl sorbate [68] [69], 1,3-cyclohexadiene [69] and methyl 9,11-octadecadienoates [69].

On the basis of kinetic and chemical data, the following sequence of reactions has been postulated for the methyl sorbate hydrogenation [24] [68].

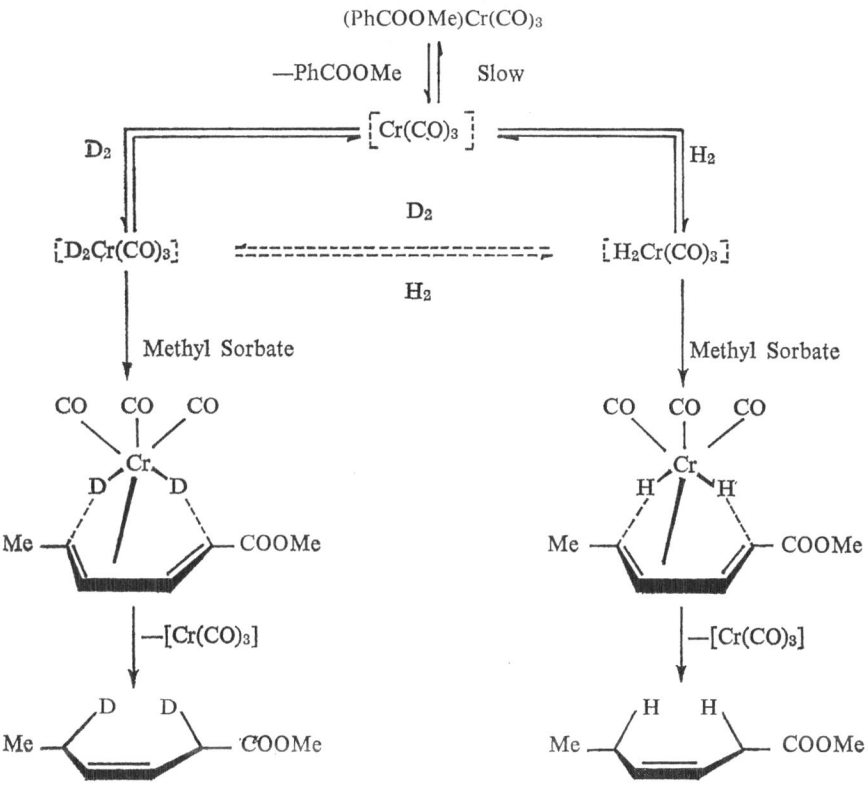

Scheme 5

The primary dissociation of π-arene-chromium carbonyls is supported by the detection, in the course of the reaction, of free aromatic ligands from the catalyst; on the other hand the reversibility of the above equation is substantiated by the inhibiting effect of added aromatics [24] [68]. Finally the existence of an induction period implies that the above dissociation is a slow step [24] [68].

Conceivably the hydrogen (or deuterium) activation is not rate determining; approximately the same rate has been obtained in the hydrogenation of methyl sorbate using hydrogen or deuterium [68]. On the other hand, preferential formation of hydrogenated with respect to deuterated monoene when using hydrogen-deuterium mixture has been found.

These results indicate that in the fast equilibrium between $H_2Cr(CO)_3$ and $D_2Cr(CO)_3$ (scheme 5) the former is thermodynamically favoured [68].

The following step, i.e. addition of the dideuteride (dihydride) to *trans-trans*-methyl-2,4-hexadienoate, is not reversible since only d_0 residual diolefins have been found in samples partially reduced with deuterium. Infrared, ^{1}H and ^{2}H n.m.r. and mass spectrometry indicate the presence, in the resulting monoene, of two deuterium atoms almost exclusively located on the methylene carbon atoms; this result substantiates the formation of the —CHD—CH=CH—CHD— group by 1,4 addition of D$_2$ to the conjugated double bond [24] [68] [69].

Table 5 [24]

HYDROGENATION OF SEVERAL DIENES (0.2M) CATALYZED BY METHYL BENZOATE--Cr(CO)$_3$ (10^{-2}M) $^{(a)}$

Substrate (% Convn.)		$k^{(b)}$ (hr^{-1})	Product(s) (% Compn.)			
1,3-Hexadiene	(100)	0.9	1-Hexene (6)	cis-2-Hexene (66)		
				trans-2-Hexene (10)		
				cis-,trans-3-Hexene (18)		
2,4-Hexadiene$^{(c)}$	(100)	2.1	cis-2-Hexene (10)	cis-3-Hexene (90)		
1,4-Hexadiene$^{(d)}$	(77)	0.22$^{(e)}$	cis-2-Hexene (23.3)	cis-,trans-3-Hexene (52.0)		
			trans-2-Hexene (24.7)			
1,5-Hexadiene	(2)	0.004$^{(e)}$	1-Hexene (50)	2-Hexene(s) (50)		
(CH$_3$ branched diene structure)	(100)	0.69	(branched hexene structure, 75)	(branched hexene structure, 17)		
			(branched hexene structure, 8)			
(CH$_3$ CH$_3$ branched diene structure)	(100)	0.64	(branched alkene structure, 82)	(branched alkane structure, 18)		
(H$_3$C–...–CH$_3$, CH$_3$ CH$_3$ diene structure)	(5)	0.007	(H$_3$C–...–CH$_3$, CH$_3$ CH$_3$ structure, 100)			
1,3-Cyclohexadiene	(100)	0.5	Cyclohexene (98)	Cyclohexane (2)		
1,3-Cyclooctadiene	(25)	0.05	Cyclooctene (100)			

(a) At 160 °C and 30 atm of H$_2$ for 6 hrs in pentane or hexane solution.
(b) Decrease in diene
(c) cis, trans (48); trans, trans (52); reaction time 2 hrs
(d) cis (28), trans (72)
(e) At 175 °C

The 100% *cis* configuration of the final 3-hexenoate has been explained by assuming 1,4 addition of $H_2(D_2)$ to a *cisoid* complexed diene system and $H_2(D_2)$ transfer (scheme 5).

Various conjugated dienes have been examined in order to determine the effect of steric configuration on the kinetics of the above 1,4 hydrogen addition [24].

The *trans,trans* diene-chromium intermediate (scheme 5) corresponds to the most favourable situation as far as stereochemical factors are concerned; indeed in this case both substituents are in the *syn* position and favourably situated for 1,4 addition. However, when *cis, trans* and *cis, cis* dienes are involved, the presence of one or both substituents in *anti* position interferes sterically with the concerted hydrogen (deuterium) addition.

It has been found that relative rates are 1.0 for *cis, cis*, 8.0 for *cis, trans* and 25.0 for *trans, trans* diene [24], in the hydrogenation of 9,11-octadecadieno-ates catalyzed by benzene-Cr(CO)₃.

The influence of the structure on the hydrogenation rate for several dienes has been studied [24] and the results are reported in table 5; 1,4-disubstitution (2,4-hexadiene) favours the reaction with respect to 4-monosubstitution (1,3-hexadiene); on the other hand the rates are dramatically decreased by the presence of several substituents on the 1- and 4-position (2,5-dimethyl-2,4-hexadiene). The presence of substituents on C-2 and C-3 of the diene system seems to decrease the rate. Finally, the facile hydrogenation of 1,3-cyclohexadiene as compared to 1,3-cyclooctadiene has been related to a more favourable conformation of the former for diene complex formation with the catalyst [24].

Generally speaking not only the kinetic of hydrogenation, but also the double bond distribution of monoenes can be affected by the steric configuration of the starting dienes: e.g. it has been reported [24] that methyl *trans*-9-, *trans*-11-octadecadienoate is reduced by benzene-Cr(CO)₃ almost exclusively to 10-octa-decenoate (37) while methyl *cis*-9-, *trans*-11-octadecadienoate is reduced to a mixture of nearly equal amounts of 9- and 10-monoenes (38):

$$\text{(37)}$$

$$\text{(38)}$$

In the latter case the closeness of C_8 hydrogen to C_{12} possibly brings about a 1,5-hydrogen shift [70].

The 1,4-addition of hydrogen to both 8,10- and 9,11-dienes explains the formation of 9- and 10- monoenes [24].

iii) *Hydrogenation of unconjugated dienes catalyzed by chromium complexes*: *isomerizing properties of the catalytic system*. In the case of unconjugated dienes, preliminary conjugation is required for hydrogenation to take place [24]: on the other hand, since π-arene-chromium carbonyls have low isomerizing properties, this condition is hardly satisfied even at high temperatures whenever more than one methylene group is situated between the two double bonds. Little or no hydrogenation of 1,5-hexadiene (table 5) and of methyl *cis* 9-, *cis* 15-octadeca-dienoate has been observed with methyl benzoate-Cr(CO)$_3$ at 175 ºC [24]. Conversely, selective hydrogenation of 1,4-dienes to monoenes takes place, provided sufficiently high temperatures are used (tables 4, 5) to allow isomeri-zation [66].

In this respect methyl benzoate- and benzene-Cr(CO)$_3$ are suitable catalysts by virtue of their thermal stability in the reaction conditions [24] [66].

The reaction path involves a preliminary slow isomerization, followed by fast 1,4 hydrogen addition to the resulting conjugated system [24]. This sequence has been demonstrated by deuterating 1,4-cyclohexadiene in the presence of methylbenzoate-Cr(CO)$_3$ as the catalyst.

The resulting cyclohexene-d_2 was found to contain 90% of the deuterium atoms located on the α-methylene carbons; thus the following reaction path can be formulated [69]:

$$\text{(39)}$$

Whenever π-arene-chromium carbonyls are used in such conditions in which they have no isomerizing properties, reduction of only 1,3-dienes in a mixture with 1,4-dienes takes place. This is the case of cycloheptatriene-Cr(CO)$_3$, which is very active in the hydrogenation of 1,3-dienes to monoenes at 120 ºC, but inert with respect to the isomerization of 1,4- to 1,3-dienes [66].

On the other hand preferential hydrogenation of 1,3-dienes in a mixture with 1,4-dienes has also been carried out by using isomerizing catalysts under the same reaction conditions in which the two substrates are individually hydro-generated. The reason for the above result can be ascribed to the preferential co-ordination on the catalyst of the conjugated diene with respect to the uncon-jugated one [24].

Competition on the chromium complex between 1,3- or 1,4-dienes and resulting monoenes is also the reason for the high obtained stereoselectivity (84-94% *cis*-monoene) [24]. In fact the isomerization of monoenes and 1,6-diene fatty esters has been observed [24] [69] as taking place to a certain extent under the hydrogenation conditions (1,3- and 1,4-diolefins being absent) together with hydrogen-deuterium exchange which involves methylene groups in α-position to the double bond. On the other hand, since all monoenes from the reduction of conjugated diolefins have a deuterium content close to 2.0, it has also been deduced [69] that the H-D exchange of monoene products is effectively inhibited by the presence of dienes.

3.2.3.2. *Cyclopentadienyl-metal carbonyls*

Bis(tricarbonylcyclopentadienylchromium) $(4 \cdot 10^{-2}$ M) has been used as catalyst in the selective hydrogenation of conjugated dienes and polyenes (3 M) at 70 °C and 90 atmospheres of hydrogen pressure (table 4) [71]. The catalytic system shows no isomerizing properties, so that the composition of the resulting monoenes corresponds to that of the primary products. Interestingly this catalyst is completely inactive with respect to both mono-olefins and polyolefins containing isolated double bounds; for this reason high selectivity to monoenes results in the reduction of conjugated systems.

Based on the nature of the resulting monoenes (table 4), it is likely that 1,4 addition of a hydridic species takes place, according to the following equation:

$$2 \, HCrCp(CO)_3 + CH_2{=}C{-}CH{=}CH_2 \rightarrow [CrCp(CO)_3]_2 + CH_3{-}C{=}CH{-}CH_3 \qquad (40)$$
$$\overset{|}{CH_3} \qquad\qquad\qquad\qquad \overset{|}{CH_3}$$

However, the occurrence of steric factors may occasionally favour 1,2-addition [71].

In contrast to the chromium complexes, the corresponding molybdenum and tungsten dimers are not cleaved by molecular hydrogen to the corresponding hydrides [72]; this is the reason why stoichiometric hydrogenation could only be obtained using preformed hydrides [72]. The nature of the reaction products is in accordance with a mechanism similar to that suggested for the chromium catalyzed hydrogenation. A decreasing activity in the order Cr > Mo > W has been found for these catalysts [72].

3.3. Noble metal complexes

Noble metal complexes have largely been used as catalysts in the homogeneous hydrogenation of several unsaturated substrates.

Suitable ligands having donor-acceptor properties such as phosphines, arsines, $SnCl_3^-$, etc. are to be used in order to stabilize the oxidation state of the metal relative to the reduction by hydrogen. In the absence of the above ligands, rather unstable active systems are obtained on a borderline between heterogeneous and homogeneous catalysts [73] [74].

A comprehensive description of noble metal catalytic systems has been given in another article of this series [1], with special reference to the mechanism of molecular hydrogen activation and to the kinetics of olefin hydrogenation.

Comparatively few papers have appeared concerning the homogeneous selective hydrogenation of polyenes to monoenes by these systems; complexes containing halogens, cyanide groups and tertiary phosphines as ligands have been used, in which the central atom was Rh(I), Ru(II), Pd(II) or Pt(II).

Special attention has been paid to the systems in which the activity of Pd(II) and Pt(II) complexes is enhanced by the addition of $SnCl_2$.

3.3.1. RHODIUM(I) COMPLEXES

The hydrogenating properties of $RhCl(PPh_3)_3$ towards the olefinic substrates have been fully described in another article of this series [1]; its activity in the selective hydrogenation of conjugated dienes to monoenes has also been reported [6] [75].

Using $RhCl(PPh_3)_3$ as catalyst (10^{-2} M), Candlin and Oldham [6] have obtained mainly cis-2-pentene as the intermediate product of the reduction of cis-1,3-pentadiene (1 M) in benzene solution at 25 °C and 1 atm hydrogen pressure (table 6). This result is in accordance with the very low isomerizing properties of the rhodium complex [76] [77]. In this case competition between the diene and monoene on the catalyst causes the reaction selectivity since kinetic factors have been shown to be unfavourable [6] [76] [78]. Separate comparative experiments have shown that 1,3-pentadiene is hydrogenated more slowly (relative rate 0.12) than both α-olefins (relative rate 1.0) and internal straight chain olefins (relative rate 0.25). However, when using a 50/50 mixture, say, of 1-pentene and 1,3-pentadiene, the monoene reduction is so strongly depressed that a lower reaction rate by a factor of 1.3 has been observed for the α-olefin with respect to the diene. Obviously the above factor will be higher when 1,3-pentadiene is hydrogenated in the presence of cis-2-pentene.

During the hydrogenation of 1,4-pentadiene in the above conditions, no isomerization of the reactant takes place and only 1-pentene is formed as intermediate (yields have not been reported).

3.3.2. RUTHENIUM(II) COMPLEXES

Highly selective hydrogenation of 1,3-pentadiene (1.2 M) to 2-pentene is carried out [79] by hydridochlorotris(triphenylphosphine)ruthenium(II) (8.3 · 10^{-4} M) in benzene solution at 25 °C and 50 cm. partial hydrogen pressure, the observed gas uptake being 54 ml min^{-1} l^{-1} (table 6).

Other authors [6] have pointed out that the hydrogenation of cis-1,3-pentadiene (1 M) catalyzed by $RuHCl(PPh_3)$ (10^{-2} M) is highly stereoselective insofar as cis-2-pentene (95% yield) is obtained; the reaction takes place at room temperature and atmospheric pressure (table 6). In this case the very good selectivity can be attributed to the negligible activity of the catalyst towards the internal olefin [79].

Interestingly, a mixture of nearly equal amounts of 1-butene and *cis*-2-butene is obtained from 1,3-butadiene in the same conditions (yields not reported) [6]. According to the above results, since $RuHCl(PPh_3)_3$ exhibits no isomerizing properties, it is reasonable to assume that reaction intermediates of type I and II are involved, in the case of C_4 and C_5 diolefins respectively.

In fact, while only *cis*-2-pentene can be obtained from complex II, formation of 1-butene or *cis*-2-butene takes place from complex I depending on the mode of opening of the allylic moiety. This mechanism is in agreement with the above experimental data, although it has been reported [80] [81] that π-allyl complexes of Co(I) and Pd(II) are less stable in the *anti* form.

$RuCl_2(PPh_3)_4$ and $RuCl_2(PPh_3)_3$ can also be used in the selective hydrogenation of dienes to monoenes; the formation of $RuHCl(PPh_3)_3$ has been proved during the reduction of α-olefins [82] [83].

3.3.3. PALLADIUM(II) AND PLATINUM(II) COMPLEXES

Palladium complexes of general formula PdX_2L_2, where $L = PPh_3$, $AsPh_3$ and $X = CN$, Cl, catalyze the hydrogenation of polyunsaturated fatty acid esters to monoenes with high selectivity [84]. The reactions were carried out on soybean oil methyl ester (39 g/l), with a $1.2 \cdot 10^{-2}$ M catalyst concentration in 3 : 2 benzene-methanol solution at 90 °C and 39 atmospheres of hydrogen pressure. The activity of the above complexes depends strongly upon the nature of the ligands and changes as follows: $PPh_3 > AsPh_3$ and $CN > Cl$. It has been observed [84] that some decomposition takes place whenever arsine complexes are employed. Using the most active catalyst, i.e. $Pd(CN)_2(PPh_3)_2$, almost quantitative conversion of polyenes to monoenes has been obtained in twelve hours, the selectivity to monoenes being 97.5% (table 6).

Interestingly, the above hydrogenations can also be carried out in nitrogen atmosphere, methanol being the hydrogen source [84].

Platinum complexes of similar formula PtX_2L_2 ($6.7 \cdot 10^{-3}$ M), where $L = AsPh_3$, $P(OPh)_3$, PPh_3 and $X = Cl$, have been used [85] in the hydrogenation of methyl linoleate (20 g/l) under the same conditions as above. Best performances have been obtained by using the arsine derivative, in which case monoenes resulted in a 47.5% yield after 5 hours, but catalyst decomposition was observed (table 6).

3.3.4. Bi-metallic systems

Many papers have recently appeared concerning the hydrogenation of unsaturated hydrocarbons catalyzed by complexes of Pt(II) and Pd(II), activated by addition of IV group metal halides, in particular SnCl₂. The mechanism of hydrogen activation for these catalytic systems has been reported in another article of this series [1], together with a full description of the role played by the cocatalysts. Both simple and liganded systems have been used, the latter containing phosphites or tertiary phosphines, arsines and stibines in addition to platinum-(or palladium-)-tin halides. A common feature of the above two systems resides in their ability to promote isomerization of terminal to internal olefins and of unconjugated dienes with only one methylene group between the two double bonds to conjugated systems; conversely, positional isomerization of internal olefins takes place to a lesser extent [86].

3.3.4.1. *Platinum*(II)-*tin halides*

The hydrogenation of polyunsaturated fatty acid esters to the corresponding monoenes catalyzed by platinum-tin halides has been reported as proceeding with high selectivity [86] [87-89].

Soybean oil (154 g/l) in acetic acid solution is easily reduced to monoenes by means of chloroplatinic acid ($6.7 \cdot 10^{-3}$ M) and tin(II) halide ($3.7 \cdot 10^{-2}$ M), almost quantitative yield being obtained after ten hours at 40 °C and 1 atmosphere of hydrogen pressure (table 6) [87]. Also the hydrogenation of single substrates has been reported: thus, for example methyl linolenate (1 M) has been reduced [89] by using chloroplatinic acid (0.2 M) and SnCl₂ (2 M) in methanol solution at 40 °C and 1 atmosphere of hydrogen pressure (table 6). In this case large amounts of octadienoates with double bonds separated by several methylene groups have been obtained.

During the hydrogenation of both methyl linoleate and linolenate, formation of conjugated double bonds has been observed, due to the extensive isomerizing properties of the catalyst on 1,4-dienic systems; for this reason it has been suggested [85] that primary conjugation is a necessary step in the reduction. Accordingly, polyolefins with double bonds separated by several methylene groups show low reactivity due to the increased difficulty in attaining conjugation [89].

The selectivity of the platinum-tin halide catalyzed hydrogenation of polyenes can be ascribed to the negligible rate of reduction of internal olefins [87] [90], although thermodynamic factors cannot be ruled out.

It is interesting to note that the activity of the catalytic system depends strongly upon the ratio Sn/Pt: at a molar ratio lower than 3 the catalyst is not

Table 6

Selective Hydrogenation of Polyenes Catalyzed by Noble Metal Complexes

Catalyst	Substrate (% Convn.)	Product(s) (% Compn.)	Conditions	Ref.
RhCl(PPh₃)₃	1,3-Pentadiene	2-Pentene	H₂ = 1 atm, 22 °C Benzene	[6]
RhCl(PPh₃)₃	1,3-Cyclooctadiene	Cyclooctene	H₂ = 1 atm, 22 °C Benzene	[6]
RhCl(PPh₃)₃	1,3-Octadiene	Octene	H₂ = 1 atm, 22 °C Benzene	[6]
RuHCl(PPh₃)₃	1,3-Pentadiene	2-Pentene	H₂ = 1 atm, 25 °C Benzene	[79]
RuHCl(PPh₃)₃	cis-1,3-Pentadiene	cis-2-Pentene (yield 95%)	H₂ = 1 atm, 22 °C 1:1 Benzene-Ethanol	[6]
RuHCl(PPh₃)₃	1,3-Butadiene	1-Butene (ca. 50), cis-2-Butene (ca. 50)	H₂ = 1 atm, 22 °C 1:1 Benzene-Ethanol	[6]
Pd(CN)₂(PPh₃)₂	Soybean Oil Methyl Ester (1) (100)	Octadecanoate (2.5), Octadecenoate(s) (97.5)	H₂ = 39 atm, 90 °C 2:3 Methanol-Benzene	[84]
PtCl₂(AsPh₃)₂	Methyl Linoleate (52.0)	Octadecenoate(s) (91.3) cis-,trans-conj., Octadecadienoate(s) (8.7)	H₂ = 39 atm, 90 °C 2:3 Methanol-Benzene	[85]
PtCl₂[P(OPh)₃]₂	Methyl Linoleate (43.6)	Octadecenoate(s) (55.3) cis-,trans-conj., Octadecadienoate(s) (44.7)	H₂ = 39 atm, 90 °C 2:3 Methanol-Benzene	[85]

(1) Soybean Oil Methyl Ester with the following composition: Octadecanoate 4.1%; Octadecenoate 24.0%; Octadecadienoate 51.6%; Octadecatrienoate 9.0%.

Table 6

SELECTIVE HYDROGENATION OF POLYENES CATALYZED BY NOBLE METAL COMPLEXES

Catalyst	Substrate (% Convn.)	Product(s) (% Compn.)	Conditions	Ref.
$K_2PtCl_4 + SnCl_2$	Methyl Linoleate (64.8)	Octadecanoate (1.9) Octadecenoate(s) (66.7) Conj. Octadecadienoates (31.4)	H_2 = 39 atm, 90 °C 2 : 3 Methanol-Benzene	[85]
$H_2PtCl_6 + SnCl_2$	Soybean Oil Methyl Ester [1]	Octadecanoate (from 5.1 to 6.3) Octadecenoate (from 24.8 to 90.2) Octadecadienoate (from 64.5 to 2.3) Octadecatrienoate (from 5.6 to 2.1)	H_2 = 1 atm, 40 °C Acetic Acid	[87]
$H_2PtCl_6 + SnCl_2 + HCl$	Methyl Linoleate (26.0)	Octadecenoates (100)	H_2 = 39 atm, 90 °C Methanol	[88]
$H_2PtCl_6 + SnCl_2$	Methyl Linolenate (95.5)	Octadecenoate(s) (6.2) Unconj. Octadecadienoate(s) (69.6) Conj. Octadecadienoate(s) (5.5) Conj. Dienetrienoate(s) (16.6) Conj. Octadecatrienoate(s) (2.1)	H_2 = 1 atm, 40 °C Methanol	[89]
$H_2PtCl_6 + SnCl_2$	Methyl Linolenate (99.4)	Octadecenoate(s) (32.6) Unconj. Octadecadienoate(s) (64.4) Conj. Octadecadienoate(s) (1.8) Conj. Dienetrienoate(s) (1.2)	H_2 = 34 atm, 65 °C 2 : 3 Methanol-Benzene	[89]
$PdCl_2(PPh_3)_2 + SnCl_2$	Soybean Oil Methyl Ester [1] (100)	Octadecenoates (100)	H_2 = 14-39 atm, 60-90 °C 2 : 3 Methanol-Benzene	[84]
$PtCl_2(PPh_3)_2 + SnCl_4$	Methyl Linoleate (49.7)	Octadecenoate(s) (59.0) Conj. Octadecadienoates (39.8) Unknown (1.2)	H_2 = 39 atm, 90 °C 2 : 3 Methanol-Benzene	[85]

(1) Soybean Oil Methyl Ester with the following composition: Octadecanoate 4.1%; other satd. compounds 11.3%; Octadecenoate 24.0%; Octadecadienoate 51.6%; Octadecatrienoate 9.0%.

Catalyst	Substrate	Products	Conditions	Ref.
$PtCl_2(PPh_3)_2 + SnCl_2$	1,5-Heptadiene	Heptene(s) (91.0) Diene and higher hydrocarbons (9.0)	$H_2 = 34$ atm, 90 °C Methylene Chloride	[92]
$PtCl_2(PPh_3)_2 + SnCl_2$	1,7-Octadiene	Octene(s) (89.0) Diene and higher hydrocarbons (11.0)	$H_2 = 34$ atm, 90 °C 2 : 3 Methanol-Benzene	[92]
$PtCl_2(PPh_3)_2 + SnCl_2$	1,3,6-Octatriene	Octene(s) (91.0) Diene and higher hydrocarbons (9.0)	$H_2 = 34$ atm, 90 °C 2 : 3 Methanol-Benzene	[92]
$PtCl_2(PPh_3)_2 + SnCl_2$	2,4,6-Octatriene	Octene(s) (90.0) Diene and higher hydrocarbons (10.0)	$H_2 = 34$ atm, 90 °C 2 : 3 Methanol-Benzene	[92]
$PtCl_2(PPh_3)_2 + SnCl_2$	1,4,9-Decatriene (100)	Decene(s) (100)	$H_2 = 41$ atm, 90 °C Methylene Chloride	[92]
$PtCl_2(PPh_3)_2 + SnCl_2$	4-Vinylcyclohexene	Monoene(s) (93.0) Diene and higher hydrocarbons (7.0)	$H_2 = 41$ atm, 90 °C Methylene Chloride	[92]
$PtCl_2(PPh_3)_2 + SnCl_2$	3-Ethylidenecyclohexene (100)	Monoene(s) (100)	$H_2 = 41$ atm, 90 °C Methylene Chloride	[92]
$PtCl_2(AsPh_3)_2 + SnCl_2$	1,5-Cyclooctadiene (100)	Cyclooctene (100)	$H_2 = 41$ atm, 105 °C Methylene Cloride	[92]
$PtCl_2(AsPh_3)_2 + SnCl_2$	Methyl Linoleate (75.2)	Octadecenoate (s) (68.6) Conj. Octadecadienoates (31.4)	$N_2 = 1$ atm, 90 °C 2 : 3 Methanol-Benzene	[85]
$PtCl_2(AsPh_3)_2 + SnCl_2$	Methyl Linoleate (80.4)	Octadecenoate(s) (82.7) Conj. Octadecadienoates (17.3)	$H_2 = 39$ atm, 90 °C 2 : 3 Methanol-Benzene	[85]
$PtCl_2[P(OPh)_3]_2 + SnCl_2$	Methyl Linoleate (94.1)	Octadecanoate (6.6) Octadecenoate(s) (93.4)	$H_2 = 39$ atm, 90 °C 2 : 3 Methanol-Benzene	[85]

stable and precipitation of metallic platinum occurs, while at values higher than 7 the hydrogenation rate gradually decreases [87]. The latter effect is conceivably due to the formation of the inactive species $[Pt(SnCl_3)_5]^{3-}$ [91].

Also, the nature of the solvent influences the activity or the catalyst; while ketones, ethers and nitrobenzene are suitable solvents [87], low activity has been observed in alcoholic media.

No intermediate has been isolated from the reaction mixtures; however in all probability a hydride complex of Pt(II) is the actual catalyst of the hydrogenation, which probably takes place through the formation of a σ-alkenyl species [1].

It is worth noting that K_2PdCl_4—$SnCl_2$ mixtures have been reported as having no catalytic activity in the hydrogenation of polyunsaturated olefins [85]; the inability of $SnCl_2$ to stabilize hydride derivatives of this metal is probably the reason of this behaviour.

3.3.4.2. *Liganded platinum(II)-and palladium(II)-tin halides*

The presence of such ligands as phosphites or tertiary phosphines, arsines and stibines in complexes of Pt(II) and Pd(II), stabilizes the catalyst; for Pd(II) derivatives this is a prerequisite for the existence of active species in the reaction conditions, while in the case of Pt(II) complexes a lower reactivity is found with respect to the unmodified systems described above.

Selective hydrogenation of polyolefins (0.3-0.6 M) and polyunsaturated fatty esters ($6.7 \cdot 10^{-2}$ M) to monoenes has been carried out in methanol-benzene or in dichloromethane solution, by using PtX_2L_2 (10^{-2} M and $6.7 \cdot 10^{-3}$ M respectively) + $SnCl_2$ (10^{-1} M and $3.3 \cdot 10^{-2}$ M respectively) as catalytic system; in the above formula X denotes a halogen atom and L represents a phosphite or a tertiary phosphine, arsine and stibine. Rather drastic conditions (90-105 °C, hydrogen pressure 34 41 atm.) were employed in order to obtain acceptable reaction rates [85] [88] [89] [92] [93] (table 6).

In some cases a complicating factor arosfrome the presence, in the reaction mixture, of precipitates which were proved to be inactive as hydrogenating agents [85].

Similarly, soybean oil methyl ester (39 g/l) has been selectively hydrogenated in the presence of $PdCl_2L_2$ ($1.2 \cdot 10^{-2}$ M, L = tertiary phosphine or arsine) + $SnCl_2$ ($8.5 \cdot 10^{-2}$ M) at 60-90 °C and 14-39 atmospheres of hydrogen pressure [84]; also in this case the formation of a precipitate has sometimes been observed.

The influence of physical and chemical parameters on the selective hydrogenation of polyenes to monoenes has been studied in detail in the case of the platinum systems.

The catalytic activity is strongly dependent upon the nature of the neutral ligands on the platinum atom; in the reaction catalyzed by $PtCl_2L_2$ + $SnCl_2$, the following order has been observed, based on both yield to monoene and conversion of substrate (methyl linoleate or 1,5-cyclooctadiene)[85] [92]:

$$P(OPh)_3 > AsPh_3 > SePh_2 > SPh_2 > PPh_3 > SbPh_3 > PBu_3 > (CH_2NH_2)_2$$

which corresponds to the increase of the σ-donor properties of the ligands [92]. It is worth noting here that the reverse order PPh₃ > AsPh₃ has been observed for the corresponding palladium system [84].

It is worth noting here that the reverse order $PPh_3 > AsPh_3$ has been observed for the corresponding palladium system [84].

Several chlorides of group IV elements have been employed as cocatalysts together with $PtCl_2(PPh_3)_2$, the system resulting in the following order of decreasing activity, based on converted starting material [85]:

$$SnCl_2 > SnCl_4 > GeCl_2 + HCl > GeCl_4 > PbCl_2 + HCl > SiCl_4 > SiHCl_3,$$

while the following order has been found with reference to the yield to monoenes:

$$SnCl_4 > SnCl_2 > GeCl_2 + HCl > GeCl_4 > SiCl_4 > SiHCl_3 > PbCl_2 + HCl.$$

The use of $PtBr_2(PPh_3)_2 + SnBr_2$ in place of the corresponding chloro compounds, results in an increase of yield to monoenes [85], while the formation of conjugated dienoates is remarkably lower [85] [94].

The tin to platinum molar ratio also affects the catalytic activity, a maximum being observed, which depends on the reaction conditions; a value of 10 is reported by Bailar et al. [92] [95] as giving best results in their experimental conditions.

The solvent has a great effect on the "liganded" platinum-tin system; in the hydrogenation of several olefins using different solvents, the following activity has been observed [92]:

$$methylene\ chloride \sim dichloroethane > acetone > tetrahydrofuran \sim$$
$$\sim 3:2\ benzene\text{-}methanol > pyridine.$$

Moreover, change of reaction media alters the rate of both isomerization and hydrogenation as well as their relative value [92].

Sometimes the solvents are the hydrogen source; e.g. when methanol is present in the reaction mixture, hydrogenation of dienes catalyzed by $PtCl_2(PPh_3)_2$ or $PtCl_2(AsPh_3)_2 + SnCl_2$ takes place even in the absence of molecular hydrogen [85] [92].

Much work has been carried out in order to investigate the influence of the substrate on the course of the reaction. An important feature of this catalytic system resides in its inability to hydrogenate internal double bonds, while high activity is shown with respect to terminal olefins [86]. Consequently, whenever internal monoenes are formed via hydrogenation of dienes or via isomerization of terminal olefins, they remain unaltered. By studying the hydrogenation of isomeric hexadienes (table 7) the following order of activity has been found [86] [96]:

$$1,5 > 1,4 > 1,3 \gg 2,4.$$

It can be deduced that the reactivity of a terminal double bond is decreased by conjugation (1,4 > 1,3); this is in accordance with the probable formation of more stable diene-catalyst intermediates.

Table 7 [86, 96]

SELECTIVE HYDROGENATION OF ISOMERIC HEXADIENES (0.2 M) CATALYZED BY $PtCl_2(PPh_3)_2$ (5 · 10^{-3} M) + $SnCl_2$ (5 · 10^{-2} M) (¹)

Diene	1-Hexene	2-Hexene	3-Hexene	Hexane	Hexadiene			
					1.5	1.4	1.3	2.4
1,5	0.8	18.0	1.0	2.5	0	17.0	2.0	58.7
1,4	0	17.5	0	0	0	5.7	3.6	73.2
1,3	0	0	7.5	0	0	0	73.1	19.4
2,4	0	0.9	0	0	0	1.3	1.8	96.0

(¹) The figures indicate the % compn. of the products. All reactions were carried out in 2:3 methanol-benzene at 90 °C and 34 atmospheres of hydrogen pressure, reaction time 3 hrs.

Conversely, inertness of isolated internal double bonds is overcome by conjugation: consequently, the isomerizing properties of the catalytic system play an important role in the hydrogenation of unconjugated internal polyolefins. The conjugation step is not rate determining since accumulation of large amounts of conjugated systems has been detected during the hydrogenation. Interestingly, bicycloheptadiene, in which the two double bonds cannot be forced into conjugation, results in a skeletal rearrangement and ethylidenecyclopentane is found as the final product instead of bicycloheptene.

From the nature of the resulting hexenes, it can be deduced that 1,2 addition of hydrogen to the conjugated system is the predominant path.

Table 8 [86, 96]

SELECTIVE HYDROGENATION OF SEVERAL CONJUGATED DIENES (0.2 M) CATALYZED BY
$PtCl_2(PPh_3)_2$
$(5 \cdot 10^{-3} M) + SnCl_2 (5 \cdot 10^{-2} M)$ [1]

Diene	Monoene (yield %)	
1,3-Butadiene [2]	2-Butene	(0.2)
1,3-Pentadiene [3]	2-Pentene	(3.0)
1,3-Hexadiene	3-Hexene	(7.5)
2,4-Hexadiene	2-Hexene	(0.9)
Methyl 2,4-Hexadienoate	Hexenoate	(9.0)
Methyl trans-9, trans-11-Octadecadienoate	Octadecenoate	(8.2)

[1] The reactions were carried out in 2 : 3 methanol-benzene at 90 °C and 34 atmospheres of hydrogen pressure, reaction time 3 hrs.

[2] Substrate concn. 0.9 M; solvent not reported.

[3] Methylene chloride as solvent.

On the other hand, the negligible formation of hexane when starting from 1,5-hexadiene can be explained on the basis of separate hydrogenation of the two terminal double bonds taking place before isomerization [86] [96].

It is of interest to note that increasing reactivity has been observed with increasing molecular weight (e.g. 1,3-hexadiene > 1,3-pentadiene [86]; 9,11-octadecadionate > 2,4-hexadiene [86]); moreover appropriate functional groups directly bonded to the conjugated system can increase its reactivity, as in the case of methyl-2,4-hexadienoate (table 8) [86]. Much work has been devoted to the isolation of intermediate species: such compounds as platinum-diolefin and hydrido-platinum-diolefin complexes have been isolated and characterized, which are probably the true intermediates of the reaction [92]. On the basis of the above results and considering that previous conjugation is a necessary step of the reaction, the following mechanism has been proposed [92].

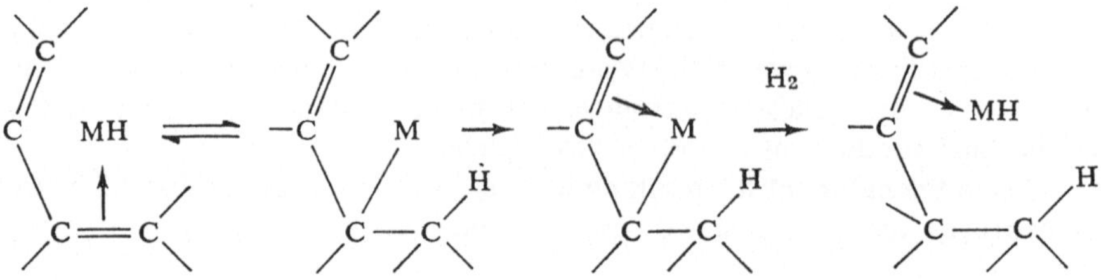

Scheme 6

Addition of diene to hydridic species could result in the formation of a σ-π complex through co-ordination of one double bond of the conjugated system to the metal [92]. The π-bond, by weakening the metal-carbon σ-bond, could make it susceptible to attack by hydrogen in a manner similar to the replacement of chloride ion in platinum hydride complex formation [92]. The 1,2-addition suggested in Scheme 6 is in accordance with the nature of the products of the reactions [86].

It is conceivable that a similar mechanism is involved in the simple platinum-tin halide catalyzed hydrogenations.

3.4. Transition metal chelates

Some Schiff-base complexes of transition metals have been employed as catalysts of moderate activity and selectivity in the hydrogenation of poly-unsaturated fatty esters to the corresponding monoenes [97].

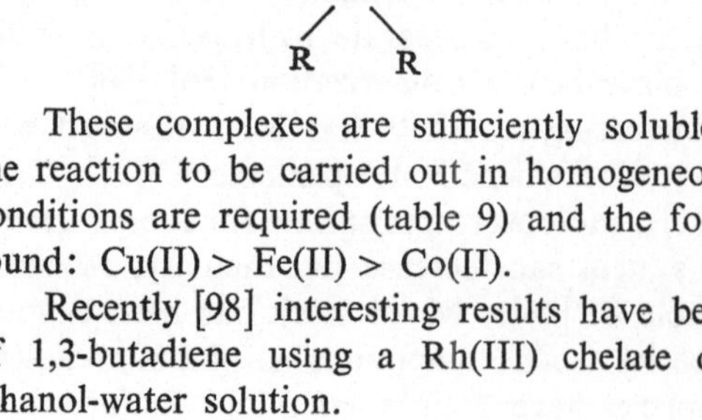

$M = Cu(II); Fe(II); Co(II)$

$R = lauryl, palmityl, stearyl$

These complexes are sufficiently soluble in the reaction media to allow the reaction to be carried out in homogeneous phase. Rather drastic reaction conditions are required (table 9) and the following order of activity has been found: $Cu(II) > Fe(II) > Co(II)$.

Recently [98] interesting results have been obtained in the hydrogenation of 1,3-butadiene using a Rh(III) chelate of formula $HRh(DMG)_2PPh_3$ in ethanol-water solution.

Table 9

SELECTIVE HYDROGENATION OF POLYENES CATALYZED BY TRANSITION METAL CHELATES

Catalyst	Substrate (% Conv.)	Product(s) (% Compn.)	Conditions	Ref.
Cu (SaProR)	Soybean oil methyl ester [1]	Octadecanoate ·(from 4.6 to 11.6) Octadecenoate (from 27.0 to 53.5) Octadecadienoate (from 58.2 to 32.2) Octadecatrienoate (from 10.2 to 2.7)	$H_2 = 150$ atm, 120 °C Hexane	[97]
Fe(SaProR)	Soybean oil methyl ester [1]	Octadecanoate (from 4.6 to 8.3) Octadecenoate (from 27.0 to 41.8) Octadecadienoate (from 58.2 to 45.6) Octadecatrienoate (from 10.2 to 4.3)	$H_2 = 150$ atm, 60 °C no solvent	[97]
RhH(DMG)$_2$PPh$_3$	1,3-Butadiene (93.5)	cis-2-Butene (49.2) trans-2-Butene (50.2) Others (0.6)	$H_2 = 1$ atm, 5 °C 9 : 1 Ethanol-Water	[98]
RhH(DMG)$_2$PPh$_3$+NaBH$_4$	1,3-Butadiene (100)	cis-2-Butene (99.5) Others (0.5)	20 °C, $N_2 = 1$ atm, 9 : 1 Ethanol-Water	[98]
Co(DMG)$_2$PyCl+NaBH$_4$	Isoprene (84)	2-Methyl-1-Butene (23.4) 3-Methyl-1-Butene (67.8) 2-Methyl-2-Butene (8.8)	$H_2 = 1$ atm, 20 °C	[99]

[1] Soybean Oil Methyl Ester with the following composition: Octadecanoate 4.1%; other satd. compounds 11.3%; Octadecanoate 24.0%; Octadecadienoate 51.6%; Octadecatrienoate 9.0%.

This chelate adds 1,3-butadiene to give a linear 2-alkenyl derivative, which has been isolated; in the presence of hydrogen this addition product undergoes hydrogenolysis in very mild conditions (about 5 °C) to produce equal amounts of cis and trans-2-butenes in almost quantitative yield. When NaBH$_4$ is used as hydrogenating agent in place of molecular hydrogen, cis-2-butene is obtained exclusively, provided sufficiently high diolefin to catalyst molar ratios are used.

A similar catalytic system i.e. Co(DMG)$_2$PyCl + NaBH$_4$ has been used recently [99] in the selective hydrogenation of isoprene in mild conditions (table 9), the main products being in this case, 2- and 3-methyl-1-butenes.

4. FIRST ROW TRANSITION METAL COMPLEXES IN THE PRESENCE OF COCATALYTIC AGENTS

4.1. Organometallic catalysts

Although the catalytic properties of Ziegler-Natta catalysts in the hydrogenation of olefins in mild conditions have been known since 1964 [100] [101], the use of these systems in the hydrogenation of polyenes have been reported only recently.

Chlorides or acetylacetonates of Fe(III), Co(II), Co(III) and Ni(II) (1.6 · · 10^{-2} M) in the presence of Al(C_2H_5)$_3$ (M/Al = 1 : 5) have been used in the hydrogenation of soybean oil methyl esters (140 g/l) in hexane solution at 150 °C and 150 atmospheres of hydrogen pressure (table 10) [102]. Good selectivity has been obtained despite the very drastic reaction conditions.

Similar systems of cobalt(II) salts (4 · 10^{-2} M) (acetylacetonate, chloride or salicylate) and Al(C_2H_5)$_3$ (M/Al = 1 : 5) have been employed in the selective hydrogenation of 1,3-butadiene and isoprene (2 M) in benzene solution [102] at 90-110 °C and 40 atmospheres of hydrogen pressure. In this case a large amount of saturated compounds has been observed in the reaction mixtures. Therefore, it appears that the nature of the substrate plays an important role in determining the selectivity of the reaction.

Modification of the above systems by addition of suitable ligands such as tertiary phosphines, phosphites, pyridine [103] and DPE, DPP, phosphorus trichloride, phosphorus oxychloride [104], results in a great improvement in selectivities (table 10).

According to Tajima and Kunioka, the hydrogenation of the resulting monoenes to saturated compounds is slowed down by the competition of the added ligands for the metal atom [103].

The preferential formation of 1-butene and cis-2-butene in the hydrogenation of 1,3-butadiene indicates that the catalytic system has very poor isomerizing properties.

Hydrogenation of diolefins such as 1,3-butadiene and isoprene has also been carried out using binary catalytic systems formed by cyclopentadienyl transition metal complexes of formula Cp$_2$MCl$_2$ (M = Ti and V) and organometallic compounds such as LiC$_4$H$_9$ or C$_6$H$_5$MgBr [105]. By using vanadium derivatives (3.5 · 10^{-2} M) and LiC$_4$H$_9$ (8.4 · 10^{-2} M) the hydrogenation of both 1,3-butadiene (1.5 M) and isoprene (1.2 M) resulted in the production of internal olefins as the main products. In both cases selectivity to monoenes was 100%, the conversion after 15 hr being 62% for 1.3-butadiene at 45 °C and 100% for isoprene at 100 °C and 60 atmospheres of hydrogen pressure. Extensive hydrogenation of 1,3-butadiene to butane has been observed in the presence of Cp$_2$TiCl$_2$—LiC$_4$H$_9$ (or C$_6$H$_5$MgBr) catalytic systems.

Interestingly the nature of the alkylating reagent is very important, no hydrogenation being observed [105] when Al(C_2H_5)$_3$ or Al(C_4H_9)$_3$ is used in place of LiC$_4$H$_9$ or C$_6$H$_5$MgBr.

Table 10

Selective Hydrogenation of Polyenes Catalyzed by First Row Transition Metal Complexes in the Presence of Cocatalytic Agents

Catalyst	Substrate (% Convn.)	Product(s) (% Compn.)	Conditions	Ref.
$Ni(acac)_2$—$Al(C_2H_5)_3$	Methyl Linoleate	Octadecenoate(s) (0.4) Octadecenoate(s) (89.0) Octadecadienoate(s) (10.6)	H_2 = 150 atm, 150 °C Hexane	[102]
$Ni(acac)_2$—$Al(C_2H_5)_3$	Methyl Linoleate	Octadecenoate(s) (74.9) Octadecadienoate(s) (17.5) Octatrienoate(s) (8.5)	H_2 = 150 atm, 150 °C Hexane	[102]
$CoCl_2$—$Al(C_2H_5)_3$	1,3-Pentadiene (80.7; 3-5% to oligomers)	Butane (75.3) 1-Butene (7.1) cis-2-Butene (12.6) trans-2-Butene (5.0)	H_2 = 40 atm, 90 °C	[103]
$CoCl_2(PPh_3)_2$—$Al(C_2H_5)_3$	1,3-Butadiene (28.0)	Butane (2.8) 1-Butene (21.1) cis-2-Butene (61.5) trans-2-Butene (14.6)	H_2 = 40 atm, 90 °C Benzene	[103]
$CoCl_2(PPh_3)_2$—$Al(C_2H_5)_3$	1,3-Butadiene (84.0)	Butane (8.2) 1-Butene (29.2) cis-2-Butene (57.4) trans-2-butene (11.2)	H_2 = 40 atm, 110 °C Benzene	[103]
$CoCl_2(PPh_3)_2$—$Al(C_2H_5)_3$	Isoprene (76.3)	3-Methyl-1-Butene (2.7) 3-Methyl-2-Butene (64.3) 2-Methyl-1-Butene (26.8) 2-Methyl-2-Butene (6.2)	H_2 = 40 atm, 110 °C Benzene	[103]

Table 10

SELECTIVE HYDROGENATION OF POLYENES CATALYZED BY FIRST ROW TRANSITION METAL COMPLEXES IN THE PRESENCE OF COCATALYTIC AGENTS

Catalyst	Substrate (% Convn.)	Product(s) (% Compn.)	Conditions	Ref.
CoCl₂(DPE)₂—Al(C₂H₅)₃	1,3-Butadiene (78.9; 5-6% to oligomers)	Butane (1.0) 1-Butene (25.9) cis-2-Butene (72.6) trans-2-Butene (0.5)	H₂ = 30 atm, 100 °C Toluene	[104]
CoCl₂—POCl₃—Al(C₂H₅)₃	Isoprene (99.8; 3% to oligomers)	2-Methyl-Butane (15.3) 3-Methyl-1-Butene (0) 2-Methyl-2-Butene (11.8) 2-Methyl-1-Butene (72.9)	H₂ = 30 atm, 100 °C	[104]
Co(SCN)₂(PPh₃)₂—Al(C₂H₅)₃	Isoprene (32.0; 2% to oligomers)	2-Methyl-Butane (1.6) 3-Methyl-1-Butene (10.1) 2-Methyl-2-Butene (58.0) 2-Methyl-1-Butene (30.3)	H₂ = 30÷50 atm, 100 °C	[104]
Cp₂VCl₂—Li(C₄H₉)	1,3-Butadiene (62.2)	1-Butene (2.3) cis-2-Butene (13.9) trans-2-Butene (83.8)	H₂ = 60 atm, 45 °C Benzene	[105]
Cp₂VCl₂—PhMgBr	1,3-Butadiene (43.1)	1-Butene (1.7) cis-2-Butene (26.1) trans-2-Butene (72.2)	H₂ = 60 atm, 45 °C Benzene	[105]
Cp₂VCl₂—Li(C₄H₉)	Isoprene (100)	2-Methyl-1-Butene (5.8) 2-Methyl-2-Butene (92.1) 3-Methyl-1-Butene (2.1)	H₂ = 60 atm, 100 °C Benzene	[105]

257

Catalyst	Substrate	(conversion)	Products (%)	Conditions	Ref.
Cp₂VCl₂—PhMgBr	Isoprene	(97.8)	2-Methyl-1-Butene (4.9) 2-Methyl-2-Butene (93.7) 3-Methyl-1-Butene (1.4)	H_2 = 60 atm, 100 °C Benzene	[105]
CpFe(CO)₂Cl—Al(C₂H₅)₃	1,3-Butadiene	(98.8)	1-Butene (2.1) cis-2-Butene (45.6) trans-2-Butene (52.3)	H_2 = 60 atm, 45 °C Benzene	[105]
CoH(DPE)₂—Al(C₂H₅)₂Cl	1,3-Butadiene	(100; 2% of oligomers)	Butane (1.4) 1-Butene (28.8) cis-2-butene (68.6) trans-2-Butene (1.2)	H_2 = 30 atm, 85-120 °C Toluene	[104]
CoH(DPE)₂	1,3-Butadiene	(42.1; 7-8% of oligomers)	Butane (7.1) 1-Butene (59.4) cis-2-Butene (27.3) trans-2-Butene (6.2)	H_2 = 50 atm, 90-100 °C Chlorophenol	[104]
CoCl₂(DPE)₂—LiAlH₄	1,3-Butadiene	(50.1; 7% of oligomers)	Butane (0.6) 1-Butene (21.4) cis-2-Butene (77.2) trans-2-Pentene (0.8)	H_2 = 30 atm, 95-98 °C Chlorophenol	[104]
CoCl₂—LiAlH₄	d-Limonene		Monoene	H_2 = 1 atm, 0 °C Tetrahydrofuran	[109]
Ni(acac)₂	Methyl Linoleate		Octadecanoate (9.0) cis-9-Monoene (46.5) cis-12-Monoene ⎱ trans-Monoenes (29.5) Isom. unconj. Dienes (15.0)	H_2 = 7 atm, 100 °C Methanol	[110]

Table 10

SELECTIVE HYDROGENATION OF POLYENES CATALYZED BY FIRST ROW TRANSITION METAL COMPLEXES IN THE PRESENCE OF COCATALYTIC AGENTS

Catalyst	Substrate (% Convn.)	Product(s) (% Compn.)	Conditions	Ref.
$Ni(acac)_2$	Methyl Linoleate	Octadecanoate (1.0); cis-Monoenes (9-, or 12-, or 15-) (10.1); trans-Monoenes (9- to 15-) (5.0); Unconj. Dienes (59.8); Partially isom. unconj. Dienes (24.2)	p H_2 = 7 atm, 100 °C Methanol	[110]
$NiI_2[P(C_6H_5)_3]_2$	Methyl Linoleate	Octadecanoate (1.1); Octadecenoate(s) (83.5); Octadecadienoate(s) (15.4)	H_2 = 39 atm, 90 °C Benzene	[111]
$NiI_2[P(C_6H_5)_3]_2$	Methyl Linoleate	Octadecenoate(s) (83.2); Octadecadienoate(s) (16.8)	H_2 = 39 atm, 90 °C Tetrahydrofuran	[111]
$NiI_2[P(C_6H_5)_3]_2$	Methyl Linoleate	Octadecanoate (1.4); Octadecenoate(s) (55.3); Octadecadienoate(s) (39.1); Conj. Octadecadienoates (4.2)	H_2 = 39 atm, 90 °C Tetrahydrofuran	[111]
$NiBr_2[P(C_6H_5)_3]_2$	Methyl Linoleate	Unknown compounds (22.0); Octadecanoate (5.8); Octadecenoate(s) (80.5); Octadecadienoate(s) (11.5)	H_2 = 39 atm, 90 °C Benzene	[111]

Cyclopentadienyl-complexes of VIII group metals, such as $CpNiC_3H_7$, $CpCo(CO)_2$ and $CpFe(CO)_2Cl$ have little or no activity in the hydrogenation of dienes; however, good selectivity and catalytic activity are obtained whenever the above complexes are employed together with $Al(C_2H_5)_3$ or C_6H_5MgBr [105]. Thus, for instance, 1,3-butadiene (1.5 M) in benzene solution is quantitatively reduced to internal butenes at 45 °C and 60 atmospheres of hydrogen pressure in the presence of $CpFe(CO)_2Cl$ ($3.5 \cdot 10^{-2}$ M) and $Al(C_2H_5)_3$ ($8.4 \cdot 10^{-2}$ M).

It is interesting to note that complexes containing two cyclopentadienyl moieties such as Cp_2M and Cp_2M^+ (M = Fe, Co, Ni) are inactive even in the presence of organometallic compounds [105].

The behaviour of $CoH(DPE)_2$ [106] is worth mentioning; this hydride has been reported to hydrogenate 1,3-butadiene and isoprene very slowly in chlorophenol as solvent. However, when $Al(C_2H_5)_2Cl$ is added to the toluene solution of $CoH(DPE)_2$ good activity and selectivity have been obtained [104].

The hydrogenation of diolefins has been suggested as proceeding through the same mechanism as that proposed by Sloan et al. [100] for mono olefins. In the present case formation of intermediate π-allyl complexes (possibly in equilibrium with σ-alkenyl forms) has generally been invoked.

It should be pointed out that metallorganic compounds may behave also as Lewis acids [107] [108], whose role is to abstract ligands from the metal with consequent formation of coordinatively unsaturated species of greater activity. In this way the role of $Al(C_2H_5)_2Cl$ in the activation of $CoH(DPE)_2$ is more easily explained.

4.2. Transition metal salts - lithium aluminum hydride systems

Iron and cobalt halides in the presence of non-transition metal hydrides give catalytic systems which are active in the hydrogenation of unsaturated substrates; the use of both simple and liganded metal halides has been reported [104] [109].

When simple salts are used, the system is very active even at 0 °C and 1 atmosphere of hydrogen pressure; however, very low selectivities are obtained in the hydrogenation of conjugated dienes to monoenes [109].

Catalytic systems showing lower activity have been obtained by using chelating phosphine complexes of Co(II) [104]; thus, for instance, $CoCl_2(DPE)_2$ (10^{-2} M) and $LiAlH_4$ ($4 \cdot 10^{-2}$ M) in toluene solution give rise to a catalytic system which is active in the hydrogenation of conjugated diolefins (10 M) at 100 °C and 1 atmosphere of hydrogen pressure.

Almost exclusive formation of 1-butene and cis-2-butene from 1,3-butadiene has been reported (table 10), thus indicating that the catalytic system has no isomerizing properties.

Since a similar composition of the hydrogenation products has been obtained by using $CoH(DPE)_2$ — $Al(C_2H_5)_2Cl$, $CoCl_2(DPE)_2$ — $Al(C_2H_5)_3$ and $CoCl_2(DPE)_2$ — $LiAlH_4$, probably the same catalytic intermediate is involved in every case.

4.3. Transition metal salts in polar media

Selective hydrogenation of soybean oil methyl esters (100 g/l) to the corresponding monoenes has been carried out by means of Ni(II), Co(III), Fe(III) and Cu(II) acetylacetonates (0.1 M) in methanol solution at 100-180 °C and 7-70 atmospheres of hydrogen pressure (table 10) [110].

The activity of acetylacetonates decreases in the order: Ni(II) > Co(III) > > Cu(II) > Fe(III): this trend seems to be related to the decreasing stability of acetylacetonates in the reaction conditions. The selectivity of the catalytic system has been found to decrease in the same order [110].

The nickel acetylacetonate catalyzed hydrogenation of methyl linoleate and linolenate gives monoenes containing 79% of the double bonds in one of the original positions, the *cis* monoenes being predominant; clearly the hydrogenation proceeds mainly on the isolated double bonds [110]. Conjugated dienes are not present in the hydrogenation products, but the formation of some *trans*-monoenes with double bonds on either side of the original position has been ascribed to a fast hydrogenation of conjugated intermediates.

Dienes with more than one methylene group between the two double bonds, have been observed to accumulate in the course of the reaction. Therefore the presence of only one methylene group between the double bonds is the condition for a fast reaction to take place, even though the reaction has been observed to proceed largely on the isolated unsaturations [110].

The use of methanol is essential: other solvents such as hydrocarbons, acetone, pyridine, chloroform and long-chain carboxylic acids have been used without success; very low activity has been observed in acetic acid or dimethylformamide. Clearly the knowledge of the role of the solvents is necessary to clarify the activation step of the catalytic systems.

In this respect, it has been suggested [110] that methanol, acetic acid and dimethylformamide are suitable solvents due to their good proton stabilizing properties or to their ability in promoting the formation of a metal acetylacetonate hydride. The assumption has been made that the first step is the substitution of one acetylacetonate ligand with methanol followed by hydrogenolysis of the solvate intermediate and formation of a metal hydride. Hydride addition to carbon-carbon double bond and further hydrogenolysis, have been assumed [110] to proceed in accordance with the mechanism of Sloan *et al.* [100]

In the above examples, addition of appropriate substances was necessary to assist the formation of active species; however it seems worth noting that such complexes as $NiX_2(PPh_3)_2$ (X = Cl, Br, I) show high hydrogenating activity (I > Br > Cl) in hydrocarbon solvents in the absence of cocatalysts [111]. Methyl linoleate has been reduced to monoene in high conversion and very high selectivity (table 10) in both benzene and tetrahydrofuran solution (90 °C and 39 atmospheres of hydrogen pressure); the hydrogenation has been reported [111] to proceed to some extent also in nitrogen atmosphere in tetrahydrofuran and in aromatic hydrocarbons.

Very fast isomerization of the starting substrate to conjugated diene has been assumed to be the first step of the reaction, followed by a fast hydrogenation of the conjugated diene to monoene; slow isomerization of the resulting mono olefin to the *trans* form also takes place under these conditions.

In this case, nothing can be said with certainty about the mechanism of hydrogenation, no intermediates having been isolated from the reaction mixtures; however, hydridic compounds have been suggested as active species. The recent isolation [112] of stable *trans*-hydridochloro-bis(tricyclohexylphosphine) nickel(II) seems to support this suggestion.

5. NON-TRANSITION METAL HYDRIDES

Lithium aluminum hydride has been previously reported to behave as a hydrogen transfer agent in the activation of some metal complexes (4.2.). The same compound has been reported [113] to catalyze selective hydrogenation of conjugated dienes to monoenes.

In a typical experiment, a tetrahydrofuran solution of LiAlH$_4$ (0.45 M) has been used in the selective hydrogenation of 1,3-pentadiene (2 M) at 190 °C

Table 11

SELECTIVE HYDROGENATION OF POLYENES CATALYZED BY NON-TRANSITION METAL HYDRIDES

Catalyst	Substrate (% Convn.)	Product(s) (% Compn.)		Conditions	Ref.
LiAlH$_4$	1,3-Cyclooctadiene (48.6)	Cyclooctene	(96.0)	H$_2$ = 56-98 atm, 190 °C	[113]
		Unidentified	(4.0)	Tetrahydrofuran	
LiAlH$_4$	1,3-Pentadiene (100)	1-Pentene	(8.2)	H$_2$ = 56-98 atm, 190 °C	[113]
		cis-2-Pentene	(10.4)	Tetrahydrofuran	
		trans-2-Pentene	(57.1)		
		Dimers	(24.3)		

and 50-100 atmospheres of hydrogen pressure. Total conversion was obtained in 6-15 min., a mixture of pentenes being formed as the main reaction product; due to the rather drastic conditions, dimers were also formed in considerable amounts (table 11) [113].

Under identical conditions, LiAlH$_4$ has been reported to add slowly to isolated double bonds (monoolefins or unconjugated polyolefins), but the resulting alkyl intermediates have no reactivity toward molecular hydrogen [114]. For this reason high selectivity to monoenes may be expected in the hydrogenation of conjugated dienes.

Deuterium tracer experiments [113] have shown that lithium aluminum hydride and molecular hydrogen each donate one hydrogen atom to the diene, as inferred from the quantitative formation of monodeuterated alkenes. This result strongly supports the following reaction path:

$$CH_2\!=\!CH\!-\!CH\!=\!CH\!-\!CH_3 + LiAlH_4 \rightarrow (n\text{-}C_5H_9)LiAlH_3 \qquad (41)$$

$$(n\text{-}C_5H_9)LiAlH_3 + H_2 \rightarrow n\text{-}C_5H_{10} + LiAlH_4 \qquad (42)$$

Since the unconverted diolefin contains no deuterium atoms, the reverse of equation (41) seems to be negligible, compared with the hydrogenolysis step.

6. SUMMARY AND CONCLUSION

In conclusion many homogeneous catalytic systems, in addition to the well known heterogeneous ones, can profitably be used for the selective hydrogenation of polyenes to monoenes.

In fact also catalysts active in the homogeneous hydrogenation of mono-olefins can be used in the selective hydrogenations of dienes; in this case, as we have pointed out, the favourable results are to be ascribed to thermodynamic factors, i. e. competition of dienes and monoenes on the catalyst. Moreover in homogeneous phase catalytic systems have also been described, which are completely inactive towards the hydrogenation of mono-olefins (for instance the cyano-complexes); clearly, from the standpoint of the selective hydrogenation, this is the most favourable case, which in very few cases can be obtained in heterogeneous phase.

Some difficulty can arise whenever the formation of a positional isomer among the possible resulting monoenes is of interest; in this case more sophisticated performances of the catalytic system are necessary; in few cases this requirement has been met (for instance with the cyano-cobaltate system by changing the concentration of the cyanide anion in the required way).

Another important feature of the selective hydrogenation resides in the possibility of directing the reaction towards the formation of particular stereoisomers (i. e. monoenes having only cis or trans configuration); in this respect, so far only chromium carbonyls have been reported to behave properly.

Unfortunately at present the factors determining both the selective hydrogenation of dienes and the preferential formation of a particular positional or steric isomer are not known with certainty; this knowledge is obviously very important since it is hoped that "tailor made" catalytic systems of suitable performances will be made in the future.

Despite the impossibility at the moment to prepare «a priori» a homogeneous completely selective hydrogenation catalyst for polyenes, it is possible to obtain, with the catalysts already known, a reasonably good selectivity when some simple rules are satisfied.

7. REFERENCES

[1] R. S. COFFEY, "Aspects Homog. Catalysis", *1*, 5, (1970) Ed. R. Ugo, Manfredi (Milan).

[2] J. KWIATEK, "Catalysis Rev.", *1*, 37, (1967), Dekker (New York)

[3] J. KWIATEK and J. K. SEYLER, "Advan. Chem. Ser.", *70*, 207, (1968).

[4] H. ITATANI, "Yuki Gosei Kagaku Kyokai Shi", *27*, 632, (1969).

[5] E. N. FRANKEL, T. L. MOUNTS, R. O. BUTTERFIELD, and H. J. DUTTON, "Advan. Chem. Ser.", *70*, 177, (1968) and references therein.

[6] J. P. CANDLIN, and A. R. OLDHAM, "Disc. Far. Soc.", *46*, 60, (1968).

[7] G. E. COATES, M. L. H. GREEN, and K. WADE, *Organometallic Compounds*, Vol. 2, p. 8, (1968), Methuen (London).

[8] J. CHATT, R. S. COFFEY, A. GOUGH, and D. T. THOMPSON, "J. Chem. Soc. A", 190, (1968).

[9] G. C. BOND, and P. B. WELLS, *Advan. Catalysis*, *15*, 155, (1964).

[10] J. J. ROONEY, and G. WEBB, "J. Catalysis", *3*, 488, (1964).

[11] J. P. CANDLIN, K. A. TAYLOR, and D. T. THOMPSON, *Reactions of transition-metal complexes*, p. 137, (1968), Elsevier (Amsterdam).

[12] R. F. HECK, "Advan. Chem. Ser.", *49*, 181, (1965).

[13] N. K. KING, and M. E. WINFIELD, "J. Amer. Chem. Soc.", *80*, 2060, (1958).

[14] B. DE VRIES, "J. Catalysis", *1*, 489 (1962).

[15] W. P. GRIFFITH, and G. WILKINSON, "J. Chem. Soc.", 2757, (1959).

[16] M. S. SPENCER, and D. A. DOWDEN, U. S. Pat. 3,009,969, (1961).

[17] J. KWIATEK, I. L. MADOR, and J. K. SEYLER, "Advan. Chem. Ser.", *37*, 201, (1968).

[18] J. KWIATEK, and J. K. SEYLER, *Proc. 8th International Conference on Coordination Chemistry, Vienna, 1964*, p. 308, (1964), Springer, (Vienna).

[19] T. SUZUKI, and T. KWAN, "Nippon Kagaku Zasshi", *86*, 713, (1965); "Chem. Abstr.", *64*, 6473*e* (1966).

[20] T. SUZUKI, and T. KWAN, "Nippon Kagaku Zasshi", *86*, 1198, (1965); "Chem. Abstr.", *64*, 11070*c* (1966).

[21] T. SUZUKI, and T. KWAN, "Nippon Kagaku Zasshi", *86*, 1341, (1965); "Chem. Abstr.", *65*, 12097*d*, (1966).

[22] A. FARCAS, U. LUCA, and O. PIRINGER, *Proc. 11th International Conference on Coordination Chemistry, Haifa, 1968*, p. 29, (1968), Elsevier (Amsterdam).

264

[23]   B. De Vries, "Koninkl. Ned. Akad. Wetenschap. Proc." Ser. B *63*, 443, (1960).

[24]   E. N. Frankel, and R. O. Butterfield, "J. Org. Chem.", *34*, 3930, (1969).

[25]   J. Kwiatek, and J. K. Seyler, "J. Organometal. Chem.", *3*, 421, (1965).

[26]   M. G. Burnett, P. J. Connolly, and C. Kemball, "J. Chem. Soc. A", 991, (1968).

[27]   J. M. Pratt, and R. J. P. Williams, "J. Chem. Soc. A", 1291, (1967).

[28]   L. I. Simandi, F. Nagy, and E. Budo, "Acta Chim. Acad. Sci. Hung.", *58*, 39, (1968).

[29]   L. I. Simandi, and F. Nagy, "Acta Chim. Acad. Sci. Hung.", *46*, 137, (1965).

[30]   L. M. Jackman, J. A. Hamilton, and J. M. Lawlor, "J. Amer. Chem. Soc.", *90*, 1914, (1968).

[31]   T. Suzuki, "Nippon Kagaku Zasshi", *88*, 440, (1967); "Chem. Abstr.", *67*, 43325*k* (1967).

[32]   W. Strohmeier, and N. Iglauer, "Z. Phys. Chem.", *61*, 29, (1968).

[33]   F. Conti, and R. Ugo, unpublished results.

[34]   A. F. Mabrouk, H. J. Dutton, and J. C. Covan, "J. Amer. Oil Chem. Soc.", *41*, 153, (1964).

[35]   K. Tarama, and T. Funabiki, "Bull. Chem. Soc. Japan", *41*, 1744, (1968).

[36]   G. Pregaglia, D. Morelli, F. Conti, G. Gregorio, and R. Ugo, "Disc. Far. Soc.", *46*, 110, (1968).

[37]   K. Takao, and S. Suzuki, Japan Pat. 12,566 (1966); "Chem. Abstr." *69*, 99888, (1968).

[38]   C. E. Wymore, "Chem. Eng. News", *46*, April, p. 52, (1968).

[39]   G. M. Schwab, and G. Mandre, "J. Catalysis", *7*, 103, (1968).

[40]   O. Piringer, A. Farcas, and U. Luca, *Proc. 9th International Conference on Coordination Chemistry, S. Moritz, 1966*, p. 202, (1966).

[41]   M. G. Burnett, "Chem. Comm.", 507, (1965).

[42]   T. Mizuta, H. Samejima, and T. Kwan, "Bull. Chem. Soc. Japan", *41*, 727, (1968).

[43]   W. H. Dennis Jr., D. H. Rosenblatt, R. R. Richmond, G. A. Finseth, and G. T. Davis, "Tetrahedron Lett.", *16*, 1821, (1968).

[44]   E. N. Frankel, H. M. Peters, E. P. Jones, and H. J. Dutton, "J. Amer. Oil Chem. Soc.", *41*, 186, (1964) and references therein.

[45]   E. N. Frankel, E. A. Emken, H. M. Peters, V. L. Davison, and R. O. Butterfield, "J. Org. Chem.", *29*, 3292, (1964).

[46]   E. N. Frankel, E. A. Emken, and V. L. Davison, "J. Org. Chem.", *30*, 2739, (1965).

[47]   E. N. Frankel, E. P. Jones, and C. A. Glass, "J. Amer. Oil Chem. Soc.", *41*, 392, (1964).

[48]   E. N. Frankel, E. A. Emken, and V. L. Davison, "J. Amer. Oil Chem. Soc.", *43*, 307, (1966) and private communication.

[49]   E. N. Frankel, N. Maoz, A. Rejoan, and M. Cais, *Proc. 3rd International Symp. Organometal. Chem., Munich, 1967*, p. 210, (1967).

[50] M. Cais, N. Maoz, and A. Rejoan, *Proc. 2nd International Symp., Venice*, p. 25, (1969).

[51] H. D. Murdoch, and E. Weiss, "Helv. Chim. Acta", *45*, 1156, (1962).

[52] H. W. Sternberg, and I. Wender, *International Conference of Coordination Chemistry, London, April 6-11, 1959*; Special Publication No. 13, p. 35, (1959), The Chemical Society (London).

[53] G. F. Pregaglia, A. Andreetta, G. F. Ferrari, and F. Conti, unpublished results.

[54] J. Tsuji, and Y. Mori, "Bull. Chem. Soc. Japan", *42*, 527, (1969).

[55] I. Ogata, and A. Misono, "J. Japan Oil Chem. Soc." (Yukagaku), *13*, 21, (1964).

[56] E. N. Frankel, E. P. Jones, V. L. Davison, E. A. Emken, and H. J. Dutton, "J. Amer. Oil Chem. Soc.", *42*, 130, (1965).

[57] I. Ogata, and A. Misono, "J. Japan Oil Chem. Soc." (Yukagaku), *14*, 16, (1965); "Chem. Abstr.", *63*, 17828, (1965).

[58] K. E. Atkins, U.S. Pat. 3,308,177, (1965); "Chem. Abstr.", *67*, 21504, (1967).

[59] G. F. Pregaglia, A. Andreetta, G. F. Ferrari, and F. Conti, unpublished results.

[60] I. Ogata, and A. Misono, "Disc. Far. Soc.", *46*, 72, (1968).

[61] G. F. Pregaglia, A. Andreetta, G. F. Ferrari, and R. Ugo, "J. Chem. Soc. D", 590, (1969).

[62] I. Ogata, "Kogyo Kagaku Zasshi", *72*, 1710, (1969).

[63] G. F. Pregaglia, A. Andreetta, and R. Ugo, "Chimica e Industria" (Milan), *50*, 1332, (1968).

[64] G. F. Pregaglia, A. Andreetta, G. F. Ferrari, and R. Ugo, unpublished results.

[65] A. Rejoan, and M. Cais, *Proc. 11th International Conference on Coordination Chemistry, Haifa, 1968*, p. 32, (1968), Elsevier (Amsterdam).

[66] E. N. Frankel, and F. L. Little, "J. Amer. Oil Chem. Soc.", *46*, 256, (1969).

[67] M. Cais, E. N. Frankel, and A. Rejoan, "Tetrahedron Lett.", *16*, 1919, (1968).

[68] E. N. Frankel, E. Selke, and C. A. Glass, "J. Amer. Chem. Soc.", *90*, 2446, (1968).

[69] E. N. Frankel, E. Selke, and C. A. Glass, "J. Org. Chem.", *34*, 3936, (1969).

[70] J. Wolinsky, B. Chollar, and M. D. Baird, "J. Amer. Chem. Soc.", *84*, 2775, (1962).

[71] A. Miyake, and H. Kondo, "Angew. Chem., Int. Ed.", *7*, 631, (1968).

[72] A. Miyake, and H. Kondo, "Angew. Chem., Int. Ed.", *7*, 880, (1968).

[73] P. N. Rylander, N. Himelstein, D. R. Steele, and J. Kreidl, "Engelhard Ind. Tech. Bull.", *3*, 61, (1962).

[74] British Pat. to Du Pont, 1,043,800, (1965).

[75] R. S. Coffey, "Chem. Comm.", 923, (1967).

[76] G. C. Bond, and R. A. Hillyard, "Disc. Far. Soc.", *46*, 20, (1968).

[77] J. A. Osborn, F. H. Jardine, J. F. Young, and G. Wilkinson, "J. Chem. Soc. A", 1711, (1966).

[78] F. H. Jardine, J. A. Osborn, and G. Wilkinson, "J. Chem. Soc. A", 1574, (1967).

[79] P. S. Hollam, D. Evans, J. A. Osborn, and G. Wilkinson, "Chem. Comm.", 305, (1967).

[80] D. W. Moore, H. B. Jonassen, T. B. Joyner, and A. J. Bertrand, "Chem. and Ind." (London), 1304, (1960).

[81] J. Lucas, S. Coren, and J. E. Blom, "J. Chem. Soc. D", 1303, (1969).

[82] D. Evans, J. A. Osborn, F. H. Jardine, and G. Wilkinson, "Nature", 208, 1203, (1965).

[83] I. Jardine, and F. J. McQuillin, "Tetrahedron Lett.", 40, 4871, (1966).

[84] H. Itatani, and J. C. Bailar, Jr., "J. Amer. Oil Chem. Soc.", 44, 147, (1967).

[85] J. C. Bailar, Jr., and H. Itatani, "J. Amer. Chem. Soc.", 89, 1592, (1967).

[86] R. W. Adams, G. E. Batley, and J. C. Bailar, Jr., "J. Amer. Chem. Soc.", 90, 6051, (1968).

[87] L. P. Van't Hof, and B. G. Linsen, "J. Catalysis", 7, 295, (1967).

[88] J. C. Bailar, Jr., and H. Itatani, "J. Amer. Oil Chem. Soc.", 43, 337, (1966).

[89] E. N. Frankel, E. A. Emken, H. Itatani, and J. C. Bailar, Jr., "J. Org. Chem.", 32, 1447, (1967).

[90] G. C. Bond, and M. Hellier, "J. Catalysis", 7, 217, (1967).

[91] R. D. Cramer, R. V. Lindsey, C. T. Prewitt, and V. G. Stolberg, "J. Amer. Chem. Soc.", 87, 658, (1965).

[92] H. A. Tayim, and J. C. Bailar, Jr., "J. Amer. Chem. Soc.", 89, 4330, (1967).

[93] J. C. Bailar, Jr., H. Itatani, M. J. Crespi, and J. Geldard, "Advan. Chem. Ser.", 62, 103, (1967).

[94] P. Abley, and F. J. McQuillin, "Disc. Far. Soc.", 46, 31, (1968).

[95] H. A. Tayim, and J. C. Bailar, Jr., "J. Amer. Chem. Soc.", 89, 3420, (1967).

[96] R. W. Adams, G. E. Batley, and J. C. Bailar, Jr., "Inorg. Nucl. Chem. Lett.", 4, 455, (1968).

[97] E. Fedeli, and G. Jacini, "J. Amer. Oil Chem. Soc.", in press.

[98] B. G. Rogachev, and M. L. Khidekel, "Izv. Akad. Nauk SSSR, Ser. Khim.", 1, 141, (1969).

[99] Y. Mizuta, and T. Kan, "J. Chem. Soc. Japan", 88, 471, (1967).

[100] M. F. Sloan, A. S. Matlock, and D. S. Breslow, "J. Amer. Chem. Soc.", 85, 4014, (1963).

[101] M. Farina, and M. Ragazzini, "Chimica e Industria" (Milan), 40, 816, (1958).

[102] Y. Tajima, and E. Kunioka, "J. Amer. Oil Chem. Soc.", 45, 478, (1968).

[103] Y. Tajima, and E. Kunioka, "J. Catalysis", 11, 83, (1968).

[104] M. Iwamoto, "Kogyo Kagaku Zasshi", 71, 1510, (1968).

[105] Y. Tajima, and E. Kunioka, "J. Org. Chem.", 33, 1689, (1968).

[106] A. Sacco, and R. Ugo, "J. Chem. Soc.", 3274, (1964).

[107] S. Otsuka, T. Kikuchi, and T. Taketomi, "J. Amer. Chem. Soc.", 85, 3709, (1963).

[108] J. L. Hérisson, Y. Chauvin, N. H. Phung, and G. Lefebvre, "Compt. Rend.", Ser. C, *269*, 661, (1969).

[109] Y. Takegami, T. Ueno, and T. Fujii, "Bull. Chem. Soc. Japan", *38*, 1279, (1956).

[110] E. A. Emken, E. N. Frankel, and R. O. Butterfield, "J. Amer. Oil Chem Soc.", *43*, 14, (1966).

[111] H. Itatani, and J. C. Bailar, Jr., "J. Amer. Chem. Soc.", *89*, 1600, (1967).

[112] M. L. H Green, and T. Saito, "J. Chem. Soc. D", 208, (1969).

[113] L. H. Slaugh, "Tetrahedron Lett.", *22*, 1741, (1966).

[114] K. Ziegler, H. G. Gellert, H. Martin, K. Hagel, and J. Schneider, "Annalen", *589*, 91, (1954).

Author index